A. Rothenberger (Editor)

Brain and Behavior in Child Psychiatry

With 44 figures and 24 tables

Springer-Verlag Berlin Heidelberg GmbH

Prof. Dr. med. Aribert Rothenberger
Kinder- und Jugendpsychiatrische Klinik
am Zentralinstitut für Seelische Gesundheit, J5
D-6800 Mannheim 1

ISBN 978-3-642-75344-2   ISBN 978-3-642-75342-8 (eBook)
DOI 10.1007/978-3-642-75342-8

Library of Congress Cataloging-in-Publication Data
Brain and behavior in child psychiatry/Aribert Rothenberger (ed.).
Includes bibliographical references.

1. Biological child psychiatry. 2. Pediatric neuropsychology. I. Rothenberger, Aribert. [DNLM: 1. Brain—physiology. 2. Child Behavior—physiology. 3. Mental Disorders—etiology. 4. Mental Disorders—in infancy & childhood. WS 350 B814] RJ486.5.B7 1990 618:92'89071—dc20 DNLM/DLC

© Springer-Verlag Berlin Heidelberg 1990
Originally published by Springer-Verlag Berlin Heidelberg New York in 1990
Softcover reprint of the hardcover 1st edition 1990

2125/3030-543210–Printed on acid-free paper

List of Contributors

Amorosa, H., Priv.-Doz., Dr.
Max-Planck-Institute of Psychiatry, Child and Adolescent Psychiatry, Kraepelinstr. 10, D-8000 München 40, FRG

Asarnow, R., M.D.
Department of Psychiatry, Division of Child Psychiatry, University of California at Los Angeles, 760 Westwood Plaza, Los Angeles, CA 90024, USA

Barthelemy, C., DR.
Inserm U316—CHRU Bretonneau, Boulevard Tonnellé, F-37044 Tours, FRANCE

Birbaumer, N., Prof. Dr.
Department of Clinical & Physiological Psychology, University of Tübingen, Gartenstr. 29, D-7400 Tübingen, FRG

Bruneau, N., Dr.
Inserm U316—CHRU Bretonneau, Boulevard Tonnellé, F-37044 Tours, FRANCE

Chiarenza, G.A., Dr.
Istituto di Neuropsichiatria Infantile, Università di Milano, Via Besta 1, I-20161 Milano, ITALY

Conners, C.K., Ph.D., Prof.
Department of Psychiatry, Duke University Medical Center, 2213 Elba Street, Durham, NC 27710, USA

Courchesne, E., Ph.D., Prof.
Neuropsychology Research Laboratory, Children's Hospital, Research Center, 8001 Frost Street, San Diego, California, 92123, USA

Curtiss, S., Ph.D.
Department of Linguistics, University of California, Los Angeles, 405 Hilgarde Avenue, Los Angeles, CA 90024, USA

Elbert, T., Priv.-Doz., Dr.
University of Tübingen, Department of Clinical & Physiological Psychology,
Gartenstr. 29, D-7400 Tübingen, FRG

Elmasian, R., Ph.D.
Neuropsychology Research Laboratory, Children's Hospital, Research Center,
8001 Frost Street, San Diego, California, 92123, USA

Esser, G., Dr., Dipl.-Psych.
Clinic for Child and Adolescent Psychiatry, Central Institute of Mental Health,
J 5, D-6800 Mannheim 1, FRG

Garreau, B., Dr.
Inserm U316—CHRU Bretonneau, Boulevard Tonnellé, F-37044 Tours,
FRANCE

Gillberg, C., Prof. Dr.
University of Göteborg, Department of Child and Adolescent Psychiatry, Child
Neuropsychiatry Centre, Box 17113, S-40261 Göteborg, SWEDEN

Gillberg, I., Dr.
University of Göteborg, Department of Child and Adolescent Psychiatry, Child
Neuropsychiatry Centre, Box 17113, S-40261 Göteborg, SWEDEN

Hebebrand, J., Dr.
University of Marburg, Clinic for Child and Adolescent Psychiatry,
Hans-Sachs-Str. 6, D-3550 Marburg, FRG

Johannsen, H.S., Prof. Dr.
University of Ulm, Department of Phoniatrics, Frauensteige 14a, D-7900
Ulm/Donau, FRG

Leckman, J.F., M.D., Prof.
Yale Child Study Center, Yale University School of Medicine, 333 Cedar Street,
New Haven, CT 06510, USA

Lehmkuhl, G., Prof. Dr.
University of Cologne, Clinic for Child and Adolescent Psychiatry,
Robert-Koch-Str. 10, D-5000 Köln, FRG

Lelord, G., Prof. Dr.
Inserm U316—CHRU Bretonneau, Boulevard Tonnellé, F-37044 Tours,
FRANCE

Lerminiaux, D., Dr.
Inserm U316—CHRU Bretonneau, Boulevard Tonnellé, F-37044 Tours,
FRANCE

Lincoln, A.J., Ph.D.
Neuropsychology Research Laboratory, Children's Hospital, Research Center,
8001 Frost Street, San Diego, California, 92123, USA

Lombroso, P.J., M.D.
Yale Child Study Center, Yale University School of Medicine, 333 Cedar Street,
New Haven, CT 06510, USA

Lutzenberger, W., Dr.
University of Tübingen, Department of Clinical & Physiological Psychology,
Gartenstr. 29, D-7400 Tübingen, FRG

Martineau, J., Dr.
Inserm U316—CHRU Bretonneau, Boulevard Tonnellé, F-37044 Tours,
FRANCE

Martinius, J. Prof. Dr.
University of Munich, Institute of Child and Adolescent Psychiatry,
Heckscher-Klinik, Heckscherstr. 4, D-8000 München 40, FRG

Muh, J.P., Prof. Dr.
Inserm U316—CHRU Bretonneau, Boulevard Tonnellé, F-37044 Tours,
FRANCE

Pauls, D.L., Ph.D., Prof.
Yale Child Study Center, Yale University School of Medicine, 333 Cedar Street,
New Haven, CT 06510, USA

Propping, P., Prof. Dr.
Institute of Human Genetics, University of Bonn, Wilhelmstr. 31, D-5300 Bonn
1, FRG

Rockstroh, B.S., Priv.-Doz., Dr.
University of Tübingen, Department of Clinical & Physiological Psychology,
Gartenstr. 29, D-7400 Tübingen, FRG

Rothenberger, A., Prof. Dr.
Clinic for Child and Adolescent Psychiatry, Central Institute of Mental Health,
J 5, D-6800 Mannheim 1, FRG

Schmidt, M.H., Prof. Dr. Dr.
Clinic for Child and Adolescent Psychiatry, Central Institute of Mental Health,
J 5, D-6800 Mannheim 1, FRG

Schulze, H., Dr. Dipl.-Psych.
University of Ulm, Department of Phoniatrics, Frauensteige 14a, D-7900
Ulm/Donau, FRG

Strandberg, R., Ph.D.
Department of Psychiatry, Division of Child Psychiatry, University of California
at Los Angeles, 760 Westwood Plaza, Los Angeles, CA 90024, USA

Tallal, P., Ph.D., Prof.
Center for Molecular and Behavioral Neuroscience, Rutgers University,
Newark, NJ 07102, USA

Tanguay, P.E., M.D., Prof.
Department of Psychiatry, Division of Child Psychiatry, University of California
at Los Angeles, 760 Westwood Plaza, Los Angeles, CA 90024, USA

Taylor, E., Dr.
University of London, Institute of Psychiatry, De Crespigny Park, Denmark
Hill, London SE5 8AF, ENGLAND

Thoma, W., Dipl.-Psych.
Clinic for Child and Adolescent Psychiatry, Central Institute of Mental Health,
J 5, D-6800 Mannheim 1, FRG

Timsit-Berthier, M., Dr.
Université de Liège, Laboratoire de Neurophysiologie Clinique, et de
Psychopathologie, Rue Trasenster 21, B-4200 Ougreé, BELGIUM

Van Engeland, H., Prof. Dr.
Department of Child and Adolescent Psychiatry, University Hospital,
Heidelberglaan 100, NL-3508 GA Utrecht, THE NETHERLANDS

Werry, J.S., M.D., Prof.
Department of Psychiatry and Behavioral Science, University of Auckland,
School of Medicine, Auckland 1, NEW ZEALAND

Contents

Sleep

Diagnostic and Therapeutic Aspects

Preface

The Brain—What Else!

All senses are connected with the brain. From sense-perception derives ... knowledge. In the brain is the sovereignty of the mind. Mind is interpreted by the brain.

Alcmaeon of Croton
(5th Century B.C.)

The ground is shifting under the traditional approaches to problems in the philosophy of mind. Earlier doctrines concerning the independence of cognition from the brain now appear untenable.

P. S. Churchland
(20th Century A.D.)

It is not objective of this volume to discuss the history and significance of neuroscience for philosophy from a developmental perspective, although this would be a rather interesting topic. Its object is the relationship between brain and behavior in children as exhibited by higher mental functions (e.g., speech and language, reasoning, perception, free will and control of motor acts, dependence of behavior on neuronal constraints, the self of the child and therapeutic activities).

Child psychiatrists commonly allude to the brain as the site of disturbance responsible for many developmental disabilities and psychopathological syndromes identifiable by observing behavior (e.g., dyslexia, delusions), neurological examination (e.g., soft signs), psychological test performance (e.g., Bender Gestalt Test), EEG (e.g., alpha-theta ratio), and CCT (e.g., pseudoatrophy). While there is nothing inherently wrong with such inferences, the fact is frequently overlooked that there is no specific set of brain-behavior relationships validating these inferences. In this respect, brain inferences on the basis of a child's performance can lead to four potential fallacies concerning the relationships between brain and behavior in children (Fletcher and Taylor 1984):

(a) Differential sensitivity fallacy (i.e., examinations sufficiently sensitive to detect a general brain dysfunction in adults do not necessarily process the same sensitivity in children)
(b) Similar skill fallacy (i.e., tests measuring "prefrontal skills" in adults do not necessarily tap similar abilities in children of different ages)
(c) Special-sign fallacy (i.e., there is no unified set of symptoms consequent to brain injury in children)
(d) Brain-behavior isomorphism fallacy (i.e., mistaking dysfunctions as a descriptor of the brain rather than as a descriptor of behavior. The same behavior does not imply identical brain organization.)

In addition, the potential interactions of psychosocial variables with the applied measurements should be taken into consideration. This not only pertains to the level of psychopathological and neuropsychological research but also comes into play at the neurophysiological level of investigation: in an ongoing multilevel (neurophysiological, neuropsychological, behavioral, psychological) longitudinal study of our group, 362 children are being examined at the ages of 3 months, 2, $4\frac{1}{2}$, and 8 years. Three grades of both organic and psychosocial risk are differentiated, and infants are assigned to one of the nine resulting experimental groups. The objective of the study is to disentangle the influences of organic and psychosocial risk factors on child development. In the 3-month-olds, who have already been investigated, remarkable influences of psychosocial risks were seen both on neurological functions and on the amplitude of auditory evoked potentials (AEP) at the cortical level. There was very low augmenting of AEP amplitude with increasing stimulus intensity in such infants. This was interpreted as a decreased ability for neuronal facilitation in infants with psychosocial risk. Whether or not this reflects environmental and/or genetical factors remains to be determined (Rothenberger et al. 1989). Significant relationships between AEP and socioeconomic status have also been reported. Children from disadvantaged families are at increased risk for malnutrition, learning disabilities, lead poisoning, and a myriad of other problems associated with poverty. In particular, an increased lead blood level seems to be related to neuronal disinhibition and attention deficit in children (Bauman et al. 1987; Winneke 1989). Thus, environmental influences on child development are not only limited to a psychodynamic level but are equally important at the genetic, neuronal, and hormonal levels.

In contrast to our increasing knowledge in this field, there is still a wide gap in our information on behavioral disturbances in children on the one hand, and the neuronal bases of these disturbances on the other hand. Therefore, more intensive combined biological and clinical research in child and adolescent psychiatry is necessary. We, as child psychiatrists, have to cooperate with the disciplines involved in fundamental neuroscience (e.g., neuropharmacology, neurochemistry, neurophysiology, genetics, nuclear medicine, molecular biology), cognitive psychology, and neurology. More child psychiatrists have to be trained to pursue relevant neuroscientific research issues in their own field. Thus, they could be able to motivate researchers working in these disciplines to join them in a fruitful cooperation regarding their common objective of solving mental health problems in children.

Within the last years, some European child psychiatrists have intensified their activities in clinically oriented biological research. This has stimulated the interest of young clinicians in this area and prompted patients and parents to offer their support for promoting this kind of research. This volume on *Brain and Behavior in Child Psychiatry* is intended to be part of this process. It not only presents the state of the art on both sides of the Atlantic, it also tries to develop an appreciation of what may be the best approaches to the subject in future research. Biological research in child and adolescent psychiatry is mostly adult-oriented. It might be very promissing to broaden the developmental perspective, including stability and change of parameters, best in longitudinal studies. Methodologically, these projects should include multiple levels of investigation and should be based on model-guided hypotheses.

So far, our knowledge of the relationship between brain and behavior in child psychiatry is scarce. Nevertheless, the efficiency of our treatments depends greatly on our knowledge of the child's brain status. For a child with a brain maturational deficit (for instance), a mismatch—too high or too low in quantity, wrong in quality—between the inner and outer worlds can be unfavorable (Rothenberger 1989). This book provides essential information that should stimulate our efforts to develop a better basis for understanding disorder-specific neuronal weaknesses, the interaction with nonspecific environmental stress factors, and the compensatory factors inside and outside a child's brain. In consequence, developmental psychopathology as well as clinical developmental neuropsychiatry should benefit from neuroscientific research integrated at the clinical level.

It was a pleasure for me to edit this book in close cooperation with the contributing authors. I wish to thank Dr. T. Thiekötter and Mr. V. Oehm from Springer-Verlag for their friendly support and the good working atmosphere. My children, Liane Tessa and Lillian Geza, helped me in a nice way to design the cover of the book. My wife, Lilo, as at similar occasions in former years, accompanied me during the production of the volume with critical understanding and gave me the time to accomplish the editorial work.

Mannheim, Spring 1990   Aribert Rothenberger

References

Bauman S, Otto D, Robinson G, Schroeder S, Barton C (1987) The relationship of late positive ERPs, age, intelligence and lead absorption in socioeconomically disadvantaged children. In: Johnson R Jr, Rohrbough JW, Parasuraman R (eds) Current trends in event-related potential research (EEG Suppl 0.) Elsevier, Amsterdam, pp 617–623

Churchland PS (1988) The significance of neuroscience for philosophy. Trends Neurosci 11:304–307

Fletcher JM, Taylor HG (1984) Neuropsychological approaches to children: towards a developmental neuropsychology. J Clin Neuropsychol 6:39–56

Rothenberger A (1989) A closer look at maturational factors. In: Schmidt MH, Remschmidt H (eds) Needs and prospects of child and adolescent psychiatry. Hogrefe and Huber, Toronto, pp 107–112

Rothenberger A, Reiser A, Esser G, Laucht M, Schmidt MH (1989) Are auditory evoked potentials (AEP) and EEG power spectra in infants influenced by organic and psychosocial risk factors? Paper presented at WAIPAD 4th World Congress, Lugano

Winneke G (1989) Subklinische Wirkungen von Blei auf das Nervensystem von Kindern. Nervenheilkunde 8:80–85

Overview

J. S. Werry

Historical Aspects

When I began my training in child psychiatry in Montreal in 1960, biological aspects were almost nonexistent, though the discovery of the neuroleptics in France, the antidepressants and benzodiazepines in Switzerland, and lithium in Australia, were bringing a revolution to adult psychiatry, not only in empirical therapeutics but also in basic research into etiology. This entailed a shift away from psychologizing and the neuroses and a return to neurobiology and the major psychoses, just as it had been in the days of Eugen Bleuler and Emil Kraepelin. (It seems to me, though, that psychiatry in Europe may have suffered less from this loss and return of the prodigal son than psychiatry in North America).

Lacking as it did much psychotic clientele and clear indications for the use of medication, child psychiatry seemed at first untouched by this biological revolution in psychiatry—except in one area which actually had little to do with adult psychiatry, that is hyperkinesis or minimal brain dysfunction or hyperactivity (as it came to be known), whose antecedents seem to have come largely from pediatrics and neurology (see contributions by Taylor and Esser and Schmidt).

Just as research in the United States by Frances Graham and her group at St. Louis and by Schulman and co-workers in Chicago (Werry 1986) in the early 1960s was showing that Strauss's idea of a specific "brain damage/behavior syndrome" was false, it took off like a rocket and for the next twenty years dominated the literature, such that even this volume finds it necessary to review and refute this thesis repeatedly throughout. It is a sad irony that it was not really until Michael Rutter took up the cause, more than a decade after the work by Graham and Schulman, that people began to take the debunking seriously.

It cannot, however, be said that just because the minimal brain dysfunction idea was false that it was a bad one or a disaster for child psychiatry. Ideas in science are in themselves neither good nor bad, only useful in stimulating research and/or practice or destructive of it. There can be no doubt that the whole concept of minimal brain dysfunction gave to child psychiatry the kind of rationale and intellectual ferment for developing a biological base which it lacked because of a sufficiency of probably-biological syndromes such as schizophrenia or mood disorder. Autism could not fulfil this role—though it has been significantly synergistic—simply because it was too rare to make an impact on the mainstream of clinical clientele in child psychiatry.

Minimal brain dysfunction, or the various subsyndromes it has become (attention deficit disorder, learning disorders, speech and language disorders, developmental motor disorders), has played a major role in the development of both measurement methodology and experimental design in child psychiatry, which have been important far beyond anything to do with biological psychiatry.

This Monograph

So where are we in 1990 with biological child psychiatry? This is the question this monograph addresses. First we must ask, what is it that this monograph can offer that others cannot? The most obvious is that it offers a window on research, knowledge, and thinking of non English-speaking European child psychiatry. The idea of producing this in English is a bold step since it will make it accessible to much more of the world's population than otherwise, and the English-speaking world, which ethnocentrically tends to ignore what does not appear in English, will have a particularly good opportunity to see what German, French, and Scandinavian child psychiatry is doing. However, it is a risky venture too, because while the demand that all write in English has been met with more than adequate comprehensibility, it is quite unreasonable to expect that idiomatic usage can be met. The English speaker, therefore, must not be repelled by some of the inevitable departures from idiomatic (none of which distorts the real meaning) because to do so would be to deprive oneself of a great deal of invaluable material.

Overview of Contents

The book has been organized into sections but there are elements of communality which are clear throughout the volume. These are:

1. There is considerably more knowledge about brain and behavior in children than most of us, even those who think ourselves experts, suspect.
2. Though this knowledge is impressive, complex, and technical, applications to clinical practice are extremely limited and previous and current attempts to do so (such as minimal brain dysfunction) are premature and simplistic.
3. Much of the relevant biological knowledge and technique requires considerable fluency in other areas of science. For this reason, child psychiatry must stay close to centers of science and find ways to work with those whose knowledge in areas such as molecular genetics or receptor physiology is sufficient to stop any ideas that we as child psychiatrists may have about brain–behavior relationships from appearing foolish.

4. While most of us lack the sophistication needed, there are among us a few who seem to be able to straddle the worlds of clinical practice and neurobiology. We must cosset them as our ambassadors if we are to find applications of biology in child psychiatry.

5. We know more about what is not (e.g., that brain damage neither inevitably nor commonly causes hyperkinesis), than what is. This may be seen as extremely discouraging especially in practice, but it also shows that at least we are making a start toward a systematic body of knowledge.

6. The brain is an incredibly complex organ, the function of which is only dimly grasped even at simple levels of function let alone at the levels of social behavior or of learning academic skills. In our hurry to know, we have tended to grasp at explanations which insult this complex organ and give false comfort to those who would deny any role to biology in children's psychopathology.

7. Human behavior is also very complex, and our ways of measuring it are even less exact than those for brain function.

8. While we all accept, at least in principle, that brain factors can influence behavior (even if we are not sure how and in what precise ways), we probably have little grasp of the extent to which psychosocial factors can influence brain function especially in development (see Preface and Chapter 1.1).

9. The brain changes and develops quite dramatically throughout childhood and thus what is found in the developed state of adulthood is not necessarily correct for child psychiatry. There is thus an urgent need to see and sustain developmental psychoneurobiology as one of our basic sciences.

10. Certain theories (such as equipotentiality of brain damage) and techniques of investigation (such as evoked cortical potentials) have been perhaps too popular and this popularity may prevent their rather meager yield from being as obvious as it should be. Further, popularity may prevent proper examination of other theories or methods of study.

Specific Contents

The contents fall loosely into seven topic areas. The first, entitled General Aspects, gives us windows on philosophy, psychobiology, neurophysiology, neurology, biochemistry, and genetics, and each offers to some degree a useful review of these areas with particular emphasis on psychiatry and child development. This section is not for the faint-hearted as trying to update oneself on most basic science relevant to psychiatry (e.g., molecular genetics) is a daunting task, though an absolutely vital one.

The second topic area looks at cognition, an area of great concern to child psychiatry and one in which we can, with some pride, point to major contributions to the science as a whole, largely because of our interest in learning problems in children with what is now called Attention Deficit Disorder or with specific learning disabilities. The biological aspects discussed here are pivotal and poorly recognized by many. The third topic covers speech and language, an area that no child psychiatrist can ignore since problems in this area are so often associated with behavioral disturbances.

The fourth section looks at the important, if numerically rather infrequent, area of childhood psychoses—autism and schizophrenia. In the latter, since it seldom develops before the age of 15 years, there has been an understandable interest in trying to pick those children who will develop schizophrenia in later life. While abnormalities in cognitive, neurodevelopmental, and, less certainly, psychophysiological function have been found to be risk factors, the problem of their failure to meet the requirement of specificity is underlined. This problem of poor specificity of biological markers/predictors is one which has so far bedevilled child (and adult) psychiatry in general as, for example, the story of soft neurological signs shows clearly.

The fifth section is concerned with minimal brain dysfunction and head injury both as separate topics in their own right and their interrelationship. Here the presentations, all from Europe, largely confirm what has been known for some time in North America and elsewhere, that there is no such thing as a brain damage/behavior syndrome, that hyperactive/clumsy children, who form such a large part of the clientele of most child psychiatric clinics, often get better as they get older but equally often do not. Also that once one eschews the concept of hyperactivity or attention deficit disorder as a homogeneous brain damage behavior syndrome, there is merit in regarding it as a valid clinical entity on an empirical basis devoid of any etiological connotations. Head injury emerges, as in studies in the United Kingdom and the United States, as a complex pathogen which results significantly often in durable psychiatric disturbance but this disturbance can take many forms and is dependent for its expression and severity on psychosocial factors. Also important is the replication of the conclusions by Fletcher and Satz (1983) that age of injury is not, as commonly believed, inversely correlated with degree of subsequent disability but bears no clear relationship to it at all.

In the sixth section, a comprehensive and informative review of sleep and its disturbances in children is presented from a developmental, neurobiological perspective, showing how this important basic function with which we, as child psychiatrists, deal constantly, is now vested with a wealth of scientific knowledge.

Finally, the all important issues of diagnosis and therapy are addressed with an understandable focus on biological methods of biofeedback, psycho-physiology, and psychopharmacology. This section helps to end the book on a rather pessimistic but realistic note of the gap between even effective methods of treatment like the stimulant drugs in attention deficit disorder and any ideas we may have about the biogenic etiology.

Conclusion

This monograph presents an impressive array of scholarship and technology, which illustrates the robust if developmentally immature state of knowledge about brain and behavior in child psychopathology. No longer can any child psychiatrist be complacent about or ignorant of this. If one is to justify one's role in child mental health as a psychiatrist, one must ask what it is that we bring specifically to the mental health team? Surely it can only be that we are physicians trained in human biology in health and disease. This should enrich, not dominate, any perspective on child psychopathology. If we lose touch with this knowledge base, we will have lost our possibility of specificity. This monograph aims to make our loss of this knowledge base unlikely.

References

Fletcher JM, Satz P (1983) Age, plasticity and equipotentiality. J Consult Clin Psychol 51:763–767
Werry JS (1986) Biological factors. In Quay HC, Werry JS (eds) Psychopathological disorders of childhood. 3rd edn, Wiley, New York, pp 294–331

General Aspects

The Interaction of Psychological and Biological Processes in Children

E. Taylor

Biological and Psychological Processes in Brain Development

History

Mental activity is part of the function of the brain; the structure of the brain is determined in part by the stimulation it has received. The relative importance of biological and psychological processes in shaping mental development has generated much passion and debate.

Prescientific psychology veered between extremes of thought. Many mediaeval world views expressed an extreme of constitutional determination. Human character developed as the unfolding of an ontogenetic programme, a homunculus, that was present as part of the father's make-up, and therefore of his father's and so on back to Adam. The new thinking of the Enlightenment formulated the contrary view. To writers such as Locke, human infants contained no mental character except the ability to learn. They were tabulae rasae, blank slates upon which experience wrote, and their later development was wholly contingent upon what happened to them.

For much of twentieth century neuropsychology, Pavlovian thinking provided the dominant scheme. Reflexes were based on the synaptic connections present at birth; the conditioning of those reflexes provided the basis for mental growth. The post-natal modification of microstructure, in this view, was the result of experience. The brain responded to its environment in a relatively passive fashion. Acquired changes in the brain then determined the future reactions of the organism to stimulation, but the equipment of the brain was essentially in place at birth.

Current Views

The advances in developmental neurobiology of the last 20 years were built on improved methods of studying the anatomical and physiological connections of brain cells. They have led to a more complex and more dynamic picture of brain development. The brain at birth is structurally immature and much of its growth takes place during childhood: not so much in terms of numbers of neurones as of their interconnections. In the first years of life there is a massive outgrowth of branches from the processes of neurones, and a massive but selective loss of many of the connections formed. This happens while the child is in touch with the outside world, and is guided by the nature of that contact as well as by genetic information. Those interconnections survive that are most fit for the purposes of communication.

Biological and psychological processes affect each other through this stage, with lasting effects. For example, abnormalities of migration of cells may have no immediate effect, but are associated with some cases of dyslexia in later childhood (Galaburda et al. 1986). Deprivation of experience in animals has profound results upon the size of the brain and its functions (Trevarthen 1986). In children, too, the early experience of the brain can have lasting effects upon its later function. The eye with a refractive error that is not corrected in the first few years may remain functionally inactive even when its optical competence is restored.

Some types of functional interconnection happen without the need for stimulation from the environment. Abilities innate to infants—such as that of recognising members of their own species—must reflect the organised formation of connections between neurones before the advent of the environmental associations that they come to recognise.

The brain is therefore changing during childhood, both actively and in reaction to the way it is stimulated. The most important kind of stimulation is of course that which comes from other brains. Nevertheless, the influence of personal relationships upon brain development is still largely unknown. At a psychological level of analysis, there is still argument about the later effects of childhood relationships. Even when personal relationships are extremely disturbed and children are grossly neglected or maltreated, normal development of intellect and personality is still possible. Documented cases in which children have been rescued after extremes of maltreatment have emphasised the capacity of children to recover if they are treated better in later childhood (Skuse 1985). However, this recovery in adjustment does not imply that the brain must have been unaffected, and should not deter research from examining the processes involved. Much potential may have been lost even in children who reach an acceptable level of function.

The neurosciences have now developed to a point where they should be able to clarify some of the mechanisms through which brain function determines structural brain development. There are already some exciting clues. The discovery that neurotransmitters also promote DNA synthesis and are therefore trophic growth factors for the brain is promising. It opens up lines of studying the cellular and molecular processes involved, and also the effects of specific types of experience upon localised structural changes.

Other mechanisms will be important. Subtractive and selective processes are susceptible to environmental effects as well as proliferative. An example comes from the carefully-studied sequences of changes when the visual input into the brain is altered by covering one eye during early development. The geniculo-striate axons that would carry the input from that eye suffer a much more drastic loss of synaptic connections than would normally be the case. Correspondingly, the axons from the open eye are much less severely pruned than in animals with two functioning eyes. This induces changes at a higher level, in cerebral organisation. The cerebral cortical changes depend upon excitatory synapses— probably involving glutamic acid as transmitter—between the terminals of the axons and the dendritic spines of the cortical cells. In addition, ascending noradrenergic nerve fibres from the locus coeruleus and cholinergic fibres from the forebrain modify the extent to which the cortex is reorganised. This allows the possibility that other kinds of experience will influence the changes in structure induced by deprivation.

A major priority for future developmental research is therefore the description of psychological influences upon brain function—including the effects of relationships as well as the physical properties of the environment, and the effects on social as well as cognitive functioning. Such a programme will be more feasible if there is development of the measures suitable for human subjects that allow for analysis at the level of physiological brain function. Improvements in the acceptability and resolution of functional imaging techniques (such as magnetic resonance imaging and mapping of evoked responses) should therefore be encouraged because of their importance to this scientific goal. At present there are several promising lines for the measurement of brain function and dysfunction in childhood but none of them is wholly satisfactory and all of them raise difficulties of interpretation of their findings (Taylor 1983). The improvements of methodology needed are not only a matter of the physical development of the instruments but will also require the collaboration of psychological scientists for the establishment of measures of psychosocial functioning with which the biological measures are to be associated.

Implications for Studying Psychopathology

Several general considerations arise from the view of brain development outlined above.

1. Injuries to the developing brain will often have wider effects than a localised deficit of function. Subsequent development of the brain will be affected in complex ways. This may imply plasticity and recovery from lesions; it may also imply harmful effects on other kinds of function. Many neuropsychological functions will need to be examined in pathological groups.
2. Since the brain operates in different ways at different ages, psychological function may be intact in a brain-damaged person at one age yet show impairment later in development. The same processes may be subserved by different parts of the brain at different stages of its development. The course of psychological disorder may therefore be quite different from the course of underlying neurological disability.
3. Impairment of an elementary neuropsychological process will not necessarily give rise to a characteristic psychiatric syndrome. Consider, for instance, the example of a specific impairment in avoidance learning. This is not purely

hypothetical: it corresponds, for instance, to the behavioral lesion produced in juvenile rats if, as foetuses, they are exposed to chloramphenicol eaten by their mother. If this happened in humans, how would we know? The translation of such a deficiency into observed behavior in natural settings would depend upon the reinforcement contingencies operating for each person.

4. Biological and psychological measures represent different levels of analysis of brain activity, not different classes of events. A physiologically recorded abnormality does not automatically mean that the effective aetiology is organic; electrophysiological and neurochemical changes can result from psychological manipulations.

5. The developmental course of different neuropsychiatric disorders suggests that psychological and biological factors may operate in diverse ways. Contrast, for example, the typical course of autism—involving persistence of characteristic cognitive and social disabilities into adult life in a way that is predicted by childhood cognitive status—with that of hyperkinetic syndrome, which seems to transform from cognitive impairment into antisocial behavior in a way that is predicted by intrafamilial relationships (Taylor 1986).

Progress in understanding the pathogenesis of neuropsychiatric disorders will therefore require the establishment of correlations between brain function and psychological performance within homogeneous subtypes of psychiatric disorder, and the longitudinal and aetiological study of cohorts defined on the basis of closely characterised neuropsychological disabilities rather than traditional psychiatric diagnoses.

Biological and Psychological Processes in Pathology

History

From the beginning of child psychiatry, practitioners have been excited by the possibility of biological causes of psychological abnormalities. They have sometimes let their excitement take them to a crude formulation of causality that presents biological and psychological causes as incompatible, competitive explanations. The truth is far removed: biological dysfunction of the brain usually produces disorder by interaction with the psychological environment. The interactions are complex, and vary with the developmental level of the child and the nature of the biological risk.

Nineteenth-century descriptions of children's disorders were dominated by the concept of moral insanity (von Gontard 1988). The rise of evolutionary thinking through the later nineteenth and early twentieth centuries led to a notion of child psychiatric disorders as the persistence of (or regression to) biologically primitive states. Indeed, "social Darwinism" identified primitive biological states with the characteristics of underdeveloped races of mankind and dreamed of programmes of eugenic breeding that would free the human stock of the taint of mental illness. Schachar (1986) has shown how the rise of hyperactivity as a concept of biologically based disorder was dependent upon the ideology and values of the paediatricians and psychiatrists who supported it— not upon scientifically persuasive findings.

The rise of psychological explanations of disorder also came in the early part of the twentieth century. For a long time, however, the approach was believed to be in opposition to biological theories. Even nowadays some authors still follow the earlier belief that biological aetiology is an exclusive alternative to psychosocial aetiology and therefore removes the burden of perceived guilt from parents.

Current Views

Psychological stressors may obviously have effects upon biological measures of brain function, and biological stressors upon psychological measures. This does not imply any necessary interaction between factors; but it is what is most commonly meant by investigators of "psychosomatic" or "somatopsychic" influences.

The *somatopsychic* strategy of investigation examines the psychological effects of biological changes, which are at their most direct for physical diseases or injuries involving the brain. In the neuropsychiatric research work of the last 20 years, empirical findings about brain damage have pointed very clearly to the need to take interactive effects into account. The epidemiological studies carried out by Rutter, Graham and Yule (1970) and their colleagues were highly influential in charting the wide range of consequences of brain injury and disease. They made it clear, for instance, that most of the psychiatric disorders in children with neurological diseases were indistinguishable in form from those afflicting children with intact brains, but occurred at a much higher rate.

The pathogenesis of disorder seems to involve both psychological and biological factors. The form of the neurological disease does to some extent determine the risk of disorder: the presence of epilepsy goes with a particularly high prevalence of psychiatric complications. Psychosocial adversity is also a factor: it has long been known that the same kinds of family problem that are thought to cause disorder in the general population are found more commonly in epileptics with psychiatric symptoms than in epileptics without such symptoms (at least since the work of Grunberg and Pond 1957). Similarly, a study of closed head injury found an association between psychological problems and psychosocial stresses within the injured group (Rutter et al. 1983). Indeed, it would be remarkable if people with abnormal brains were free of the stresses that upset other people, or immune to their effects!

In short, psychological and biological mechanisms of pathogenesis are not antithetic; both types of factor operate together in most cases. By itself, however, this conclusion is not very helpful in understanding the pathogenesis of disorder. Rather, it opens up the possibility of studying the specific mechanisms of interaction in different circumstances. This kind of study is likely to be a central part of the programme of neuropsychiatry over the next few years.

The *psychosomatic* strategy of investigation has had a more chequered course. There was a rapid and premature acceptance of theories (such as that of Alexander) that postulated specific intrapsychic conflicts to be the cause of specific physical illnesses or symptoms. The consequent paradigm of studies was the comparison of different physical illnesses—such as asthma, migraine, or ulcerative colitis—for evidence of specific emotional concomitants. While useful information has been gathered in this way, and the frequency of psychological problems in the chronically ill should now be recognised by every paediatrician, the main theoretical purpose has not proved to be scientifically productive. It may even have been harmful in distorting the role of the psychiatrist in paediatric liaison and falsely suggesting that the main role of psychiatric treatment is to remove the physical evidence of disorder rather than to promote psychological adjustment. The main heir to this tradition is probably the study of so-called type A personality as a risk factor for cardiovascular disease. This research has the merit of examining the consequences of a psychological deviation rather than merely the correlates of physical disease. Even so, it carries the risk of missing important correlates of stress if it focuses solely upon type A characteristics rather than upon measures drawn from general theories of personality and temperament.

We know far too little about the physical consequences of life events and other stressors in childhood. Tennes et al. (1986) and Lundberg (1986) have described interesting correlations between catecholamine and cortisol levels on the one hand and classroom behaviours and the stress of school attendance on the other. Since it is likely that both the situation and the child's personality have independent and interacting effects, the framework of life event research is likely to be a good one for this sort of psychophysiological work. The great development of measures of immune system functioning in recent years should also allow for knowledge to be developed about the effects of stress; and indeed the means now exist by which one may investigate the effects of allergic and other immune disorders upon the physical and mental functioning of the brain.

It will be clear that a good deal of knowledge has been gathered about psychosomatic and somatopsychic associations. Future research has a harder task with which to grapple: the analysis of the different possible means by which such factors may work together.

Mechanisms of Interaction

Additive Effects of Psychological and Biological Factors

The simplest way that physical and psychological stresses can work together is by exerting independent effects. Each stressor then works in much the same way as it would if it occurred in isolation. Two risk factors are worse than one, and the effects of each risk sum together.

This is not an interaction at all in the statistical sense, nor in the sense used below of a specific vulnerability to the operation of other factors. Nevertheless, it is probably a very common way that risk factors work together. Brown et al. (1981) showed it with some elegance for children who had suffered a severe closed head injury. The injury led to increased rates of psychiatric disability both in those with and those without coexisting family adversity. Family adversity also led to increased rates of disorder—and this was in some contrast to the findings for cognitive impairment, where the severity of the injury was the main determinant of IQ and family factors played a minor role. The head injury raised the prevalence of psychiatric disorders by a similar degree in those who did and those who did not have evidence of family and social problems.

Since psychological and biological processes can both produce similar effects independently, multivariate statistics are often necessary to demonstrate the action of either or both. The risk from exposure to environmental lead has often been examined in this way, with multiple regression analyses used to determine whether measures of body lead still contribute to behavior problems after psychosocial factors have been allowed for. In all such analyses, the association between lead levels and disturbed behaviour diminishes when other factors are considered; in many, some association remains nevertheless and suggests a causative role.

Transaction Between Biological and Psychological Adversity

Another way to think of the interaction between biological and psychological stressors is to remember that either of these may raise the prevalence of the other. This is probably clearest for the circumstances in which social disadvantage creates a biological risk. Perinatal morbidity and mortality have long been known to be more common in families of low socioeconomic status. The reasons are probably several: Poor maternal nutrition and low stature play a part and indicate the action of factors back to the mother's childhood and before. Prenatal and perinatal care is worse for lower social classes in many countries, and this will contribute to higher rates of brain damage. Finally, parents living in poverty often acquire attitudes of cynicism or hostility that act as a further barrier to their using those services that are available.

Another clear instance of psychosocial factors causing organic dysfunction is that of acquired trauma to the brain. There are many causes of head injury; they include the actions of parents and other adults (Craft et al. 1972) and a range of psychosocial adversities that can contribute to the risk of accidents (Langley 1984).

Psychological factors may also influence the physical constitution of the brain directly, as already considered. For instance, investigation of learned helplessness in adults with depression and in animals subjected to unavoidable stress has indicated that experimental psychological manipulation can alter the turnover of neurotransmitters in the brain—and the chemical change induced could then be a cause of behavioural abnormality.

Psychological causes of biological risk are relatively clear; it is easier to overlook the extent to which biological risk factors raise the prevalence of psychosocial stressors. This pathway must, nevertheless, be relatively common. Family genetic and twin studies have found, for example, that there is a genetic contribution to the occurrence of life events (Scarr and MacArtney 1985). The interpretation must be that individuals determine their own lives; that constitutional qualities of individuals determine the way that they are treated by others and the nature of the events that occur to them.

Stigma is one, but only one of the pathways through which the constitution exerts its transaction with the psychological environment. It is a matter of everyday experience that people with damaged brains are regarded with fear and distrust by many of their "normal" fellows. I believe that epilepsy is especially likely to evoke this response, and that this may account for some of the very high rates of psychiatric disorder in this condition. Research has not yet charted all the facets of this, facets that will probably vary between cultures. Many mothers taken from the general population say that they believe epilepsy to be contagious, and many more refuse to let their own child play with somebody who has epilepsy. Research needs to specify the reaction of society more clearly in order to allow public education to be sensibly focussed. It also needs to investigate the circumstances in which children with brain damage acquire a negative image of themselves from the attitudes of others, and those in which they resist.

Even when brain damage is not labelled as such and is not obvious (as in the case of uncomplicated epilepsy before it is diagnosed), still the reactions of others can be harmful. One instance comes from observational studies of the interactions between the mothers of handicapped toddlers and their offspring. There seems to be a relative dearth of interaction by comparison with mother–child relationships in neurologically normal groups. This type of finding is very suggestive for the pathogenesis of psychiatric disorder in certain cases. Over-protection is a less studied but clinically important type of reaction. Both point the way to more extensive and deeper studies: for instance, of the effect of counselling or of training programmes, and of whether experimental intervention would have any influence upon the later social adjustment of the children or upon the well-being of the parents. There is a real need for descriptive and short-term predictive studies in this area, not only because of the developmental interest of the interactions but also because of the possibility of prevention that arises when this type of intrafamilial risk can be detected and reduced.

Transactions such as these between psychological and biological factors could in theory be protective as well as harmful, and I hope that future research will address this possibility. In some parents the stress of a handicapped child produces a passionate tenderness. Ounsted (1955) rather cruelly called this "hyperpaedophilia" and suggested it to be a morbid process. Its effect, however, may be highly functional for handicapped children, who are thereby spared some of the rejections to which they would otherwise be prone.

Potentiation of Life Stress by Biological Impairment

Apart from their transactional role in increasing the prevalence of other stressors, psychological and biological factors can work together in a true interaction. Either factor may increase the sensitivity of the organism to the other type of factor.

A brain-injured person may show a qualitatively increased reaction to a life stress. This is often stated or implied by clinicians as the dominating mechanism, but research evidence is scanty. This is a key issue for research to explore in as wide a range as possible of stressors and vulnerability.

In some studies an interactive effect does indeed appear. The follow-up studies from the island of Kauai provide a good example (Werner and Smith 1977). Children who had survived potentially damaging events perinatally were contrasted in later childhood with their peers. Where there was no psychosocial disadvantage, there was no increase in the rate of psychiatric disorder in those who had been at risk in infancy. Where there was psychosocial disadvantage, the rate was of course higher in both the cases and the controls, but the rate in the cases of perinatal disadvantage was higher than in the controls. The implication is that brain damage operates primarily by increasing vulnerability to psychosocial disadvantage.

The differences of findings between these types of study may well reflect the differences in the type of brain insult studied. The severe head injuries studied by Brown et al. (1981) had given rise to obvious neurological consequences with prolonged unconsciousness and residual amnesia. The perinatal adversity studied in Kauai was a risk for brain damage, not evidence that the brain had actually been damaged. The longer follow-up period in the Kauai work allowed for fuller recovery from any damage that had happened, and it is possible that recovery was more rapid and more complete in this earlier period of development. For all these reasons, it is likely that the direct effects of neurological damage would be greater in the trauma study, and therefore more likely to give rise to some behavioral consequences even in the absence of family or social difficulties.

More Complex Interactions

In theory, biological damage to the brain might offer protection against the effect of psychological stressors. In the past, for example, it has been suggested that the intellectually handicapped are less likely to react to stresses with an identifiably depressive reaction. Few perhaps would care to repeat the assertion today, but it should be useful nonetheless to examine the possibility that depressive reactions take a different form in the neurologically, as in the chronologically, immature.

A temperamental quality such as hyperactivity is in part determined biologically. Its risk potential has been repeatedly stressed. In certain circumstances, however, its presence may be a protection. "Demanding" and "difficult" behavior is functional when resources are very limited. Under conditions of drought and starvation, a "difficult" temperament may even enhance survival. Schaffer's study of institution-reared children (1966) did not come to the usual conclusion that overactivity was a risk factor. Rather, the overactive children were less prone to psychiatric disability at follow-up. It is not, of course, clear that in such a population hyperactivity has the same aetiological significance as it does in those growing up in families: it may signify resistance to the pathological forces causing inertness and lack of interest. Nevertheless, the finding emphasises the complexity of the relationship between constitution and environment. It is plausible to think that under conditions of great deprivation the most demanding and active children will extract a greater share of the limited resources of food or of affection. When resources are adequate then the same behavior may be a source of irritation to caretakers and correspondingly poison personal relationships.

So far I have considered only the way that biological factors change the impact of psychological factors, but of course the converse may also hold. The previous temperament of the child predicts the degree to which acquired head injury will lead on to psychiatric symptoms. This, however, is most readily explained by the mechanism, already invoked, in which neurological dysfunction enhances the operation of the same pathogenetic factors as apply to the neurologically normal; and there is no evidence for the physical changes in the brain being conditioned in pre-existing psychological experience.

I have also considered potentiation as a quantitative effect, but there are also circumstances in which the whole meaning of a life event is altered by the previous biological status of the child. Puberty, for example, is a biological event that normally makes for greater independence in development. In the handicapped, its meaning is reversed; it is an occasion for increased protection and restrictiveness on the part of the parents as they seek to protect the handicapped person from the dangers to which their sexual maturity exposes them.

Finally, these patterns of interaction may themselves be modulated by external agents. The gender of the child, for example, is a powerful determinant of how far a risk factor will lead on to disability.

Issues for Future Research

Research into the aetiology of neuropsychiatric disorders will have to be planned in the light of a multiplicity of risks. Without this planning, serious obstacles would arise. Consider, for example, a simple case-control study of a condition such as attention deficit–hyperactivity disorder, which is supposed to have biological underpinnings but can probably result from psychosocial causes as well. Any biological cause is then very difficult to identify, for any true characteristic of a brain disturbance is diluted by the presence of many cases in which alternative aetiologies are operating. The result might well be uninter-pretable collections of unreplicated or nonspecific findings.

This problem might be overcome in several ways. In theory, all cases caused by psychological factors could be excluded. In practice, this is not yet feasible because of lack of knowledge about details of psychosocial processes. It is certainly possible to exclude cases with obvious breakdowns of parental care, and many investigators already do this; but it is likely that psychosocial causes may be subtle, and it is certainly not practical to exclude all cases with any hint of relationship problems. Another research route is through multivariate analyses and, as considered above, this will be increasingly necessary, but not sufficient. The interaction between biological and psychological factors may very well be different at different points in their range; and this could well invalidate conclusions from multiple regression analyses. Another research line could be a more efficient subtyping of disorders—for instance, to obtain a group with a high likelihood of biological causation.

Valid subtyping of disorders is a difficult process (Taylor 1988). Fortunately, possible solutions are appearing. At the level of localisation of brain damage, the many recent developments in imaging the structure and function of the brain offer major hope for the future. At the level of syndrome differentiation, clinical research has been clarifying the nature of the conditions to be recognised and the criteria to be adopted in recognition. The process has gone farthest for autism (Rutter 1985), progressive disintegrative disorders (Corbett et al. 1976) and Tourette's disorder (Corbett and Turpin 1985); hyperkinetic disorder is following in the tradition (Taylor 1988). There is obviously a long way to go in describing brain function and behavioural disorder, but progress has been made. The usual design defines syndromal contrast groups and applies brain measures as dependent variables. There is also a need for measures of brain structure and function to be applied in order to secure homogeneous subgroups of localised brain disorder, so that within these groups the degree of disorder can be related to the degree of disturbance of basic psychological processes.

What is the mechanism of increased vulnerability to psychiatric disorder in those with brain damage? It cannot be simply that psychosocial stressors become commoner and more severe, for the increase of rate attributable to brain damage is marked even when one considers only those in whom psychosocial stress is evident.

The sharpest comparison would be between groups of brain-damaged and neurologically normal children who have encountered the same types of defined stress, and between subgroups of children with compromised brains who have encountered different forms of stress. Some ingenuity and opportunism may be needed here to establish suitable comparison groups, but they should be illuminating.

Measures of psychosocial interaction are not important solely because of the need to control for unwanted nuisance variables in understanding brain–behavior relationships. Rather, since psychosocial processes are part of the pathogenesis of disorder, they need to be more closely understood in their own right. There is accordingly a strong research need to quantify the types of family, school, and peer group interactions that are responsible. The power of doing this has been pointed up by the research into the role of expressed emotion in those with schizophrenia. Not only do high levels of expressed criticism (and over-involvement) predict relapse, but modifying the family attitudes confers protection against relapse. Of course, there may well not be anything very specific about the process. Expressed emotion is a marker for risk, not the risk factor itself. Surely anybody would do worse if they were living in an emotional atmosphere of rejection and hatred? And surely any disabled person will do better if the people looking after him understand and accept his condition? But the quantification of the idea allowed its testing, and discouraged the notion that schizophrenia is simply an illness pursuing its own course with little relation to other aspects of life.

Treatment trials will help to clarify what processes cause disorder. If changing the process changes the course, understanding is advanced. Investigators will need to apply measures of the supposed process of treatment, as well as the outcome, and examine the agreement between the two. The place of an active therapy will become clearer when change in the immediate target of treatment proves to correlate with change in desired social adjustment.

The research implications of taking interactive processes as the major subject of study have been suggested through this article, and many investigators are already carrying out such work. It will mean a bringing together of rather different kinds of expertise in brain physiology and psychosocial risk, and a correspondingly collaborative style of research work. There are also clinical and training implications. Neuropsychiatry will need to be a comprehensive psychiatry for those with disturbed brains. Psychological interventions may be as important as physical ones for the prevention and treatment of neuropsychiatric disabilities. Clinical professionals will need a broad training and services will need to call on and coordinate a wide range of specialists.

References

Brown G, Chadwick O, Shaffer D, Rutter M, Traub M (1981) A prospective study of children with head injuries: III. Psychiatric sequelae. Psychol Med 11:63–78

Corbett J, Harris R, Taylor E, Trimble M (1977) Progressive disintegrative psychosis of childhood. J Child Psychol Psychiatry 18:211–219

Corbett J, Turpin G (1985) Tics and Tourette's syndrome. In: Rutter M, Hersov L (eds) Child and adolescent psychiatry: modern approaches, 2nd edn. Blackwell, Oxford, pp 516–525

Craft AW, Shaw DA, Cartlidge NE (1972) Head injuries in children. B Med J 3:200–203

Galaburda AM, Sherman GF, Rosen GD, Aboitiz F, Geschwind N (1985) Developmental dyslexia: four consecutive patients with cortical anomalies. Ann Neurol 18:222–232

Grunberg F, Pond DA (1957) Conduct disorders in epileptic children. J Neurol Neurosurg Psychiatry 20:65–68

Langley J (1984) Injury control: psychosocial considerations. J Child Psychol Psychiatry 25:349–356

Lundberg U (1986) Stress and Type A behavior in children. J Am Acad Child Psychiatry 25:771–778

Ounsted C (1955) The hyperkinetic syndrome in epileptic children. Lancet 2:303–311

Rutter M (1985) Infantile autism and other pervasive developmental disorders. In: Rutter M, Hersov L (eds) Child and adolescent psychiatry: modern approaches. 2nd edn. Blackwell, Oxford, pp 545–566

Rutter M, Chadwick O, Shaffer D (1983) Head injury. In: Rutter M (ed) Developmental neuropsychiatry. Guilford, New York, pp 83–111

Rutter M, Graham P, Yule W (1970) A neuropsychiatric study in childhood. (Clinics in developmental medicine 35/36), Heinemann, London

Scarr S, MacArtney K (1985) How people make their own environments: a theory of genotype-environment effects. Child Dev 54:424–435

Schachar R (1986) Hyperkinetic syndrome: historical development of the concept. In: Taylor E (ed) The overactive child. (Clinics in developmental medicine 97), Blackwells, Oxford, pp 19–40

Schaffer HR (1966) Activity level as a constitutional determinant of infantile reaction to deprivation. Child Dev 37:595–602

Skuse D (1985) Extreme deprivation in early childhood: A reply. J Child Psychol Psychiatry 26:827–828

Taylor E (1983) Measurement issues and approaches. In: Rutter M (ed) Developmental neuropsychiatry. Guilford, New York, pp 239–255

Taylor E (1986) The causes and development of hyperactive behaviour. In: Taylor E (ed) The overactive child. (Clinics in developmental medicine 97), Blackwells, Oxford, pp 118–160

Taylor E (1988) Attention deficit and conduct disorder syndromes. In: Rutter M, Tuma AH, Lann IS (eds) Assessment and diagnosis in child psychopathology. Guilford, New York, pp 377–407

Tennes K, Kreye M, Avitable N, Wells R (1986) Behavioral correlates of excreted catecholamines and cortisol in second-grade children. J Am Acad Child Psychiatry 25:764–770

Trevarthen J (1986) Neuroembryology and the development of perceptual mechanisms. In: Falkner F, Tanner JM (eds) Human growth. 2nd edn. Academic, New York, vol. 2, pp 361–383

Von Gontard A (1988) The development of child psychiatry in 19th century Britain. J Child Psychol Psychiatry 29:569–588

Werner E, Smith R (1977) Kauai's children come of age. University of Hawaii Press, Honolulu

Considerations for the Study of Event-Related Brain Potentials and Developmental Psychopathology

A. J. Lincoln, E. Courchesne, and R. Elmasian

Introduction

The focus of this chapter is on neurophysiological indices of developmental psychopathology. The scope of the discussion will be limited to neurophysiology as measured by the event-related brain potential (ERP) and will not include consideration of other neurophysiological techniques which have also been employed in the study of cognition, development, and psychopathology. The discussion will be largely theoretical insofar as general issues related to the study of ERPs and developmental psychopathology will be discussed. Two early childhood disorders, infantile autism and receptive developmental language disorder (R-DLD), will be discussed as a way of exemplifying pertinent issues.

Autism and Receptive Developmental Language Disorder

Autism and R-DLD are both developmental disorders that seriously compromise a child's capacity to function effectively in his or her environment. Autism is a more pervasive and significantly handicapping disorder than R-DLD. Unlike children with R-DLD, children with autism have both severe social (Kanner 1943) and intellectual deficits (Rutter 1978; Lincoln et al. 1988). However, both disorders involve delayed and deviant speech and language development (Cohen et al. 1976; Menyuk 1978). In autism, the pragmatic foundations of language (i.e., reciprocity, prosody, gaze, etc.) are delayed and impaired but in R-DLD they remain essentially intact. Furthermore, unlike R-DLD children, autistic children demonstrate deficits in imitation, the use of gesture for communicative intent, and representational play. They often manifest immediate and delayed echolalia, pronoun confusion, and atypical word usage. Thus, though both disorders present clinically with early delays in speech and language development, and even though this sometimes makes differential diagnosis difficult during early childhood because both disorders are currently diagnosed exclusively according to behavioral criteria (American Psychiatric Association 1987), they nevertheless are fundamentally different in the qualitative manifestation of psychopathology and this difference typically becomes more apparent across early development.

Some theorists interpret the differences in the manifestation of the language impairment in these two disorders as being due to a qualitatively similar dysfunction that is more severe in autistic than in R-DLD children (Churchill 1972). Others have suggested different underlying neuropathologies (Cohen et al. 1976). Descriptive behavioral analyses alone may not be sufficient to draw either conclusion, or to establish which neurophysiological mechanisms are impaired in these disorders. A methodology that employs both behavioral and neurobiological techniques would be useful in further understanding each disorder. In addition, because both disorders are developmental disorders, there is also the need to better understand the relationship of their respective brain–behavior interactions with those observed in normal development.

While it is essential to understand the problems of (a) the neurophysiological mechanisms impaired in persons with autism and R-DLD and (b) how their neurophysiology varies from normal development, there is also the need to better understand the neurophysiology as it pertains to the time course of each disorder (Garber 1984; Rutter 1983). Unlike young nonretarded autistic children who have severe delays and impairments in language and social development, older non-retarded adolescents and adults with autism are generally able to communicate, although in an abnormal manner. They are able to read, compute mathematical problems, and perform a variety of tasks. Such tasks include those found on measures used to test intelligence (Lincoln et al. 1988), independence of most self-help skills, employment, and use of public transportation. For example, one autistic individual was described by his parents as being completely echolalic and avoidant of people at 6 years of age. He was enrolled in a private school for autistic children. However, by his 18th birthday he could engage in basic conversation, tested in an average range on measures of nonverbal intelligence (Lincoln et al. 1988, see Table 3), used public transportation independently, and repaired bicycles.

Children with early-onset, severe R-DLD are sometimes misdiagnosed as having autism. At young ages the R-DLD child's capacity both to understand and to communicate with words is extremely impaired. Pragmatic functions are generally better preserved while phonological, syntactic, and semantic functions are impaired. However, children with R-DLD may also demonstrate improved speech and language ability as they grow older (Cohen et al. 1976; Paul et al. 1983; Stark et al. 1984). One girl with R-DLD who participated in our research was described by her mother as being virtually noncommunicative by 5 years of age. She was enrolled in classes for children with severe disorders of language. By junior high she was placed in classes for teenagers with learning handicaps, and when she graduated high school she had been mainstreamed into some nonacademic courses. Although speech and language testing placed her as having a language age at about half her chronological age, she nevertheless was able to carry on conversations, follow verbal directions, speak coherently, and read. After graduating she became employed as a nanny.

Thus, it appears that autistic disorders and R-DLD may be different or less pervasive during later periods of development than during early development.

The possibility then arises that there are also changes in the impaired aspects of neurophysiology over the course of development in each disorder.

Brain–Behavior Relationships and Childhood Psychopathology

The study of childhood psychopathology and the understanding of brain–behavior relationships in childhood psychopathology are more complicated than the study of adult psychopathology. The reason for this involves, in part, the complex relationship and issues of continuity between etiological factors and symptoms associated with childhood psychopathology and their similarities to and differences from etiological factors and symptoms in adult psychopathology (Garber 1984). At a more general level it also involves the impact of maturation on brain–behavior relationships (Taylor 1984; Rutter 1983).

We know little about the continuity between childhood and adult psychopathology (Garber 1984). For example, are the neurophysiological factors related to symptomatology in autistic children the same as the neurophysiological factors related to sypmtomatology in autistic adults? Would we expect to find similar abnormal neurophysiological mechanisms in 10-year-olds and 40-year-olds with autism. Similar abnormalities might suggest continuity of the physiology related to the psychopathology, while differences might suggest different physiological mechanisms impacting each group of individuals or physiological mechanisms which change across time.

There is a growing body of knowledge related to the developmental time course of ERP components, thus making them reasonable candidates for investigating evidence of neurophysiological patterns of continuity. The ERP technique allows one to examine the dynamic, moment-to-moment activity of numerous neural systems ranging from sensory to cognitive to motor. There are systematic changes in the physiology of the ERP as the brain matures from infancy to adulthood (Courchesne 1978, 1979, in press; Kurtzberg et al. 1983; Mullis et al. 1985). Some components of the ERP change substantially over this span of development while others remain relatively stable. These involve changes in the amplitude and latency of some components (N1, P2, Nc, P3b, and Slow Wave), while the amplitude and latency of other components remain amazingly stable (A/Pcz/300 and ABER waves I-V).

The problem of continuity in psychopathology is even broader and extends to current issues about continuity in normal development. Cognitive stage theory is a discontinuous model of cognitive development (Brainerd 1978). This has been suggested to explain why behavioral measures of intellectual ability are poorly correlated with measures taken during childhood and adulthood (McCall 1981; McCall et al. 1972; Sigman et al. 1981). Although the infant, child, and adult all manifest intelligent behavior, the operational definitions and behavioral measurements of intellectual ability are radically different over this time period.

Another important consideration with respect to childhood psychopathology is the idea of the developmental lesion. In this case the term "lesion" refers to one or more biological process felt to interfere with normal brain development and ultimately related to the development of psychopathology. The lesion can be analogous to the concept of a genotype—present, but only manifesting its phenotypic impact at some phase of development (Nonneman 1984; Goldman 1976). The lesion can also potentially subtly influence the course of physiological processes. It is known that some brain lesions that occur early in development result in aberrant pathways which would not normally have developed (Steward 1984).

The locus, extent, and time of onset of a lesion are also factors which might influence the behavioral manifestation of cognitive pathology (Schneider 1979). For example, lesions to the caudate nucleus and mid-dorsal thalamus result in similar impairment in both infant and juvenile monkeys (Goldman 1974; Nonneman et al. 1984). In contrast to these subcortical lesions, infant monkeys with dorsolateral cortex lesions demonstrate early sparing of function, but later in development become impaired (Nonneman et al. 1984). Subtotal lesions produced more sparing of function in infant rats than in adults rats (Simons and Finger 1984). It has also been demonstrated that infant and adult rats that have complete destruction of specific cortical areas (sensorimotor cortex) are equally impaired on a tactile battery of tests and are impaired in the acquisition of various learned behaviors (Simons and Finger 1984).

It is also possible that the effects of a lesion will only become apparent when combined with other internal or external factors. For example, abnormally developed vermal lobules VI and VII, which Courchesne et al. (1988) recently reported as being hypoplastic in the autistic child, may remain dormant until combined with other factors, upon which the observed symptomatology is produced. On the other hand, it is possible that the abnormality of the cerebellar vermis is never dormant, but instead slowly influences brain organization and the otherwise delicate homeostasis of neurophysiological and neurochemical relationships. Over time this influence may produce a deviation from a normal pattern that is large enough to be behaviorally observed and categorized as disordered functioning. There might also be critical periods during development when various factors have to be present for the disorder to become "fully" behaviorally expressed.

It is also possible that a lesion creates secondary and tertiary effects that extend well beyond the primary ones (Rutter 1983). Remote neural loss due to a primary lesion would be one example. It is also possible, for example, that the detection of an abnormal physiological brain process (e.g., Nc) that relates to aberrant information processing (e.g., arousal and orienting) might in an adult have little impact on the adult's current functioning and only really matter when the adult is pushed to the limit in some type of information processing experiment. We thus could find evidence of the lesion, but except for special circumstances that lesion is no longer really very important. On the other hand, the same lesion may have had a profound impact on the individual's functioning at younger ages when the brain was more dependent on physiological processes being normal to promote effective brain organization and development.

It is also possible that the impact of the lesion on physiological processes early in development may have contributed to the rest of the otherwise normal brain processing abnormal information, and thereby led to severe secondary or tertiary effects on learning. For example, the abnormality found in the cerebellar vermis and hemispheres of autistic adolescents and adults (Courchesne et al. 1988; Murakami et al. 1989) may well have had an impact on the development of brain stem and cortical regulation of attention, orienting, and arousal. In this regard the manifest pathology is the result not only of a lesion that causes abnormal brain physiology and processing, but also of abnormal learning that over time interferes with the on-going development of cognitive associative processes and thought.

ERPs and Developmental Psychopathology

In the study of psychopathology and particularly developmental or childhood psychopathology, knowledge about ERPs with respect to their (a) developmental time course, (b) associated sensory, cognitive, and behavioral states, and (c) scalp topography makes it possible to developmentally assess related behavioral and pathophysiological processes. These include:

1. The assessment of brain maturation in terms of developmental changes of specific ERP components (see discussion below)
2. The assessment of normality and pathology of sensory information processing within and between modalities
3. The assessment of sensory involvement versus high-order cognitive processes (i.e., as manifested in "exogenous," stimulus-bound components, versus less stimulus-dependent, endogenous ERPs)
4. The assessment through cross-validation of concurrent behavioral measures which can be employed as a way to operationalize the sensory or cognitive functions which the ERPs are experimentally designed to measure (e.g., target detection, attention)
5. The normality or pathology of the topographical distribution of ERP components

For the purpose of this review, we will focus on two components of the ERP, the P3b and the Nc. The P3b has been studied extensively over the past 20 years and much has been learned about its physiology and its relationship to cognitive events (Pritchard 1981; Hillyard and Picton 1987). The P3b is an endogenous physiological event insofar as it is not dependent on the physical characteristics of the stimulus per se. The P3b is most typically elicited by the recognition of a low-probability or an unexpected event occurring within the context of high-probability background events. In adults, during auditory conditions, its peak amplitude occurs beteen 280 and 400 ms after stimulus onset. Its latency becomes progressively shorter from early childhood until the middle of adolescence (Courchesne 1983). The amplitude of P3b in standard "odd-ball" target detection tasks (e.g., detection of the less frequent or deviant event) is largest at Pz.

The Nc component, a frontally located, negative potential, has not been as thoroughly studied as the P3b. Like the P3b, the Nc can be elicited by a low-probability event that occurs in the context of high-probability background events, even if the low-probability event is the omission of a stimulus (Courchesne et al. 1989). Thus, it also appears to be an endogenous component of the ERP. The latency of the peak amplitude of Nc overlaps that of P3b in adults. Its amplitude and latency also change with development (Courchesne 1977, 1983; Kurtzberg 1985; Karrer and Ackles 1987; review: Courchesne 1983), with a different time course from that of P3b. The Nc emerges in infancy prior to the appearance of P3b (see below).

Although the Nc and P3b occur under similar stimulus conditions, they differ in many ways. Furthermore, in factor analytic studies we found evidence that the P3b and Nc waves had their highest factor loadings on different factors (Lincoln et al. 1987). We also found that in autistic adolescents and adults the Nc is extremely small or absent, while the P3b of these individuals is present, at the appropriate latency, although somewhat reduced in amplitude (Courchesne et al. 1989).

The Nc wave is discernible in infants as young as 6 months and can also be measured in adults. However, its amplitude and latency decrease with increasing age (Courchesne et al. 1987). The P3b, on the other hand, is not apparent until about 3 years of age, and its amplitude and latency also decrease as the individual gets older (Courchesne et al. 1987). The labeling of these components as Nc or P3b over the course of development does convey an assumption of continuity in the face of substantial changes in both their amplitude and latency. ERP studies of developmental change have been primarily cross-sectional, and there are few longitudinal studies (Courchesne and Yeung-Courchesne 1988). There have been no studies that report intrasubject correlations of P3b or Nc from the same experiment employed across two or more ages. Thus, if there is continuity of Nc and P3b generators, they nevertheless have different operational definitions in terms of latency over the span of development. Although it is an open question as to whether Nc and P3b are continuous or discontinuous physiological events, the existing data are supportive of the contention that they are continuous developmental processes; but this issue needs further research.

By using ERPs we can evaluate whether differences in physiology that we find in autistic and R-DLD children reflect abnormal patterns of development, and whether such differences are more uniquely associated with each disorder. Furthermore, by comparing older and younger individuals with the same disorder, we can find whether there is evidence of continuity in abnormal aspects of the neurophysiology. Finding such continuity would be important in understanding the neurobiological basis of the disorder.

There is a paucity of ERP studies of these two disorders. There are a few studies of information processing in autism and only one in R-DLD (Review: Courchesne and Yeung-Courchesne 1987). There are no published developmental ERP studies of these two disorders. Thus, there is a significant gap in our understanding of developmental patterns of physiology in both autism and R-DLD.

ERP Findings in Autism

Adolescent and adult high-functioning individuals with autism have consistently been found to have smaller P3b amplitudes in response to auditory than to visual odd-ball target stimuli in our laboratory and in others (Novick et al. 1979, 1980; Dawson et al. 1988; Ciesielski et al. 1990; Courchesne et al. 1984, 1985b, 1989). There is no published information on whether this is the case for high-functioning children with autism. The P3b elicited by omitted target events are also smaller than normal in adolescents and adults with autism. In this paradigm subjects are asked to press a button when they detect that a stimulus has been omitted from a continuous sequence of stimuli. Thus, they are responding to the absence of a stimulus, and the associated P3b reflects purely endogenous processes associated with the decision of stimulus omission. This finding suggests that the abnormally small P3b amplitude is not the result of abnormal functioning of sensory mechanisms per se. The frontal-central A/Pcz/300 in response to novel auditory stimuli was also smaller in amplitude than in normals (Courchesne et al. 1984). This might suggest impaired arousal or orienting mechanisms. In autistic subjects the latency of P3b and A/Pcz/300 did not differ from normal.

Perhaps the most striking finding in our ERP studies of autism is the abnormally small and often absent Nc component (Ciesielski et al. 1990; Courchesne et al. 1985b, 1989) which occurs 300–600 ms after a target or novel stimulus and is largest in amplitude at frontal electrode sites (see Fig. 3). This finding is particularly interesting since the Nc is an attention-related component that appears early during the course of human development (Courchesne et al. 1981; Karrer and Ackles 1987; Kurtzberg et al. 1983). The abnormal reduction of Nc in adolescents and adults with autism might signify a neurophysiological generator or generators that have been abnormal since birth. Since Nc may be related to mechanisms of arousal and attention (Courchesne 1977, 1978, in press; Ciesielski et al. 1990), it is possible that an early impairment of such mechanisms may have a cumulative impact on cognitive and language development (Courchesne 1987; Ornitz 1983; Rimland 1964). The presence of the Nc abnormality in adolescence and adulthood might be a sign of continuity of pathophysiology between infancy and adulthood. P3b, on the other hand, has a somewhat later developmental onset than does Nc. Thus, it might be less likely to show abnormal physiological continuity since its onset follows the onset of symptoms associated with autism. The P3b abnormality could be a secondary phenomenon.

ERP Findings in R-DLD

Event-related brain potentials of adolescents with R-DLD were studied in our laboratory under experimental paradigms identical to those we have used to study autistic individuals. We have completed two experiments, one which is a standard odd-ball target detection paradigm, and another which manipulates levels of probability and interstimulus interval. In both sets of experiments we found large auditory P2 and auditory P3b amplitudes in R-DLD adolescents, in contrast to the abnormally small P3b amplitude in our autistic subjects (Courchesne et al. 1989; Adams et al. 1987, unpublished). We also found that the auditory P3b response to an omitted target stimulus was larger in the R-DLD subjects than in either autistic or normal control individuals. This latter finding suggests that the larger amplitude of P3b in R-DLD was not due to the acoustic characteristics of the stimuli. This P3b effect was found only during auditory tasks, not visual ones. We also found that Nc was present and of equivalent amplitude in R-DLD adolescents when compared with age-matched normal controls. Finally, we have not observed any latency differences in R-DLD between autistic and normal control subjects (see Fig. 3).

Developmental Changes in ERPs of Autistic and R-DLD Children

We have recently completed a preliminary study on non-retarded children with autism and R-DLD that employed methodology identical to the Courchesne et al. (1989) study mentioned above. In this pilot study, we did not include an omitted stimulus and only required our subjects to press a button in response to either an auditory target ($P = 10\%$; 1000 or 2000 Hz) or a visual target ($P = 10\%$; red or blue slide). The auditory and visual conditions were presented separately. In the auditory conditions target stimuli occurred in a context of 90% standard stimuli (1000 or 2000 Hz depending on the target frequency). In the visual condition target stimuli occurred in a context of 90% standard stimuli (red or blue slide depending on the color of the target). Conditions and stimuli were counterbalanced across subjects.

In this pilot experiment, we examined the ERPs of six autistic children, six dysphasic children, and eight normal children. These children ranged in age from 7 years 11 months to 12 years 11 months. There were no mean age differences among the three groups. All of the autistic and dysphasic children fully met the diagnostic criteria for either autism or R-DLD according to the DSM-III-R

(American Psychiatric Association 1987). All autistic children had been seriously socially, linguistically, and cognitively impaired early in their development (by 30 months). They did not have normal (a) peer relationships, (b) social discourse skills, (c) interaction with parents or siblings, or (d) communication skills. Their communicative skills were atypical and included pragmatic impairment, immediate and delayed echolalia, pronoun confusion, and perseverative insistence in talking about idiosyncratic and highly egocentric topics. These autistic children all experienced distress over trivial changes in their environment, and some expressed repetitive and stereotypic interactions and behavior with people or objects. All of them had a current and past history of normal hearing examinations. None had any other diagnosed psychiatric, medical, or neurological disorder. All were in public school classes for autistic or severely communicatively disabled youngsters, and all lived at home and were clients of a state agency for developmentally disabled youngsters because of their diagnosis of autism.

The R-DLD children all demonstrated delayed speech and language development, but were normal in their social, cognitive, and motor development. In all cases these children were observed to be experiencing significant speech and language delay by 4 years of age. The delayed speech and language development was not related to a hearing disorder or expressive speech disorder. None of the children came from bilingual environments. None had any other diagnosed psychiatric, medical, or neurological disorder. All of the dysphasic children were in public school classes for severely communicatively disabled youngsters and lived at home. The normal control children were all in regular public school classes. None had a history of language or hearing problem. None of the children came from bilingual environments. None had any other diagnosed psychiatric, medical, or neurological disorder.

The children in all three groups had Wechsler Intelligence Scale for Children—Revised (Wechsler 1974) performance IQs greater than 70. Thus, they could all be considered to be nonretarded. However, the three groups were not equal in performance IQ (autistic mean = 88.3, SD = 12.8; dysphasic mean = 95.7, SD = 9.5; and normal control = 109.7, SD = 10.9; $F = 6.7$, $P = 0.007$). In spite of the differences among the three groups in the level of their performance IQs, they nevertheless were not significantly different from one another in their RT performance on the ERP task in either the visual or auditory conditions (visual: $F = 1.9$, $P = 0.175$; auditory: $F = 1.6$, $P = 0.219$). Thus, each of the three groups of children behaviorally performed at a comparable level in speed of target detection, suggesting that their differences in IQ did not interfere behaviorally with respect to task performance.

It is interesting to note, however, that the wave shape of Nc was very similar across studies involving children of a similar age (see Figs. 1, 2; 12-year-olds; Courchesne 1983) and the normal children in the present study. We also find it interesting that the component latency of Nc and P3b appears to be similar in children of similar ages who were previously studied (see Figs. 1, 2; 12 year-olds; Courchesne 1983) and those children in the present study. Amplitude differences,

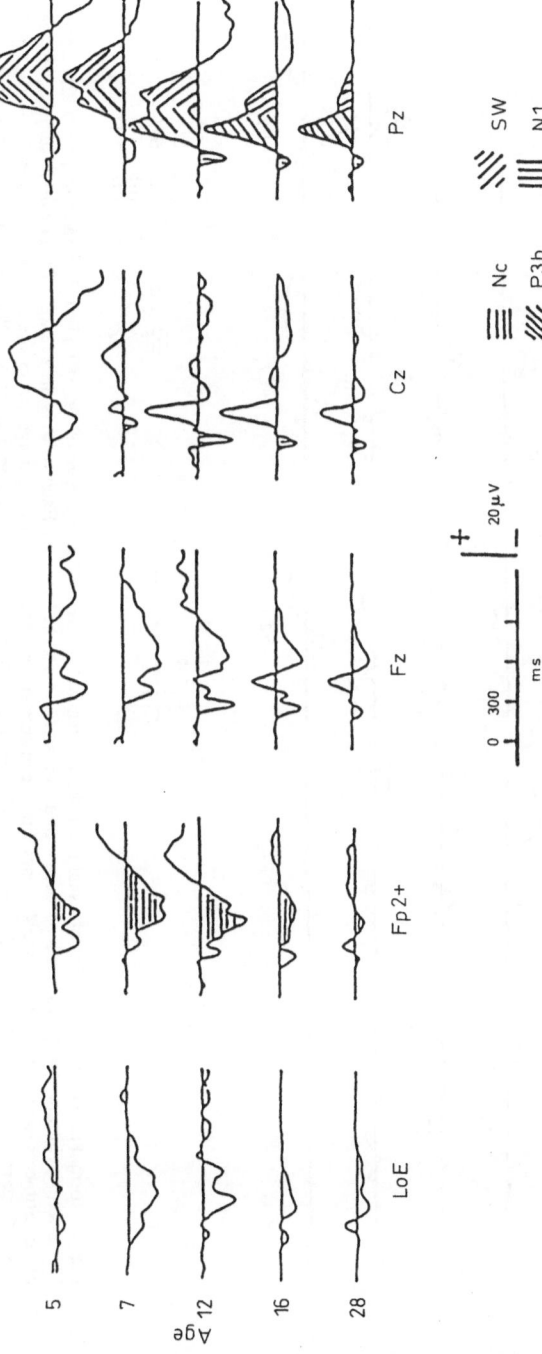

Fig. 1. Normal development patterns of auditory Nc, P3b, slow wave (*SW*), and N1 components of the ERP elicited by random ($P = 0.12$) targets (the word, YOU) contained within a series of background stimuli ($P = 0.76$; the word, ME). Each trace is the average of ten or more subjects per age group. Subjects ranged in age from 4 to 44 years. The mean ages and number of subjects per age group are: age 5 ($\bar{x} = 4.8$ years, n = 10); age 7 ($\bar{x} = 7.6$ years, n = 13); age 12 ($\bar{x} = 12.1$ years, n = 12); age 16 ($\bar{x} = 16.6$ years, n = 14); and age 28 ($\bar{x} = 28.8$ years, n = 13). (Adapted from Courchesne 1983)

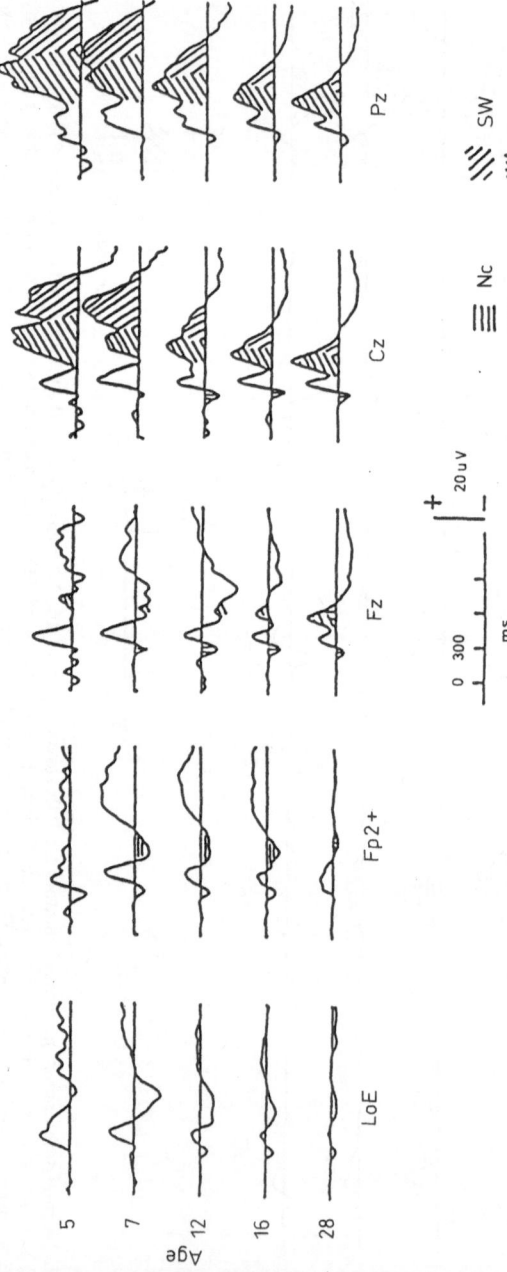

Fig. 2. Normal development patterns of visual Nc, P3b, slow wave (*SW*), and N1 component of the ERP elicited by random ($P = 0.12$) targets (the letter, A) contained within a series of background stimuli ($P = 0.76$; the letter, B). Each trace is the average of ten or more subjects per age group. Subjects ranged in age from 4 to 44 years. The mean ages and number of subjects per age group are: age 5 ($\bar{x} = 4.8$ years, n = 10); age 7 ($\bar{x} = 7.6$ years, n = 13); age 12 ($\bar{x} = 12.1$ years, n = 12); age 16 ($\bar{x} = 16.6$ years, n = 14); and age 28 ($\bar{x} = 28.8$ years, n = 13). (Adapted from Courchesne 1983)

on the other hand, might be more reflective of differences in the subject populations or task demands (i.e., diagnosis, age variability, IQ variability, specific task demands, stimuli, and experimental design).

However, in spite of similar behavioral performance and similar ERP latencies, we found significantly different ERP amplitudes among the three groups of children. Compared with both our R-DLD and normal control subjects, there were smaller Nc amplitudes in the autistic subjects at frontal electrode sites (Fz) in response to auditory but not visual targets (autistic mean = $-0.7\,\mu$V, SD = 3.9; R-DLD mean = $-9.3\,\mu$V, SD = 4.3; and normal control mean = $-7.5\,\mu$V, SD = 5.1; $F = 8.4$, $P = 0.003$; see Fig. 4).

Though there were significant amplitude differences for Nc, the latencies of Nc were quite similar across the three groups, and fell within a window previously reported for children of these ages (see Figs. 1, 2). This effect of Nc is analogous to the Nc effect we found in older adolescents and adults with autism and R-DLD,

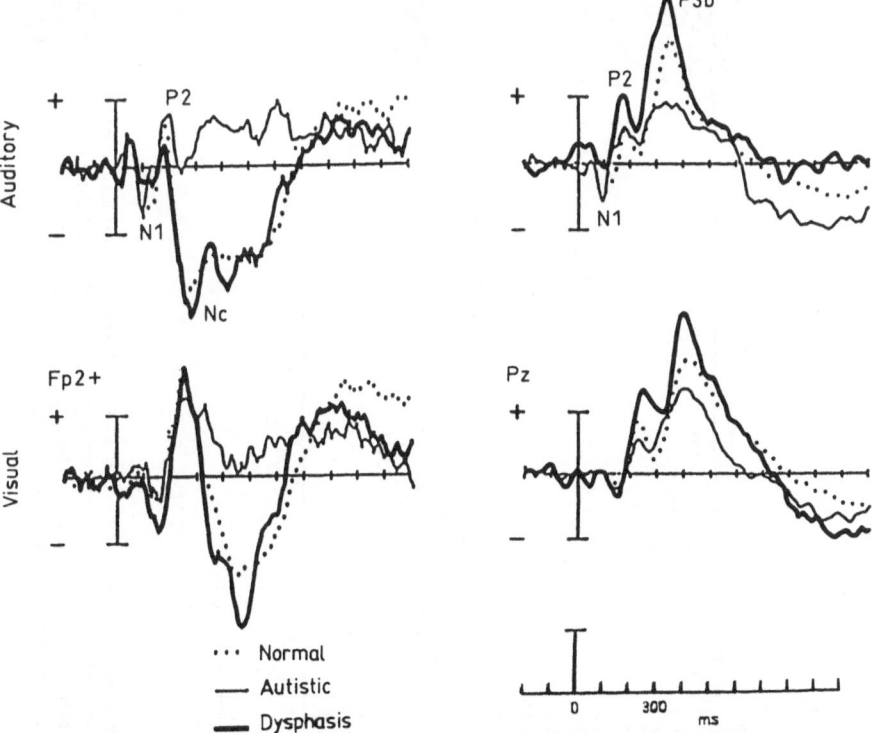

Fig. 3. The averaged auditory and visual ERPs of dysphasic, autistic, and normal groups of adolescents and young adults as they detected target events. Auditory target stimuli were either a randomly presented 1000- or 2000 Hz tone ($P = 0.10$) contained within a series of standard 1000- or 2000-Hz tones ($P = 0.90$). Auditory and visual tracings at Fp2 + are scaled to $5\,\mu$V each side of the baseline. Auditory and visual tracings shown at Pz are scaled to $10\,\mu$V each side of the baseline. Note the absent Nc component at Fp2 + and the smaller than normal amplitude of P3b in the autistic subjects. (Adapted from Courchesne et al. 1989)

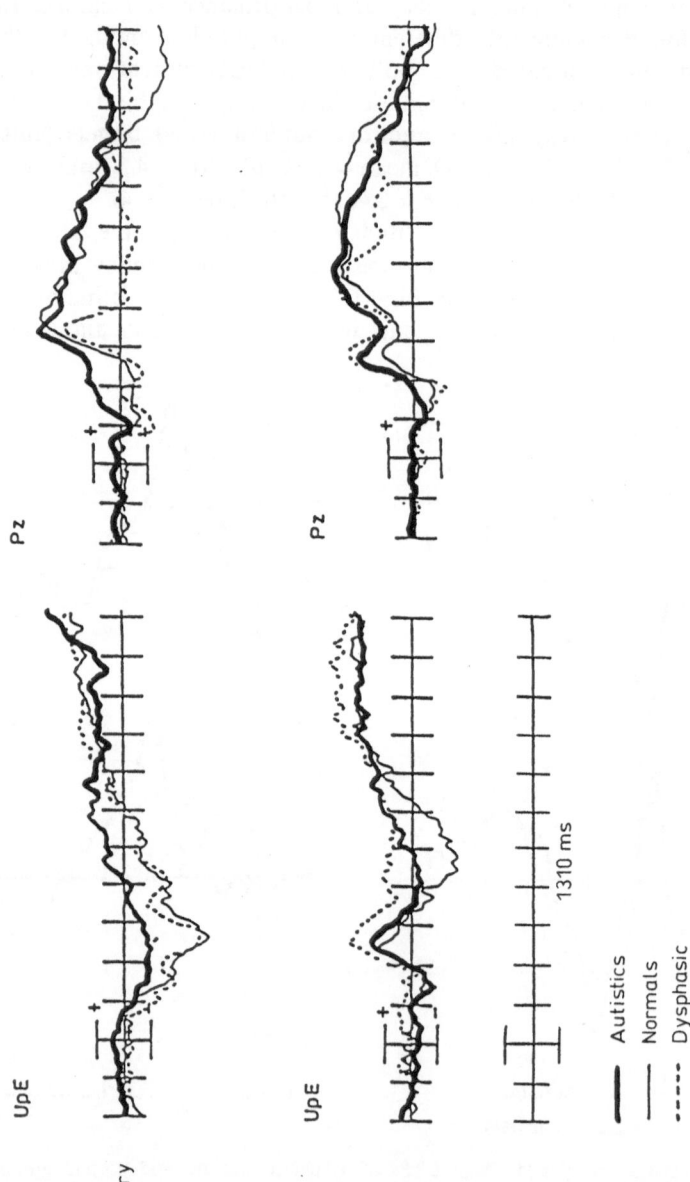

Fig. 4. The averaged auditory and visual ERPs of dysphasic, autistic, and normal groups of children as they detected target events. Auditory target stimuli were either a randomly presented 1000- or 2000-Hz tone ($P = 0.10$) contained within a series of standard 1000- or 2000-Hz tones ($P = 0.90$). Note the small Nc component at UpE (Ep2 +) and the normal amplitude of P3b in the autistic subjects. Each division represents 100 ms. Auditory and visual tracings are scaled to 6 μV each side of the baseline

e.g., small auditory Nc amplitude and normal latency in autism (Courchesne et al., in press). This finding suggests that Nc may be a candidate for studying the continuity in the pathophysiology of autism.

However, this preliminary study did not find the effect of P3b in either the autistic or the R-DLD children. The three groups did not significantly differ among each other in their P3b amplitudes at parietal electrode sites (auditory: $F = 1.6$; $P = 0.215$; visual: $F = 0.323$, $P = 0.733$; see Fig. 4). The latency of P3b was similar across the three groups, and fell within a window expected for children of these ages (see Figs. 1, 2).

The absence of a smaller P3b amplitude in the autistic group and of a larger P3b amplitude in the R-DLD group raises the question of whether the cognitive functioning reflected by P3b might change during development in these disorders. The latency of P3b appears to be more stable across this developmental time span than the amplitude. This is an area that needs more research.

It appears possible that the behaviors related to the formation of a diagnosis (autism or R-DLD) are related to amplitude differences of the Nc among the children we studied, while the latency of the Nc and associated RTs are not affected in a similar way.

Conclusions

In this chapter we have attempted to discuss and exemplify the complexities involved in the study of brain–behavior relationships as they pertain to cognitive aspects of childhood psychopathology. The methodological issues addressed include: (a) the use of ERPs to assess development and psychopathology; (b) the issue of developmental continuity in cognition and psychopathology: (c) the effect of a "lesion" on development, brain maturation, and function; and (d) the long-term effects of abnormal brain functioning on learning and the development of cognitive abilities. It will be difficult for the neuroscientist wishing to establish brain–behaviour relationships and/or continuity in brain–behavior relation-ships for some types of psychiatric disorder. The difficulty is that the psychiatric disorder is operationally defined by behavioral criteria which might not become evident until some particular time point in development, and in some cases only long after critical intervening factors have taken place. It is also possible that the behavioral criteria used to assess psychiatric/cognitive impairments are not effective enough, and therefore the effects of the lesion are never really measured (Steward 1984). How does one know which subjects should be studied to evaluate whether or not the premorbid neurophysiology or "lesion" was ever present?

We report evidence that children with autism appear to show some neurophysiological differences from normal and R-DLD children, and that these differences are also found in adolescents and adults with autism. Both younger and older autistic individuals have smaller Nc waves than younger and older

R-DLD or normal individuals. This suggests continuity in the pathophysiology of Nc between older and younger individuals with autism. Such continuity is interesting since Nc is the earliest endogenous component of the ERP observed. The auditory P3b, on the other hand, appears normal in younger autistic individuals, but is reduced in amplitude in older autistic individuals. The opposite appears true for individuals with R-DLD. Younger R-DLD individuals have small auditory P3b waves and in older R-DLD individuals they are of higher amplitude. Thus, the amplitude of the auditory P3b does not appear continuous throughout this span of development. There may, therefore, be different mechanisms affecting the physiology of these two ERP components, in spite of the fact that they are both elicited under similar experimental conditions.

The specific brain systems responsible for generating ERP components such as P3b and Nc are not known. However, Courchesne (1987) has speculated that developmental changes in the latency of Nc may be related to myelination of thalamocortical tracts. Amplitude changes may be related to synaptogenesis and increasing specificity of synaptic connections. In adolescents and adults with autism, for example, there is an absent or very small Nc, and P3b is typically about a third of the amplitude of normals. The latency of these components, nevertheless, appears to be similar to that in normals. Thus, there may be normal myelination of the generators and systems related to the generators of Nc and P3b, while synaptogenesis related to the generators of these components may (a) never have developed and be hypoplastic, (b) be overly specific, with fewer than normal appropriate neuronal connections, or (c) have atrophied. In a disorder such as autism, it is possible that brain generators which should produce an Nc early in development were never properly functioning, and that behavioral features associated with this physiological event (arousal and orienting) were also faulty (Courchesne 1987).

Further research is needed to determine whether the abnormally small Nc of the autistic children we studied can be found in even younger autistic children. Further research is also needed to determine when in development the P3b of autistic individuals begins to differ in amplitude from that of normals.

These results lead to the question of whether the generators of the Nc have been functioning abnormally in the autistic individual since birth. The abnormal functioning of Nc generators in autistic individuals is perhaps associated with interference in effective information processing. Over time such interference could have a cumulative effect on the development of symptoms and cognitive abilities associated with autism. These findings represent a preliminary, first step in establishing a relationship between the abnormal physiology of the Nc in younger and older individuals with autism.

Acknowledgments. This work was supported by NINCDS grant R01-NS26814-01 awarded to Alan Lincoln. Special thanks to Mark Allen for his help in recruiting subjects and assisting with experiments. The senior author wishes also to thank the San Diego Regional Center for the Developmentally Disabled and the San Diego Unified School District for their continued assistance and support. Finally the senior author wishes to thank Linda Lincoln, Martha Hillyard, and Stephen Hillyard for their helpful review and comments during the preparation of this manuscript.

References

Adams J, Courchesne E, Elmasian RO, Lincoln AJ (1987) Increased amplitude of the P2 and P3 components in adolescents with developmental dysphasia. In: Johnson R, Rohrbaugh JW, Parasuraman R (eds) Current trends in event-related brain potential research (EEG Suppl 40). Elsevier, Amsterdam, pp 577–583

Adams J, Courchesne E, Elmasian RO, Lincoln AJ, Akshoomoff-Haist N (in preparation) Examination of evoked potentials and behavioral performance under varying task demands in non-retarded autistic young adults

American Psychiatric Association (1987). Diagnostic and statistical manual-III (revised). American Psychiatric Association, Washington

Brainerd C (1978) The stage question in cognitive developmental theory. Behav Brain Sci 2:173–213

Churchill DW (1972) The relation of infantile autism and early childhood schizophrenia to developmental language disorders of childhood. J Autism Child Schizophr 2:182–197

Ciesielski K, Courchesne E, Elmasian R (1990) Effects of focused selective attention tasks on event-related potentials in autistic and normal individuals. Electroencephalogr Clin Neurophysiol 75:207–220

Cohen DJ, Caparulo B, Bennett & Shaywitz B (1976) Primary childhood aphesia and childhood autism. Am Acad Child Psychiatry 15:604–645

Courchesne E (1977) Event-related brain potentials: a comparison between children and adults. Science 197:589–592

Courchesne E (1978) Neurophysiological correlates of cognitive development: changes in long-latency event-related potentials from childhood to adulthood. Electroencephalogr Clin Neurophysiol 45:468–482

Courchesne E (1979) From infancy to adulthood: the neurophysiological correlates of cognition. In: Desmedt JE (ed) Cognitive components in cerebral event-related potentials and selective attention: progress in clinical neurophysiology, vol. 6 Cognitive components in event-related potentials. Karger, Basel, pp 224–242

Courchesne E (1983) Cognitive components of the event-related brain potential: changes associated with development. In: Tutorials in ERP research: endogenous components Gaillard AWK, Ritter W (eds) North-Holland, Amsterdam, pp 329–344

Courchesne E (1987) A neurophysiological view of autism. In: Schopler E, Mesibov GB (eds) Neurobiological issues in autism. Plenum, New York, pp 285–324

Courchesne E (1989) Chronology of postnatal human brain development: event-related potentials, positron emission tomography and synaptogenesis studies. In: Rohrbaugh JW, Parasuraman R, Johnson R (eds) Event-related brain potentials: issues and interdisciplinary vantages. Oxford, New York (in press)

Courchesne E, Yeung-Courchesne R (1988) Event-related brain potentials. In: Rutter M, Tuma H, Lann I (eds) Assessment and diagnosis in child psychopathology. Guilford, New York, pp 264–299

Courchesne E, Ganz L, Norcia A (1981) Event-related brain potentials to human faces in infants. Child Dev 52:804–811

Courchesne E, Kilman BA, Galambos R, Lincoln AJ (1984) Autism: processing of novel auditory information assessed by event-related brain potentials. Electroencephalogr Clin Neurophysiol Evoked Potentials 59:238–248

Courchesne E, Courchesne RY, Hicks G, Lincoln AJ (1985a) Functioning of the brainstem auditory pathway in non-retarded autistic individuals. Electroencephalogr Clin Neurophysiol 61:491–501

Courchesne E, Lincoln AJ, Kilman BA, Galambos R (1985b) Event-related brain potential correlates of the processing of novel visual and auditory information in autism. J Autism Dev Disord 15 (1):55–76

Courchesne E, Elmasian R, Yeung-Courchesne (1987) Electrophysiological correlates of cognitive processing: P3b and Nc, basic, clinical and developmental research. In: Halliday AM, Butler SR, Paul R (eds) A textbook of clinical neurophysiology. Wiley, London, pp 645–676

Courchesne E, Yeung-Courchesne R, Press G, Hesselink J, Jernigan T (1988) Hypoplasia of cerebellar vermal lobules VI and VII in autism. N Engl J Med 318:1349–1354

Courchesne E, Lincoln A, Yeung-Courchesne R, Elmasian R, Grillon C (1989) Pathophysiologic findings in non-retarded autism and receptive developmental language disorder. J Autism Dev Disord 19:1–17

Dawson G, Finely C, Phillips P, Galpert L, Lewy A (1988) Reduced P3 amplitude of the event-related brain potential: its relationship to language ability in autism. J Autism Dev Disord 18 (4):493–504

Garber J (1984) Classification of childhood psychopathology: a developmental perspective. Child Dev 55:30–48

Goldman PS (1974) An alternative to developmental plasticity: heterology of CNS structures in infants and adults. In: Stein DG, Rosen JJ, Butters N (eds) Plasticity and recovery of functions in the central nervous system. Academic, New York, pp 149–173

Goldman PS, Lewis ME (1978) Developmental biology of brain damage and experience. In: Cotman CW (ed) Neuronal plasticity. Raven, New York, pp 291–310

Hillyard SA, Picton TW (1987) Electrophysiology of cognition. In: Plum F (ed) Handbook of physiology sect I. The nervous system vol 5. Higher functions of the brain. Part 2. American Physiological Society, Bethesda MA, pp 519–584

Kanner L (1943) Autistic disturbances of affective contact. Nerv Child 2:217–250

Karrer R, Ackles PK (1987) Visual event-related potentials of infants during a modified oddball procedure. In: Johnson R, Rohrbaugh JW, Parasuraman R (eds) Current trends in event-related brain potential research (EEG Suppl 40). Elsevier, Amsterdam, pp 603–608

Kurtzberg D (1985) Late auditory evoked potentials and speech sound discrimination in newborns. Presented at the meetings of the Society for Research in Child Development, Toronto

Kurtzberg D, Vaughan H, Courchesne E, Friedman D, Harter M, Putnam L (1983) Developmental aspects of event-related brain potentials. In: Karrer R, Cohen J, Teuting P (eds) Brain and information: event-related potentials. Ann NY Acad Sci 425:300–318

Lincoln AJ, Courchesne E, Elmasian R (1987) Hypothesis testing with principal components analysis: the dissociation of P3b and Nc. In: Johnson R, Rohrbaugh JW, Parasuraman R (eds) Current trends in event-related brain potential research (EEG Suppl 40). Elsevier, Amsterdam, pp 211–219

Lincoln AJ, Courchesne E, Kilman B, Elmasian R, Allen M (1988) A study of intellectual abilities in high-functioning people with autism. J Autism Dev Disord 18 (4):505–524

McCall R (1981) Early predictors of later IQ: the search continues. Intelligence 5:141–147

McCall R, Hogarty P, Hurlburt N (1972) Transitions in infant sensorimotor development and the prediction of childhood IQ. Am Psychol 27:728–748

Menyuk P (1978) Language: what's wrong and why. In: Rutter M, Schopler E (eds) Autism: a reappraisal of concepts and treatment. Plenum, New York, pp 105–116

Mullis RJ, Holcomb PJ, Diner BC, Dykman RA (1985) The effects of aging on the P3 component of the visual event-related potential. Electroencephalogr Clin Neurophysiol 62:141–149

Murakami J, Courchesne E, Press G, Yeung-Courchesne R, Hesselink J (1989) Reduced cerebellar hemisphere size and its relationship to vermal hypoplasia in autism. Arch Neurol 46:689–694

Nonneman AJ, Corwin JV, Sahley CL, Vicedomini JP (1984) Functional development of the prefrontal system. In: Finger S, Almli CR (eds) Early brain damage vol. 2 Academic, New York, pp 139–153

Novick B, Kurtzberg D, Vaughan HG (1979) An electrophysiologic indication of defective information storage in childhood autism. Psychiatry Res 1:101–108

Novick B, Vaughan Jr. HG, Kurtzberg D, & Simson R (1980) An electrophysiologic indication of auditory processing defects in autism. Psychiatry Res 3:107–114

Ornitz E (1983) Neuroanatomical basis of early infantile autism. Int J Neurosci 19:85–124

Paul R, Cohen DJ, Caparulo BK (1983) A longitudinal study of patients with severe developmental disorders of language learning. J Am Acad Child Psychiatry 22:525–534

Prichard WS (1981) Psychophysiology of P300. Psychol Bull 89:506–540

Rimland B (1964) Infantile autism. Appleton-Century-Crofts, New York, pp 285–324

Rutter M (1978) Language disorder and infantile autism. In: Rutter M, Schopler E (eds) Autism: a reappraisal of concepts and treatment. Plenum, New York, pp 85–104

Rutter M (1983) Issues and prospects in developmental neuropsychiatry. In: Rutter M (ed) Developmental neuropsychiatry. Guilford, New York, pp 577–598

Schneider GE (1979) Is it really better to have your brain lesion early? A revision of the "Kennard Principle." Neuropsychologia 17:557–582

Sigman M, Cohen S, Forsythe A (1981) The relation of early infant measures to later development. In: Sigman M, Friedman S (eds) Preterm birth and psychological development. Academic, New York, pp 313–327

Simons D, Finger S (1984) Some factors affecting behavior after brain damage early in life. In: Finger S, Almli CR (eds) Early brain damage vol. 2 Academic, New York, pp 327–347

Stark RE, Bernstein LE, Condino R (1984) Four-year follow-up study of language impaired children. Ann Dyslexia 34:49–68

Steward O (1984) Lesion-induced neuroplasticity and the sparing or recovery of function following early brain damage. In: Almli CR, Finger S (eds) Early Brain Damage vol. 1 Academic, New York, pp 59–77

Taylor HG (1984) Early brain injury and cognitive development. In: Almli CR, Finger S (eds) Early brain damage vol 1, Academic, New York, pp 325–345

Wechsler D (1974) The Wechsler Intelligence Scale for Children-Revised. Psychological Corporation, New York

The Role of the Frontal Lobes in Child Psychiatric Disorders*

A. Rothenberger

General Considerations on Frontal Brain Functions

A large number of clinical and experimental studies primarily devoted to the investigation of the frontal cortex and its functional properties have been conducted over the last decade. This surge of interest has not been restricted to animal experiments and brain damage studies carried out in departments of neurology and neurosurgery; efforts have also been invested in adult psychiatry, a research field which has become increasingly concerned with the functional role of the frontal lobes in various psychopathological phenomena (Weinberger 1988; Malloy 1987).

Much of this work has received its impetus from the fact that frontal lobes are associated with the highest levels of cortical function, including those inherent in intellectual activity such as goal-oriented behavior and self-directed behavioral planning. Implications and the present state of research in this field have been reported by Stuss and Benson (1986, 1987), Fuster (1989), and Perecman (1987). In order to describe the organization of frontal lobes and to offer a comprehensive view of their neuroanatomical and neurofunctional structural bases, Perecman resorts to the developmental model of cerebral morphology proposed by Sanides (1964, 1971, 1972). According to this model, the frontal cortex is conceived of as having a dual developmental origin, namely the hippocampus as well as the olfactory cortex. Dorsally, the hippocampus differentiates itself through the cingulate gyrus and becomes responsible for the development of the cortex on the medial and dorsolateral surfaces of the frontal and prefrontal areas respectively. The hippocampus also lies at the origin of the dorsal premotor and precentral cortical regions. This system seems mostly involved in processing the spatial attributes of sensory information. Ventrally, the olfactory cortex differentiates into insular cortex and affects the formation of the orbitofrontal cortex, the ventrolateral surface extending up to the principal sulcus as well as the ventral premotor and precentral areas. This system seems mostly associated with emotional aspects of sensory information.

---

*This work was partly supported by the German Research Society (Deutsche Forschungsgemein-schaft, Sonderforschungsbereich 258, Universität Heidelberg).

Besides the frontodorsolateral cortex and the frontoorbital cortex, the tripartition of the frontal cortex includes the frontomesial cortex with the supplementary motor area (SMA). The latter is very important for timing in sequential motor tasks and seems to determine the right moment for starting the movement (Deecke et al. 1985).

As one attempts to characterize in general terms the frontocortical functions, it becomes obvious that they are activated especially when the cognitive tasks to be accomplished reach a high level of complexity. The first important step in performing such tasks is the elaboration of a strategy to face the situation ("what to do"), as has already been described in earlier reports on adults with frontal lobe damage (Häfner 1955, 1957; Jouandet and Gazzaniga 1979). The task is then carried out by means of the various cognitive, verbal, and motor resources of the given subject ("how to do it" and "when to do it"). The outcome is compared to the initial planning construct, an operation which implies memory processing. Therefore, the last critical step requires (a) the recognition of differences between the original plan and its psychological reality, (b) the subsequent correction of these differences, and (c) the necessary adaptation to the new state, whose adequacy lies in the capacity to process feedback information correctly (Fig. 1). The task can be adequately performed only when the whole sequence of operations takes place in the right order. A so-called short circuit, that is to say an "endless repetition" in parts of the programming loops, leads to a perseveration phenomenon which manifests itself behaviorally as inability to modify an ongoing action and deal with new information in order to adapt in an optimal manner.

Perseveration behavior has been observed in patients with superior fronto-cortical lesions when solving the Wisconsin Card Sorting Test (Heaton 1981). It was virtually impossible for them to change their response strategies spontaneously although they could even name their own error (Milner and Petrides 1984). An impressive example was noted in studies of adult monkeys with

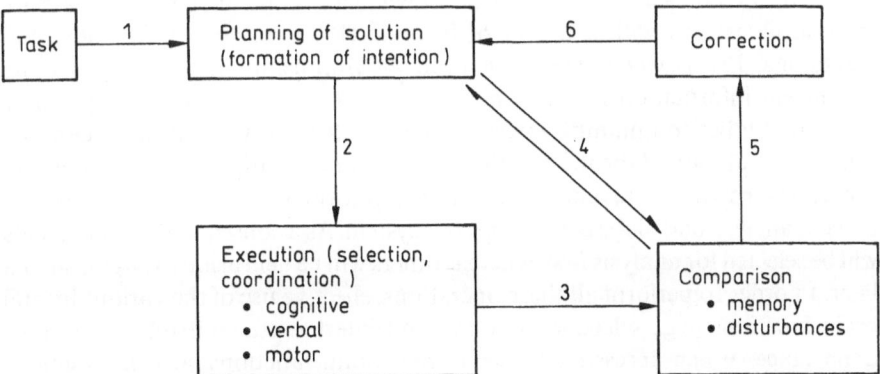

Fig. 1. Frontal brain function in coping with a given task. Adequate performance can be expected only when the whole sequence of operations is executed in the correct sequence. Short circuits or interruptions in a loop (perseveration) endanger the proper solution and learning processes

prefrontal lesions, which were observed to behave like 71/2 to 9 month old human infants when submitted to an Object Permanency Test. Human infants as well as monkeys with bilateral prefrontal lesions showed perseveration by searching for the target object in the same location even though they could see that it had been placed elsewhere by the experimenter. Looking further at the characterizing features of patients with frontal lesions, it is legitimate to assume that such behavior represents a special instance of a more general defect. In this respect, one needs only to mention deficits such as memory problems, set-shifting difficulty, errors in dealing with test instructions, difficulty in learning maze tasks (weakness in differentiating between relevant and irrelevant correction stimuli/external cues), or the abundance of impulsiveness errors in maze tests (even though the right path can be perceived from the beginning onward). It is typical of frontal lobe patients to exhibit a dissociation between ability to verbalize the demands of a given task and aptitude to employ such linguistic knowledge as a guiding tool to solve the test. A further example is seen when patients with frontal lobe damage—as opposed to patients with other localized lesions—start working with paper and pencil without awaiting any instructions and apparently in a nonpurposeful manner (with no objective in mind). All these cases seem to point to a lack of inhibition and control in these patients which is also reportedly found in their everyday behavior (Milner and Petrides 1984).

This implies that subjects with frontal brain dysfunctions have trouble compensating the pathological aspects of their behavior caused by defects affecting sites other than the frontal brain per se. As previously mentioned, the frontal cortex represents the brain area primarily involved in organizing and regulating cognitive tasks, motor behavior, and emotional responses. Hence, frontal lobes are heavily involved in a possible compensation of deficits. In this regard, a recent study (Rothenberger 1984) documented that the tics observed in children as a result of an involuntary subcortical motor dysregulation can be temporarily compensated by means of voluntary frontocortical control mechanisms. It was found that low cognitive load of interference is closely related to efficient frontocortical regulation of motor behavior. Thus, our empirical data are in good agreement with a theory of Pribram (1973, quoted by Jouandet and Gazzaniga 1979), which states that an environmental situation providing insufficient information for determining adequate behavior normally prompts the frontal lobes to minimize interference while they ensure a flexible, context-oriented regulation of the information flow. In other words, processing time and processing capacity are allocated different priorities by the frontal lobes to prevent an overload of the central nervous system. As a consequence, some events will be selected for analysis first, whereas others will be held aside to be considered later. In order to perform all these operations, effective use of the various frontal brain functions (e.g., selective attention and interference control) is of utmost importance. When accessible, other frontal brain functions such as cognitive flexibility (shifting aptitude) and capacity for goal-oriented planning (purposive movement) are surely also brought into action to assist in compensating behavioral deficits.

Development of Frontal Brain Functions

The frontal cortex is the most recently developed part of the brain in humans and primates. In humans myelinization proceeds promptly between 4 and 13 years of age. In this process, functionally equipped neuronal pathways appear to be laid out quite early, but actually become functional only after considerable time delays, when the myelinogenetic cycles of the respective neural systems reach their term (Yakovlev and Lecours 1967; Goldman and Alexander 1977; Reines and Goldman 1980). Consequently, caudate nucleus and mediodorsal thalamic lesions lead to a similar symptomatology in young and adolescent monkeys. As opposed to these subcortical lesions, however, dorsolateral lesions of the frontal cortex do not initially cause any functional disturbances in very young monkeys. Only later, when they grow up, i.e., at puberty, do they show signs of functional deficits ("sleeper effect"). This clinical manifestation of the disorder thus corresponds to a point of time at which genetic programs responsible for the function under consideration normally become activated in order to permit an adequate developmental course. This opinion was also expressed by Golden (1981), who described those cognitive capacities mediated by primary frontal brain regions as developing during adolescence and therefore stated that deficits following early frontal lesions first occur at about 12–15 years of age and do not cause any manifest cognitive impairment before this period.

Unfortunately, virtually no work is as yet available which could provide information on the development and extent of pattern organization of frontocortical functions in children and adolescents. Considering the lack of empirical evidence in this respect, the data of Passler et al. (1985), and especially the findings of Chelune and Baer (1986), become all the more important. The latter showed that 6-year-old children achieve a level of performance on the Wisconsin Card Sorting Test (WCST, a test sensitive to frontal brain functions) similar to that achieved by adults with localized frontal lesions. Moreover, it is relevant that these authors observed a certain WCST performance plateau among 8 year-olds; further, at 10 years of age, children submitted to the same test reached values that were not significantly different from those obtained by normal adults.

Frontocortical Regulatory Mechanisms in Psychiatrically Disordered Children

Preliminary Remarks

As mentioned at the beginning of the present paper, the frontal lobes are closely related in their functions to other brain areas, especially the limbic and vegetative nervous systems. Most of the projections are reciprocal ones, which implies that the prefrontal cortex not only records, compares, integrates, and converges impinging information but also exerts, in turn, some efferent neurophysiological influence on afferent systems. Keeping this functional organization in mind, I shall next report on the nature of frontocortical mechanisms in psychiatric disorders such as multiple tics, attention-deficit hyperactivity disorder (ADHD) and anorexia nervosa.

Tic Disorders and Frontocortical Control Mechanisms

Previous studies by myself and my co-workers (Rothenberger 1984, 1988a; Rothenberger and Kemmerling 1982; Rothenberger et al. 1986b) concerned the electrical brain activity registered prior to a voluntary movement in children presenting a tic disorder [in our case Gilles de la Tourette syndrome (TS) or chronic multiple tics (CMT)] and/or ADHD (according to DSM III-R). Using a neurophysiological approach, the objective of the studies was to attempt to clarify the pathogenetic relationships prevailing in tic disorders. Tics were first perceived as a motor disinhibitory phenomenon which could affect voluntary motor execution. Our main interest was to find out, on the basis of brain electrical activity, whether the preparation for a voluntary movement in children with CMT and TS differs considerably from that in healthy controls. Finally, another question considered was whether pharmacological treatment with blockers of dopamine receptors (the current treatment of choice in tic disorders) in fact exerts an influence on movement-related electrical brain activity.

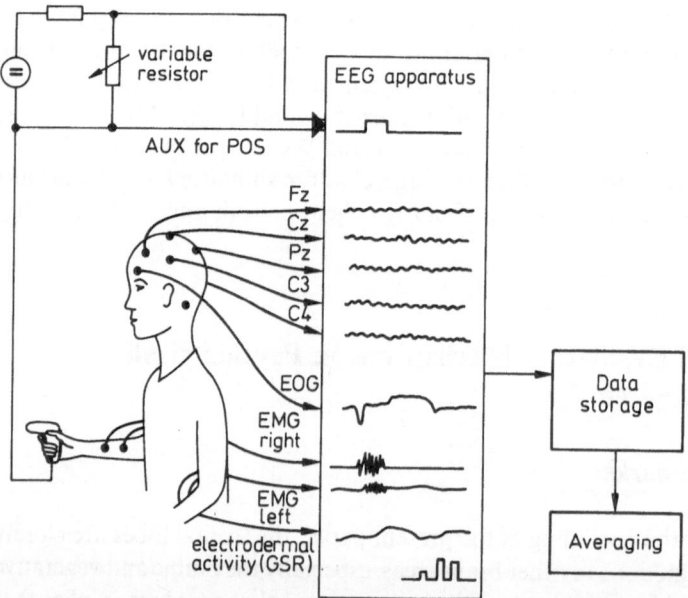

Fig. 2. Schematic diagram of the experimental set-up used in the main study

Since it turns out that about 50% of children with tic disorders also suffer from ADHD, not only were the children with tics ($n = 10$; $\bar{x} = 14.1$ years) compared with healthy controls ($n = 10$; $\bar{x} = 13.9$ years) in a pilot study, but both normal children ($n = 22$; $\bar{x} = 12.6$ years) and children with tics ($n = 22$; $\bar{x} = 12.0$ years) were compared with hyperkinetic children ($n = 22$; $\bar{x} = 12.1$ years) in a subsequent study. Six children with tics could be investigated before as well as during drug intake, a further eleven during medication only. Polygraphic electrophysiological recordings were carried out. As a motor task, the children were required to flex the index finger of their right hand every 15 s (Fig. 2). The Bereitschaftspotential (BP) was recorded over frontal (Fz), central (Cz, C3, C4), and parietal (Pz) areas, the electrodes being placed according to the international 10–20 system. The relationship between Fz and Cz leads, which is reported here, showed the most interesting results. The BP is a slowly developing negative brain potential which is closely linked to motor preparation and initiation of a voluntary movement. In normal children and adults, this negativity shows highest amplitudes (Fig. 3) over central recording sites (motor area). Aside from the above-mentioned motor task, the children also had to perform various neuropsychological tests to control for attentional performance, cognitive interference (Stroop Color-Word Test, Bäumler 1985) and task involvement factors. Intelligence was within the normal range. For group comparisons, non-parametric statistics were used.

Relationship between Voluntary Control and Frontal Brain Activity

Over frontocentral areas, differences in BP amplitudes were found between groups of examined children (Fig. 4). Normal children developed a significantly higher BP amplitude over Cz than over Fz, whereas other groups of children had a much reduced difference in BP amplitudes. This difference was not only smaller in patients, but in the case of children with tics the BP amplitude was actually higher over Fz than over Cz. The results support the hypothesis that, prior to voluntary motion, children with hypermotor behavior (TIC, ADHD) shift their neuronal activity along the midline from a central brain region to a more frontal one (since the total sum of frontocentral brain activity remained about the same). While recording the electrical brain activity along the midline preceding a movement, the relative contribution of frontal areas (Cz − Fz) was observed to be progressively larger going from the normal (NORM) to the hyperkinetic (ADHD) group, followed by children with tics (TIC) and children with tics as well as ADHD (TIC − ADHD − 1). Therefore, it can be plausibly deduced that an increase in hypermotor behavior (especially when tics and ADHD coexist) necessarily calls for more neuronal activity from frontal areas in order to perform a voluntary act adequately. Considering the studies of Deecke et al. (1985) and Lang et al. (1983), which demonstrated in adults higher levels of frontal brain activity prior to movement with increasing complexity of the motor tasks to be performed, it seems legitimate to conclude that the motor task required from our subjects was much more difficult for hypermotor children than for normal controls.

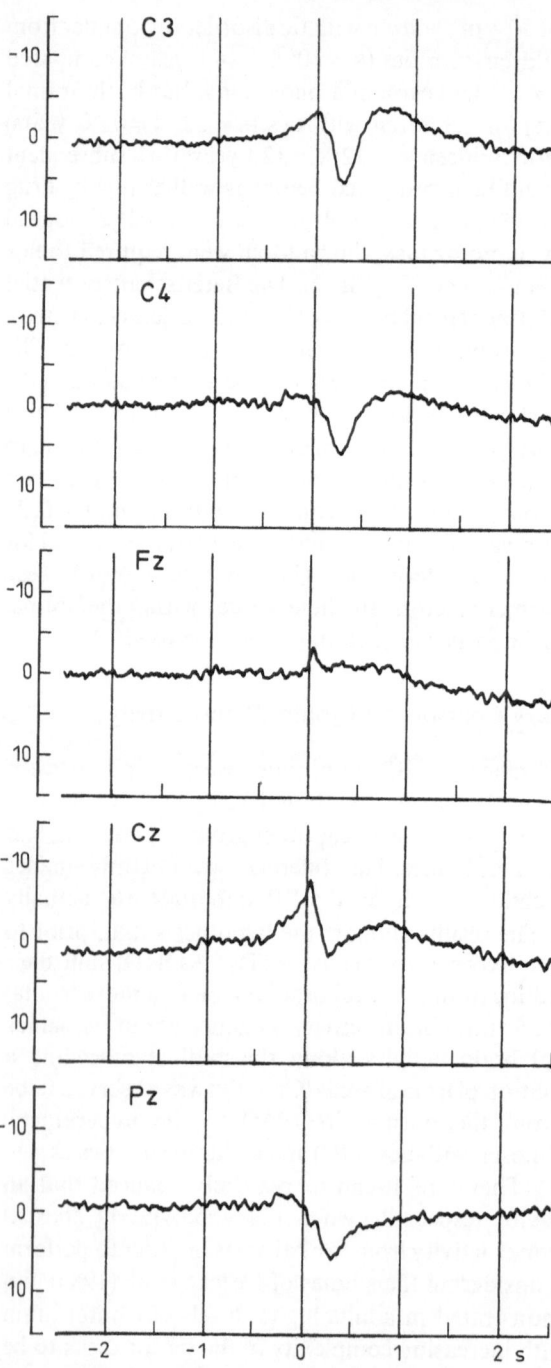

Fig. 3. Grand averages of the movement-related potentials recorded in the normal group (NORM) of children in the main study. The BP amplitude over central brain areas is, as in adults, larger contralateral to the movement (here, C3) than ipsilateral to it (C4) and is also greater over the vertex (Cz) than frontally (Fz). The movement onset is at time point zero

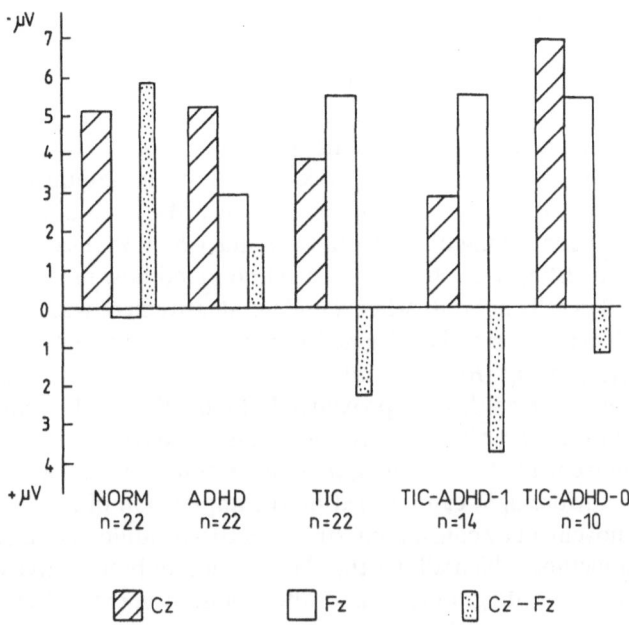

Fig. 4. Median values of the BP amplitudes at Cz and Fz as well as medians of the difference in BP amplitudes between the two sites (Cz − Fz) within various groups of subjects tested in the main study. Relative to the BP amplitude over Cz, the BP amplitude over Fz increases from the group NORM to the groups ADHD, TIC-ADHD-0, TIC, and finally TIC-ADHD-1

A lack of anterior–posterior differentiation in slow brain potentials has also been documented by Karrer and Ivins (1976) who resorted to a CNV paradigm when comparing mentally retarded adolescents with age-matched healthy controls. In respect to the subject currently under discussion, this finding is quite informative since the topography of the late CNV components (i.e., the negativity just preceding the imperative stimulus) is known normally to show a frontocentral differentiation over the midline (highest amplitude over the vertex) which is similar to the BP one (Rockstroh et al. 1982). Karrer and Ivins (1976) interpreted their results as evidence for a necessity of mentally retarded children to activate larger populations of neurons in order to perform the given task. Furthermore, according to Karrer and Warren (1979), it appears that children with mental retardation need to recruit larger frontocentral areas prior to a voluntary button press. In an analogous manner, the increased frontal brain activity in the examined hypermotor children (TIC, ADHD) would be an expression of their need to invest more resources in voluntary control than normal children in order to achieve an adequate performance at a given task.

Functional Significance of the SMA

In this respect, one very important area of the frontal lobes has yet to be considered, namely the SMA, which is located in the region of the vertex. As far as is known from cyto- and myeloarchitectonic studies, the SMA is a rather young structure which has developed from the phylogenetically old limbic system (Sanides 1964) and therefore is connected to the cingulate gyrus. Electrical stimulation of the SMA in human subjects usually elicits complex synergetic movements of the contralateral side of the body and its extremities; it often also leads to vocalization and occasionally to ipsilateral movements (Penfield and Jasper 1954). It is also known that the SMA takes part in the cognitive planning and controlling of sequential voluntary motion (Deecke et al. 1985). Relative to the resting state, the rate of cerebral blood flow in this region (rCBF) increased bilaterally not only when a unilateral movement was being made, but also when the subject simply imagined himself performing an action without in fact carrying it out (Roland et al. 1980). Because of its low temporal resolution (one measurement lasts approximately 1 min), the rCBF method cannot tell whether the brain activity which is being measured developed before or after the imagined movement. To overcome these limitations, one must resort to other methods.

More light can be shed on the implication of the SMA in planning a given movement by considering studies of BPs in adults. The analysis of BPs represents a method which allows the classification of brain activity into temporal phases (i.e., preceding, during, and after the movement). Thus, preparatory electrical brain activity (BP) is observed over the vertex even when a movement is intended but is not actually executed (Libet et al. 1983). Consequently, the SMA can be regarded as a possible BP generator prior to movement (Eccles 1982). Thus our BP results are at least partly related to SMA activity.

Control of Interference

In the motor interference task of our study (voluntary vs involuntary motion), other parts of the frontal lobes also seem to be involved (Stamm and Kreder 1979; Delgado 1979; Krauthamer 1979) by being assigned a major regulatory function. The motor interference task was paralleled and significantly correlated with the cognitive interference task, i.e., the more electrical activity had to be developed by the frontal lobes to carry out the motor task, the more time was needed to perform the Stroop Color-Word Test. This probably reflects a problem of capacity of the frontal lobes to minimize interference. The importance of the frontocortical system in suppressing interfering stimulus events in the brain has already been emphasized by Bartus and Levere (1977, cited by Skinner 1978). As a matter of fact, these authors showed that the behavior of monkeys with dorsolateral ablations of the frontal cortex was similarly affected by relevant and irrelevant stimuli during serial presentation of stimuli in a discrimination task. Transposed into the current topic of interest and in relation to our finding of a centrofrontal midline shift of electrical brain activity before movement, such results permit the assumption that hypermotor children rely more on their frontocortical system than do healthy children of a comparable age in order to inhibit irrelevant stimuli. Unfortunately, children with ADHD have been observed not to achieve such inhibition so well, lending support to the hypothesis of a poorer performance of this frontocortical system in hyperkinetic children.

Drugs and Localization of Disturbance

As for children with tics (but without ADHD), it is assumed that there exists a hypersensitivity of dopamine receptors located in the striatum and/or a disturbance of the enkephalinergic neurons located in the striatum and projecting to the pallidum (Albin et al. 1989). This results in a low level of motor

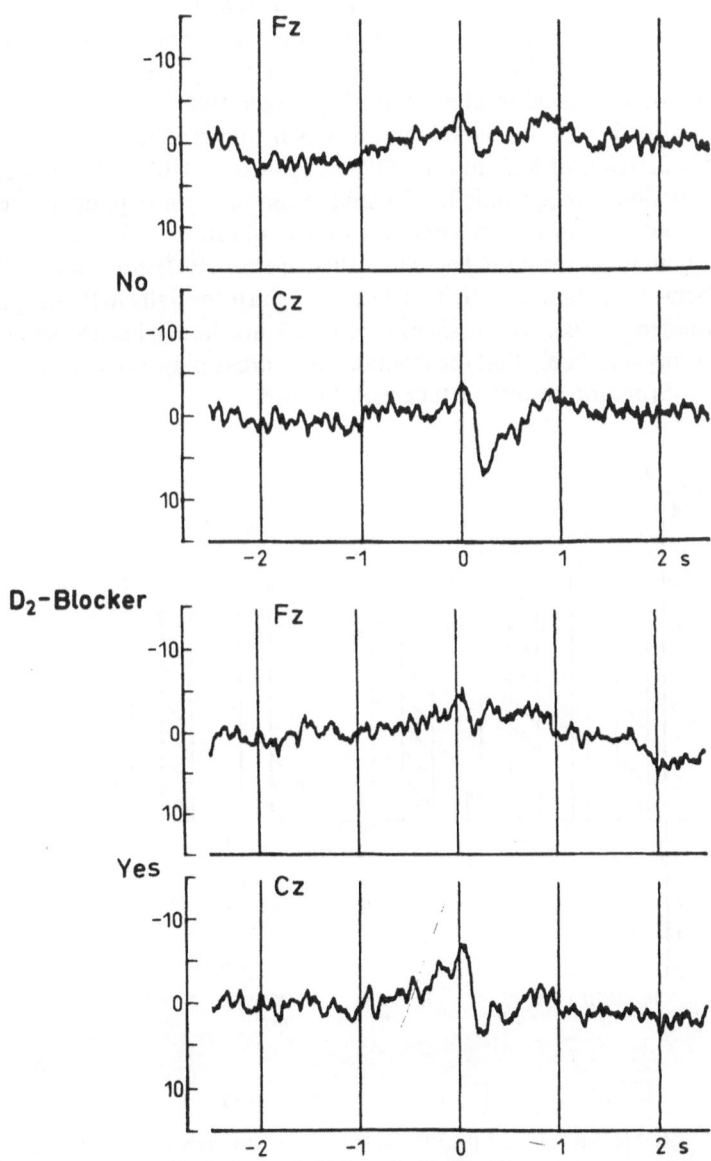

Fig. 5a. Grand averages of movement-related potentials of six children with tics before and during medication with a dopamine receptor blocker. During treatment, there is an enhanced anterior-posterior differentiation (BP amplitudes at Cz higher than at Fz)

inhibitory function of this structure in tic children who, in turn, are able to compensate this deficit by frontocortical brain activity. From this point of view, our BP results indicate that potent blocking agents of dopamine receptors exert their action principally at the striatal level and therefore "relieve" the inhibitory function of the frontocortical system. This is reflected by a shift of the BP amplitude maximum from frontal to the central location (i.e., towards a normalization of BP topography) observed in children with tics treated with blockers of dopamine receptors (Fig. 5a); this finding indicates that less frontocortical electrical brain activity is being needed to solve the task under pharmacological treatment (Rothenberger 1988a; Fig. 5b).

Further to the contribution which our results can make to the issue of localization of function in children with tics, it had been hoped that positron emission tomographic (PET) studies among adults patients with tic disorders would lead to decisive findings on the question of localizing the disturbance. Preliminary results (Chase et al. 1984; Foster 1983) indicated that the correlation between metabolic activity in basal ganglia (especially in the corpus striatum) and the cerebral cortex is higher in these patients than in healthy subjects. Although it seems very likely that the frontocortical areas play a particular role therein, firm confirmation awaits further investigation.

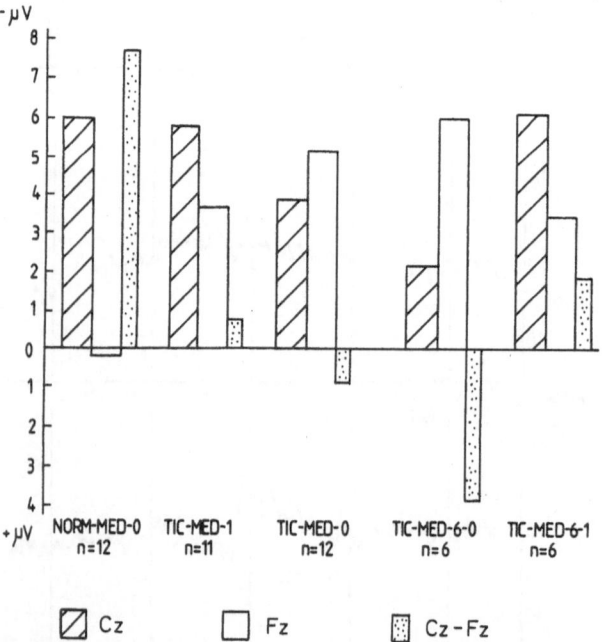

Fig. 5b. Median values of the BP amplitudes recorded over Cz and Fz as well as medians of the difference in BP amplitudes between the two sites (Cz − Fz) within various groups of subjects tested in the main study. In children with tics under neuropharmacological treatment (TIC-MED-1, TIC-MED-6-1), the BP amplitude is seen to be smaller over Fz than Cz; this approaches the situation prevailing in normal children

Disinhibition–Inhibition Model

In conclusion, a neurophysiology-oriented pathogenetic model for motor tic disorders can be employed to deal with brain structures as well as brain functions and to describe part of their intricate interconnection (Fig. 6; Rothenberger 1984). One should start from the fact that neurons in subcortical structures (e.g., basal ganglia) generate spontaneous discharges. Such spontaneous activity could, by chance, initiate motor programs that exist at this level in the form of motor subroutines to be readily used (Eccles 1982). These potential processes are normally prevented from taking place by, inter alia, inhibitory subcortical feedback loops. However, in tic disorders, it is possible that insufficient self-inhibition of spontaneous subcortical discharges leads to the execution of these motor subroutines, which express themselves as undesired tics (cf. Obeso et al. 1982).

When, for the reasons mentioned above, a child can no longer resort to the subcortical self-control of spontaneous firings, he/she is forced to bring into action other regulatory means of his/her brain in order to suppress the tics. The results we obtained lead us to assume that primarily frontocentral midline brain areas are activated for this purpose. Special attention should be drawn here to the SMA and the frontodorsal cortex. The SMA is known to act as a kind of "reference library for motor programs" and is responsible for compiling and activating these subcortically stored motor programs. The last relay station to be activated by the SMA is the primary motor cortex, which ensures the realization

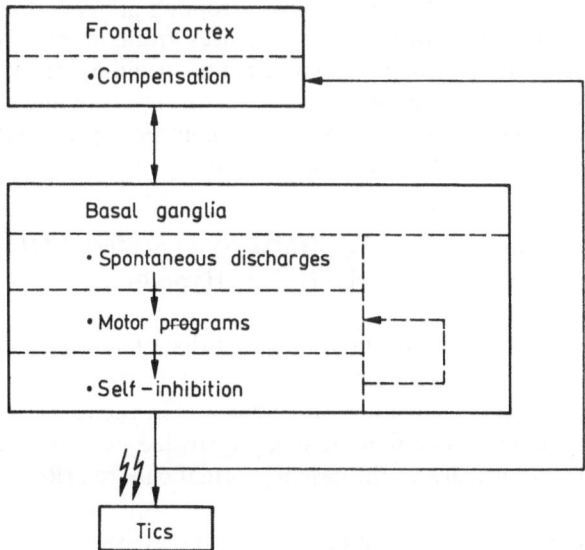

Fig. 6. Pathogenetic model (disinhibition–inhibition model) of tic disorders. It is assumed that the insufficient inhibition of subcortical spontaneous neuronal discharges must be compensated at the cortical level in order to prevent the manifestation of tics and the occurrence of nonfluent speech. *Forked arrows*, disturbance of subcortical self-inhibitory process

of highly complex activities by sending signals to the appropriate peripheral muscles (Eccles 1982).

When children with tics want to carry out a voluntary movement properly, it is to be assumed that these regulatory centers (SMA and dorsolateral frontal cortex) will have to bear a heavy processing load which can, as we stated above, be reduced by dopamine blocking agents. On the one hand, the motor program of tic children must be selected, saved, initiated, and monitored during the execution of a voluntary act by the SMA, to ensure that the motor subroutines necessary to achieve this multicomponent task do not come into conflict with unwanted motor subroutines initiated by spontaneous subcortical neuronal firing. In addition, other frontal brain areas apart from the SMA must be activated, as demonstrated by Roland et al. (1980) when studying complex sequential movements.

The assumption that children with tics suffer from a dysfunction of the midbrain (Devinsky 1983) has to be considered with caution, since the "high responsiveness to stimuli" observed in some children could be related to an additional ADHD and not to the tic disorder itself. Nevertheless, the frontal cortex is likely to regulate and integrate behavior, not only in its motor aspects (Delgado 1979) but also at the sensory level (Skinner 1978). If there is an increased activation of the mesencephalic reticular formation in children with tics, this structure interacts with the frontocortical system to act upon the thalamic reticular nucleus. The latter, in turn, influences the sensory nuclei of the thalamus. Thus, the frontal cortex appears to play a major role in determining the entry of sensory information (Skinner 1978). For this reason, the frontal cortex has to be strongly activated in order to prevent incoming peripheral sensory stimuli and spontaneous subcortical motor impulses from interfering with the planning and execution of voluntary movements, including complex sequential motor tasks such as speech performance. Thus, the frequent finding of nonfluent speech in tic children could arise from intervening tics or alternatively from perturbing but uninhibited afferent sensory input.

Attention-Deficit Hyperactivity Disorder (ADHD) and the Frontal Lobe Deficit Hypothesis

In the mentioned study, we found that children with ADHD needed to develop more electrical brain activity over the frontal areas than healthy children to perform the same task. In other words, the regulatory functions of the frontal brain had to be more strongly activated prior to the intended movement than would usually be the case in normal children (Rothenberger 1988b; Fig. 7).

Cerebral Circulation and Brain Metabolism

Besides electrophysiological methods, various computer-based nuclear medicine imaging techniques have become available, allowing closer inspection of the neurobiological background of ADHD. Following inhalation or intravenous

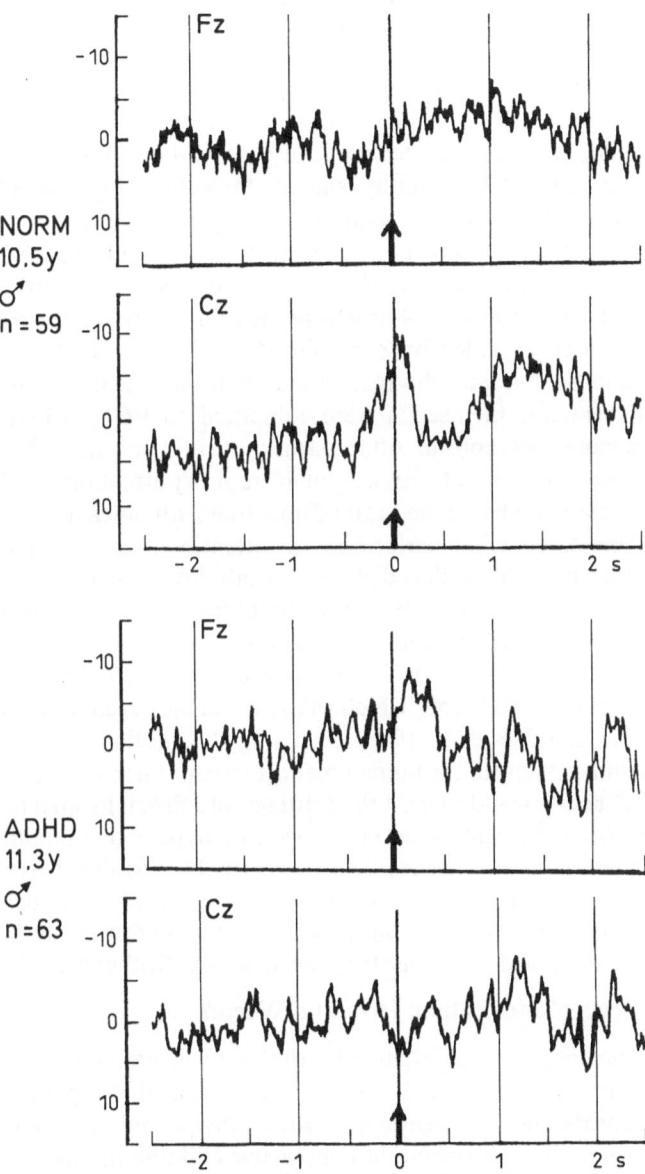

Fig. 7. Representation of movement-related electrical brain activity over the midline frontal (Fz) and central (Cz) recording sites. The *vertical arrow* indicates the time at which the movement begins. The curves at the *top* are from a normal 10.5-year-old boy (59 averaged 5 s long EEG segments); those at the *bottom* are derived from a hyperactive 11.3-year-old boy (63 averaged EEG segments). Looking at the mean amplitudes in a range of 0.5 s prior to the movement (*on the left of the arrow*), the amplitude of the BPs appears to be clearly larger over Cz than over Fz in the normal boy. However, the BP amplitude is slightly larger over Fz than over Cz in the hyperactive boy. This means in relative terms that the latter had to develop more electrical brain activity over anterior brain areas and had to invest more energy in the task than the normal boy

injection of radioactive markers, the cerebral blood supply as well as various metabolic processes in virtually all brain areas can be visualized, be it under resting conditions or during the performance of cognitive tasks, with or without medication. To our knowledge, the only such study of a group of hyperactive children was conducted by Lou et al. (1984). Comparing hyperactive children with age-matched healthy controls, the authors reported a lower blood perfusion (that is to say, a lower metabolic activity) at the level of the frontal lobes in each of 11 examined hyperactive patients, while seven subjects also showed a concurrent reduced circulation in the caudate nucleus area. Further evidence pointing at the important involvement of the frontal lobes in the pathogenesis of ADHD has been provided by Zametkin et al. (1986), who studied the parents of such children. Parents who manifested attention deficits and who were diagnosed as hyperactive themselves were submitted to PET to investigate their cerebral glucose metabolism. After intravenously receiving [^{18}F] 2-fluoro-2-deoxy-D-glucose, the parents carried out an auditory attentional task during the period of glucose uptake by the brain. Compared with normal adults, these parents made more errors while performing the attentional task; the levels of labeled glucose also witnessed a reduced glucose metabolism in all brain areas. The lower glucose uptake was particularly marked in the frontal areas of the brain. In this context of functional localization, one should also consider the evidence of a reduced capacity for subcortical dopamine uptake observed at the level of the corpus striatum in F344 rats, which were used in the development of an animal model of ADHD (Unis et al. 1986). Applying the caution necessary when transposing animal results to the human realm, one can still integrate these various sources of information and suspect the existence of a deficit located in subcortical as well as in frontocortical areas in the case of hyperactive children. However, further investigation is needed before it will be possible to establish relationships between the different dysfunctions involved and determine the relative contribution of the mesocortical dopaminergic system as well as the frontocortical compensatory mechanisms (Rothenberger 1984).

Effects of Medication: How and Where?

More insight could be gained in the above-mentioned direction by studying more thoroughly the effects of medication on the organism and its behavior. Considering that treatment with methylphenidate failed to improve the low frontal blood perfusion of hyperactive children in any systematic manner even though positive behavioral changes had been observed (Lou et al. 1984) along with improved blood perfusion in centrencephalic regions, the question must be raised as to whether the substrate of the deficit corresponds to the substrate upon which the medication acted therapeutically. It seems rather unlikely that a pharmacological agent would exert its influence exclusively at the site of neuronal dysfunction, returning it to normal operation and thereafter permitting the reestablishment of normal behavior. A much more probable interpretation would be that the brain initially develops spontaneous compensatory mechanisms attempting to deal with the aberrant processes which have established themselves. These compensatory mechanisms could subsequently be promoted by means of a pharmacological or psychological treatment which eventually results in disappearance of the behavioral signs of the deficit despite persistence of the underlying neurobiological correlates. In this regard, it is worth mentioning the case of neuronal compensatory mechanisms in children with acquired aphasia (Rothenberger 1986a). The brain structure which may be central to therapeutic response in hyperactive children seems to be the nucleus accumbens, a subcortical region of the forebrain, as suggested by the following pharmaco-dynamic considerations.

In the Federal Republic of Germany, methylphenidate is the medication of choice in the treatment of hyperactive children, along with amphetamines and pipamperone. Despite the differences in their biochemical mechanisms, these drugs do not differ widely in their clinical effects on attention and motor activity. Analogous to their impact on the energy metabolism of the brain, methylphenidate and amphetamine also display a selective pattern in influencing the glucose metabolism. At low amphetamine doses in animal studies (0.2 and 0.5 mg/kg body weight, comparable doses to those used in children), an increase in metabolic activity has been observed mainly in one particular brain structure, namely the above-mentioned nucleus accumbens. The similar distribution of regional changes in glucose uptake in the presence of methylphenidate and

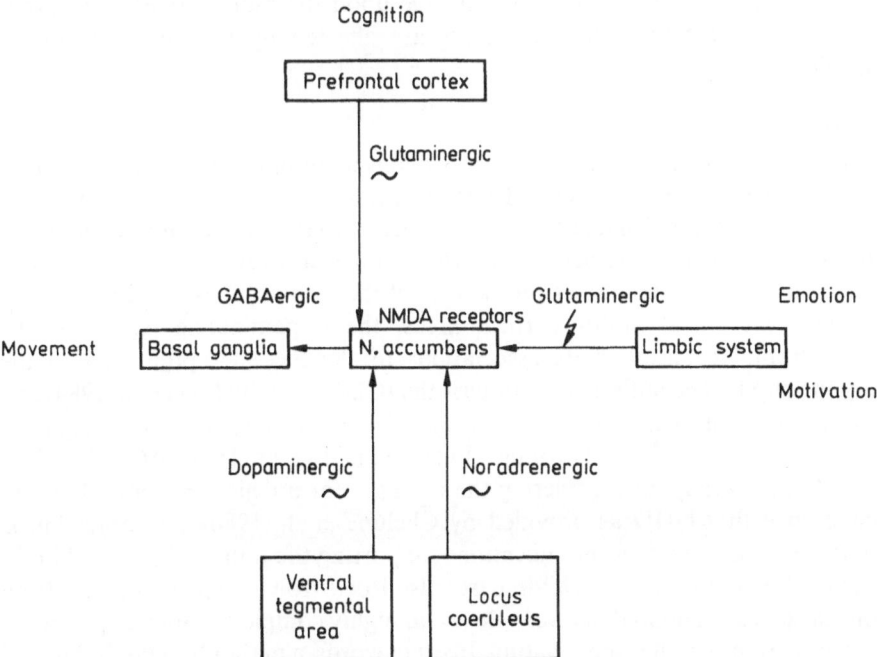

Fig. 8. The nucleus accumbens illustrated as the central target of the psychopharmacological action exerted by potent drugs used to treat ADHD. ∼, modulation; *forked arrow*, excitation

amphetamine suggests an important role of the nucleus accumbens in the therapeutic response of hyperactive children to psychostimulants.

The nucleus accumbens receives glutaminergic neuronal projections from various limbic structures, is significantly affected by dopaminergic inputs from the ventral tegmental area, and is innervated by glutaminergic afferents originating from the prefrontal association cortex. On the other hand, it sends fibres to the basal ganglia (Fig. 8). Hence, one could speculate that, by virtue of its anatomical connections, the nucleus accumbens represents a kind of functional intermediate between limbic and motor brain mechanisms (Porrino and Lucignani 1987). In this respect, the so-called NMDA receptors (having L-glutamate as neurotransmitter), which are indeed found in the entire brain but predominate in the nucleus accumbens and the hippocampus, may play an important role. Neurotransmitters such as dopamine and norepinephrine possibly only act as modulators on these NMDA receptors; the NMDA receptors themselves could actually represent behaviorally effective factors in terms of compensatory mechanisms (Cotman and Iversen 1987; Cotman et al. 1987). This would explain the similar behavioral effects observed after treating children with ADHD with methylphenidate, pemoline, blocking agents of dopamine receptors like pipamperone, α-adrenergic agonists such as clonidine (Unis et al. 1986), and tricyclic antidepressives such as imipramine. The nucleus accumbens appears, then, to represent a possible locus of action at which the therapeutic processes could take place and a site where cognitive and motivational influences from the cortex and the limbic structures converge upon the motor system. Therefore, both cognitive and motor symptoms are affected by the above-mentioned medications.

Cortical–Subcortical Interaction

For a summary of the evidence on the involvement of frontal lobe functions in hyperactive children, one should refer to the review article by Zametkin and Rapoport (1987), which reflects the increasing interest of the scientific community in this topic. Their overview begins with the general review of Mattes (1980), which introduces hypotheses on the role of the frontal lobes, the anterior and medical precentral motor cortical areas, the dopaminergic pathways, the orbitofrontal cortex and its associations to the medial septum and to the hippocampus. The authors then discuss the rCBF study by Lou et al. (1984) and the findings of Nasrallah et al. (1986). The latter reported minor signs of frontocortical atrophy in 24 young adults who had suffered from ADHD in their childhood. Finally, in considering the evidence for a deficit in frontal lobes in children with ADHD as provided by Chelune et al. (1986), an impression is obtained of the most recent indications supporting the notion of a frontal brain dysfunction in hyperactive children (see also Rockstroh et al., this volume). Here one must bear in mind that a very close and highly complex relationship prevails between frontal cortex and striatum. In other words, a pathophysiological model of this deficit should include inhibitory frontocortical activity acting upon subcortical striatal structures.

The data issued from the animal studies of Tassin (1980) and the results of Lou et al. (1984) with respect to the cerebral blood flow of hyperactive children suggest deficits in the mesocortical dopaminergic system of such children. Therefore, it becomes understandable that hyperactive children cannot resort sufficiently to their frontocortical neuronal network in order to compensate their subcortical dysfunction. This may also hold for autistic individuals (see Bruneau et al. this volume). Further, in a study on focused selective attention in autism, Ciesielski et al. (in press) reported that autistic persons, when compared with normal subjects, showed no evidence of a modulation of frontally distributed selective attention event-related responses. Nevertheless, a more detailed and systematic investigational approach would easily show the attentional differences between ADHD and autism.

Finally, additional insight into the intricate cortical-subcortical mechanisms underlying ADHD may be gained from the work of Stamm and Kreder (1979), who viewed the constellation of psychological disorders of hyperactive children (i.e., inadequate attention, improper control of impulsiveness, and poor control of response sequences) as being analogous to the deficits demonstrated in patients with known lesions of the anterior, dorsolateral, and polar regions of the frontal lobes. The dysfunction of the frontal cortex postulated by the authors could lead to an inadequate control of the inhibitory function mediated by the subcortically located caudate nucleus and, consequently, to disturbances in the inhibitory control of the subcortical psychological processes and motor responses exerted by the caudate. Future research, therefore, should focus on two aspects, namely (a) the cortical-subcortical mechanisms (as reported by e.g., Olton 1989), and (b) the differentiation of frontal cortex dysfunction in psychopathologically separate entities (e.g., in ADHD versus autism).

Anorexia Nervosa: Is There a Central Nervous Disconnection Syndrome?

Surprisingly, the available methods for investigating the electrical brain activity have rarely been used in patients with anorexia nervosa (Pirke and Ploog 1986), although they offer the advantages of good temporal resolution and provide a means for differentiating subcortical and cortical neuronal activities within the framework of evoked-potential studies.

Standard EEG and Body Weight

In the first step, we were therefore prompted to analyze the standard EEGs recorded from normal adults, healthy 13-year-olds, adolescent patients with anorexia nervosa, and children and adolescents with other psychiatric disorders (Table 1). Nearly 60% of the anorectic patients were found to have an atypical EEG. The observed abnormalities were mostly independent of body weight. Only the occurrence of paroxysmal dysrhythmia decreased in correspondence with an increase in body weight. What distinguishes our findings from previous reports is

Table 1. Abnormalities present in the standard EEG (given as percentages)

Group	Vertex transients		Paroxysmal dysrhythmia		Overall assessment	
	Non-periodic	Quasi-periodic	Present	Rare	Normal	Abnormal
Adults (n = 98; N)	8	6	1	0	94	6
13-year-olds (n = 100; N)	15	4	1	0	86	14
Psychiatrically disordered (n = 100; PCA)	29	0	10	3	80	20
Anorexia nervosa (n = 100; PCA)	71	58	22	8	44	56

PCA, groups of psychiatrically disordered children and adolescents; N, groups of healthy controls

that in the group of anorectic patients we were able to detect particular graphoelements (so-called vertex transients) which occurred much more frequently than in other groups of subjects and which may be interpreted as signs of disinhibition. With increasing disinhibition these vertex transients can develop into 3–4/s sharp-slow waves. The origin of these graphoelements has been mainly attributed to subcortical areas, but also partly to cortical regions (Niedermeyer and Lopes da Silva 1982). A similar waveform known as vertex wave has been recorded under conditions of reduced vigilance and, hence, low levels of cortical neuronal inhibition. No description of the phenomenon has so far been reported in alert patients. If it is correct that patients with anorexia nervosa typically have an elevated level of vigilance, as proposed by Herholz et al. (1987), but nevertheless show such graphoelements frequently in their standard EEGs, these vertex transients may be perceived as a stable indicator for an electrical vulnerability of the brain in the sense of a developmental deviation. However, the finding is not specific to the given psychopathology and can be observed in other child psychiatric disorders as well.

Evoked Potentials and Body Weight

The spontaneous electrical activity of the brain was relatively independent of changes in the body weight of the examined anorectic patients. Therefore, in a second step, we thought that a stimulus-related processing of auditory information by means of auditory evoked potentials (AEPs), (i.e., a task which directly addresses the performance of a central nervous subsystem relevant for everyday situations) could possibly better represent the general influence of starving on electrical brain activity. In addition, we were interested to test whether a transitory frontocortical disturbance exists in adolescents with anorexia nervosa, since flexibility of behavioral control, which is closely related to frontal lobe functioning, is severely disturbed in these patients.

With this objective in mind, we resorted to the method of AEPs and their increase (augmenting) in amplitude with higher levels of stimulus intensity (i.e., when louder tones are presented; Bruneau et al. 1985; Garreau 1985). Accordingly, we recorded the AEPs subcortically as well as at the cortical level in order to test the active modulation of sensory stimuli by the patients along the auditory pathways (Fig. 9). Moreover, it was intended to investigate whether a functional relationship could be established between the cortical pseudoatrophy documented in the CCT scans of the anorectic patients and the cortical sources of the recorded AEP waves.

We examined two groups of anorectic patients, each including only two male adolescents. One group of 27 patients with anorexia nervosa investigated once only had body weights of 32–51 kg ($\bar{x} = 42$ kg). Patients with lower body weights had just started their treatment, whereas those with higher body weights were already showing some progress in response to therapy. A second group of 12 anorectic adolescents was examined twice: once during the first week of treatment (low body weight, $\bar{x} = 39$ kg) and again after having reached and maintained the target weight for 2 weeks ($\bar{x} = 48$ kg). In most cases, the weight set as the therapeutic goal lay in the normal range for the age group in question. All patients fulfilled DSM III criteria for anorexia nervosa and had normal hearing. As a control group for these 12 anorectic patients, 12 children (ten girls and two boys) diagnosed as having a "specific emotional deficit" (corresponding to ICD 313, normal body weight, no eating disorder) were examined once.

All anorectic patients were submitted to a CCT scan and recording of AEPs at each examination. In the control group with psychiatric disorders only AEPs were recorded. Additionally, CCT scans were available from a similar control group of 12 teenagers with emotional disorders. All subjects had their body weight measured each time they were tested. When looking at the CCT scans, we counted the total number of sulci seen in the temporal, frontal, and parietal sections (Kohlmeyer et al. 1983). Early and late AEPs were recorded over the vertex in response to a lower (55/50 dBHL) and a higher (75/70 dBHL) intensity of sound. Amplitudes were measured with the use of cursors to define the range to be considered (Fig. 9). Nonparametric tests were used in the statistical analysis of our data.

We could replicate the fact that in low body weight anorectic patients, there can be a large number of cortical grooves, readily visualized on the CCT scan, which decrease significantly with increasing body weight, while in the psychopathological control group there were no relevant CCT findings. No significant correlation was found between AEP and CCT data. Thus, organic evidence other than the volume changes of the cortical folds needs to be found in order to explain our AEP results.

Our study indicated that anorectic patients, compared with psychopathological controls, had difficulties modulating the auditory stimuli adequately (even after weight gain). This applied particularly to the subcortical level of their central nervous pathways: looking at the brainstem AEP amplitudes Vp–Vn, this was reflected by a very small augmenting or even reducing pattern of responses to louder stimuli in anorectic patients compared with controls. When one considers that an active processing of stimuli in terms of a biofeedback already occurs at the

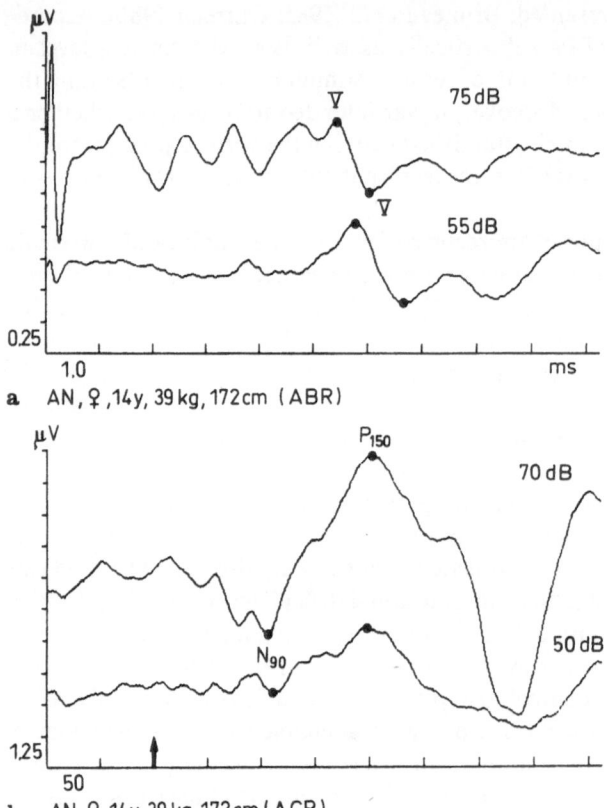

a   AN, ♀, 14y, 39 kg, 172 cm (ABR)

b   AN, ♀, 14y, 39 kg, 172 cm (ACR)

Fig. 9a, b. AEPs in response to stimuli of various intensities. The curves are from a 14-year-old anorectic girl with low body weight. **a** Cortical responses to auditory stimuli recorded over Cz/Fz (stimulus rate = 0.8/s). **b** Auditory evoked brainstem potentials recorded from Cz/M1 (stimulus rate = 11.3/s). The AEP amplitude increases with increasing stimulus intensity at the cortical level. In contrast, the amplitude at the brainstem level decreases under these conditions

level of the brain stem (Finley 1984), one could assume that the small increments in amplitudes recorded subcortically with increasing stimulus intensity account for a poor modulation capacity in anorectic patients. In contrast to our expectations, at the cortical level (N90–P150) no significant differences in AEP amplitude changes to louder stimuli were observed between anorectic patients and controls. Nevertheless, the increase in amplitude at cortical responses (in contrast to subcortical AEP) tended to be higher (i.e., clear augmenting) in both groups of anorectic patients when their body weight was low.

The Neuronal Decoupling Hypothesis

Thus, low weight anorectic patients showed diverging patterns with respect to cortical versus subcortical AEP responses (i.e., cortical augmenting vs subcortical reducing of AEP amplitudes). This suggests a functional deficit arising from a decoupling of cortical and subcortical neuronal systems, possibly at the level of the thalamus. Our control group of psychiatrically disordered children as well as healthy subjects can be reported to be augmenters for early as well as late AEPs (Garreau 1985; Rothenberger et al. 1987). Our interpretation of the AEP results may be seen in parallel to the topographical mapping of late AEP (P300) of Malloy (1987) in obsessive-compulsive patients. He suggests a frontocortical-subcortical disconnection syndrome with a disturbance at the orbitomedial part of the frontal cortex. Since anorectic patients often show obsessive-compulsive behavior, this line of research might prove fruitful.

Additionally, the PET studies by Herholz et al. (1987) have indeed shown that the metabolic rates do not change significantly in the frontal lobe regions of anorectic patients, despite the observation of increased metabolic activity in the caudate nucleus when these patients have low body weights. Given that damage to both areas (the frontal lobe and the caudate nucleus) may lead to the same sort of functional disorders, it follows that our experimental approach for examining a possible frontocortical-subcortical dysfunction in anorectic patients should be pursued in future research.

General Conclusions

As disorder-specific behavioral abnormalities start manifesting themselves (e.g., tic disorder: muscle twitches; ADHD: general motor restlessness; anorexia nervosa: obsessive thoughts), both children and parents will mobilize all possible resources to counteract any of the behavioral abnormalities. The child can make use of an important internal mental capacity, the activation of existing frontal brain functions, in attempts to compensate disturbing behavioral elements. As could be shown, children with tics succeed in utilizing this possibility spontaneously to suppress their tics for a certain amount of time. This seems to be a good prerequisite for a behavioral therapeutic approach to the disorder. Children with ADHD, on the other hand, have to struggle with a deficit of frontal lobe function and therefore they depend on external support to control their behavior. Anorectic patients, especially those with a high degree of obsessive-compulsive symptomatology, seem to be handicapped by a disturbance of frontocortical-subcortical neuronal interaction.

Thus, any treatment efforts (psychological and pharmacological) should include the aim to help psychiatrically disordered children regain and maximize the effective use of their frontal brain functions for new objectives, thereby assisting in gaining better results in control of symptoms and task performance, be it motor, emotional, cognitive, or social in nature.

Applied to the above-mentioned deficits (tic disorder, ADHD, and anorexia nervosa), this signifies that, given a similar level of severity in the manifestation of the disorder, a patient with good functioning of frontal lobes and capable of using the full potential of the latter should, in the long run, have more success in dealing with his disturbance than another patient who neither has these possibilities nor is able to acquire them by means of therapy. Furthermore, it seems legitimate to ask whether someone with good frontal brain performance may not be able to influence his psychopathological symptoms favorably on a long-term basis via self-control, considering the latter as being primarily regulated by frontal lobes. One might easily imagine that frontal lobes could also activate compensatory mechanisms of and in other cerebral areas (i.e., new organization in the cybernetics of the cerebral network for maintaining normal functions; Rothenberger et al. 1989).

Overall, we may assume that further research on frontal brain functions and child psychiatric disorders will provide precious knowledge that will assist clinicians in their daily supervision of the children under consideration and facilitate reliable prognosis.

References

Albin RL, Young AB, Penney JB (1989) The functional anatomy of basal ganglia disorders. Trends Neurosci 12:366–375

Bäumler G (1985) Farbe-Wort-Interferenztest (FWIT). Hogrefe, Göttingen

Bruneau N, Roux S, Garreau B, Lelord G (1985) Frontal auditory evoked potentials and augmenting-reducing. Electroencephalogr Clin Neurophysiol 62:364–371

Bruneau N, Garreau B, Barthélémy C, Martineau J, Muh J, Lelord G (1990) Clinical, electro-physiological, and biochemical markers and monoaminergic hypotheses in autism, this volume, pp 217–234

Chase TN, Foster NL, Fedio P, Brooks R, Mansi L, Kessler R, DiChiro G (1984) Gilles de la Tourette syndrome: Studies with the fluorine-18-labeled fluorodeoxyglucose positron emission tomographic method. Ann Neurol (Suppl) 15:175

Chelune GJ, Baer RA (1986) Developmental norms for the Wisconsin Card Sorting Test. J Clin Exp Neuropsychol 8:219–228

Chelune GJ, Ferguson W, Koon R, Dickey TO (1986) Frontal lobe disinhibition in attention deficit disorder. Child Psychiatry Hum Dev 16:221–234

Ciesielski KT, Courchesne E, Elmasian R (in press) Effects of focused selective attention tasks on event-related potentials in autistic and normal individuals. Electroencephalogr Clin Neurophysiol

Cotman CW, Iversen LL (1987) Excitatory amino acids in the brain – focus in NMDA receptors. Trends Neurosci 10:263–265

Cotman CW, Monaghan DT, Ottersen OP, Storm-Mathisen J (1987) Anatomical organization of excitatory amino acids receptors and their pathways. Trends Neurosci 10:273–280

Deecke L, Kornhuber HH, Lang W, Lang M, Schreiber H (1985) Timing function of the frontal cortex in sequential motor and learning tasks. Hum Neurobiol 4:143–154

Delgado JMR (1979) Inhibitory functions in the neostriatum. In: Divac I, Öberg RGE (eds) The neostriatum. Pergamon, Oxford, pp 241–261

Devinsky O (1983) Neuroanatomy of Gilles de la Tourette's syndrome. Possible midbrain involvement. Arch Neurol 40:508–514

Eccles JC (1982) The initiation of voluntary movements by the supplementary motor area. Arch Psychiat Nervenkr 231:423–441

Finley WW (1984) Biofeedback of very early potentials from the brainstem. In: Elbert T, Rockstroh B, Lutzenberger W, Birbaumer N (eds) Self-regulation of the brain and behavior. Springer, Berlin Heidelberg New York Tokyo, pp 143–163

Foster N (1983) NIH TS PET Scan project. Tourette Syndrome Association Newsletter X (Summer):1

Fuster JM (1989) The prefrontal cortex. 2nd edn. Raven, New York

Garreau B (1985) Étude des activités évoqués du tronc cérébral et de la région frontale chez 1' enfant autistique. Thesis, Université P. et M. Curie, Paris

Golden CJ (1981) The Luria-Nebraska Children's Battery: theory and formulation. In: Hynd GW, Obrzut JE (eds) Neuropsychological assessment and the school-age child. Grune and Stratton, New York, pp 277–302

Goldman PS, Alexander GE (1977) Maturation of prefrontal cortex in the monkey revealed by local reversible cryogenic depression. Nature (London) 267:613–615

Häfner HH (1955) Störung des Plan- und Entwurfvermögens bei Stirnhirnläsionen. Arch Psychiatr Z Neurol 193:569–582

Häfner HH (1957) Psychopathologie des Stirnhirns 1939–1955. Fortschr Neurol Psychiatr 25:205–252

Heaton RK (1981) Wisconsin Card Sorting Test Manual. FL: Psychological Assessment Resources, Odessa, FL

Herholz K, Krieg JC, Emrich HM, Pawlik G, Beil C, Pirke KM, Pahl JJ, Wagner R, Wienhard K, Ploog D, Heiss WD (1987) Regional cerebral glucose metabolism in anorexia nervosa measured by positron emission tomography. Biol Psychiatry 22:43–51

Jouandet M, Gazzaniga MS (1979) The frontal lobes. In: Gazzaniga MS (ed) Handbook of behavioral neurobiology. vol. 2 Neuropsychology, Plenum, New York, pp 25–59

Karrer R, Ivins J (1976) Steady potentials accompanying perception and response in mentally retarded and normal children. In: Karrer R (ed) Developmental psychophysiology of mental retardation. Thomas, Springfield, pp 361–417

Karrer R, Warren C (1979) Functional organization of the brain in the mentally retarded: evidence from event-related potentials. In: Obiols J, Ballus C, Gonzales Monclus E, Pujol J (eds) Biological psychiatry today. Elsevier, Amsterdam, pp 1350–1355

Kohlmeyer K, Lehmkuhl G, Poustka F (1983) Computed tomography of anorexia nervosa. Am J Neuroradiol 4:437–438

Krauthamer GM (1979) Sensory functions of the neostriatum. In: Divac I, Öberg RGE (eds) The neostriatum. Pergamon, Oxford, pp 263–289

Lang W, Lang M, Kornhuber A, Kornhuber HH (1983) Human visuomotor learning and cerebral potentials with inverted and distorted hand tracking. Preliminary poster reports, EPIC VII, Florence, p 89

Libet B, Wright EW, Gleason CA (1983) Preparation- or intention-to-act, in relation to pre-event potentials recorded at the vertex. Electroencephalogr Clin Neurophysiol 56, 367–372

Lou HC, Henriksen L, Bruhn P (1984) Focal cerebral hypoperfusion in children with dysphasia and/or attention deficit disorder. Arch Neurol 41:825–829

Malloy P (1987) Frontal lobe dysfunction in obsessive-compulsive disorder. In: Perecman E (ed) The frontal lobes revisited. IRBN Press, New York, pp 207–223

Mattes JA (1980) The role of frontal lobe dysfunction in childhood hyperkinesis. Compr Psychiatry 21:358–369

Milner B, Petrides M (1984) Behavioural effects of frontal lobe lesions in man. Trends Neurosci 7:403–407

Nasrallah HA, Loney J, Olson SC, McCalley-Whitters M, Kramer J, Jacoby CG (1986) Cortical atrophy in young adults with a history of hyperactivity in childhood. Psychiatry Res 17:241–246

Niedermeyer E, Lopes da Silva F (eds) (1982) Electroencephalography. Urban und Schwarzenberg, Munich

Obeso JA, Rothwell JC, Marsden CD (1982) The neurophysiology of Tourette syndrome. In: Friedhoff AJ, Chase TN (eds) Gilles de la Tourette syndrome. Raven, New York, pp 105–114 (Advances in neurology, vol 35)

Olton DS (1989) Frontal cortex, timing, and memory. Neuropsychologia 27:121–130

Passler MA, Isaac W, Hynd GW (1985) Neuropsychological development of behaviour attributed to frontal lobe functioning in children. Dev Neuropsychology 1:349–370

Penfield W, Jasper H (1954) Epilepsy and the functional anatomy of the brain. Brown, Boston

Perecman E (ed) (1987) The frontal lobes revisited. IRBN Press, New York

Pirke KM, Ploog D (1986) Psychobiology of anorexia nervosa. In: Wurtman RJ, Wurtman JJ (eds) Nutrition and the brain, vol 7, Raven, New York, pp 167–198

Porrino LJ, Lucignani G (1987) Different patterns of local brain energy metabolism associated with high and low doses of methylphenidate. Relevance to its action in hyperactive children. Biol Psychiatry 22:126–138

Reines S, Goldman JM (1980) The development of the brain. Thomas, Springfield

Rockstroh B, Elbert T, Birbaumer N, Lutzenberger W (1982) Slow brain potentials and behavior, Urban and Schwarzenberg, Munich

Rockstroh B, Elbert T, Lutzenberger W, Birbaumer N (1990) Biofeedback: evaluation and therapy in children with attentional dysfunctions. This volume, pp. 345–357

Roland PE, Larsen B, Lassen NA, Skinhoj E (1980) Supplementary motor area and other cortical areas in organization of voluntary movements in man. J Neurophysiol 43:118–136

Rothenberger A, Kemmerling S (1982) Bereitschaftspotential in children with multiple tics and Gilles de la Tourette syndrome. In: Rothenberger A (ed) Event-related potentials in children. Elsevier, Amsterdam, pp 257–270 (Developments in neurology, vol 6)

Rothenberger A (1984) Bewegungsbezogene Veränderungen elektrischer Hirnaktivität bei Kindern mit multiplen Tics und Gilles de la Tourette Syndrom. Postdoctoral thesis, University of Heidelberg

Rothenberger A (1986a) Aphasie bei Kindern. Fortschr Neurol Psychiatr 54:92–98

Rothenberger A, Kemmerling S, Schenk GK, Zerbin D, Voss M (1986b) Movement-related potentials in children with hypermotoric behaviour. In: McCallum WC, Zappoli R, Denoth F (eds) Cerebral psychophysiology: studies in even-related potentials (EEG Suppl 38), Elsevier, Amsterdam, pp 496–498

Rothenberger A, Reiser A, Grote I, Woerner W (1987) Modulation of sensory input in infants at different psychosocial and organic risk. In: Kutas M, Renault B (eds) ICON IV, Conference Proceedings, Paris-Dourdan, pp 85–88

Rothenberger A (1988a) Treatment of tic-disorders with dopamine receptor blockers. Less frontal lobe activity necessary to control the tics. Psychopharmacology 96 (Suppl):153

Rothenberger A (1988b) Klassifikation und neurobiologischer Hintergrund des hyperkinetischen Syndroms. In: Franke U (ed) Das aggressive und hyperkinetische Kind in der Therapie. Springer, Berlin Heidelberg New York Tokyo, pp 5–26

Rothenberger A, Woerner W, Dumais-Huber C, Eisert H-G, Etchepareborda M, Niemeyer J, Stratmann F, Schmidt MH (1989) Zentralnervöse Kontrollmechanismen und kinder-psychiatrische Störungen. 21st Meeting of the Deutsche Gesellschaft für Kinder- und Jugend-psychiatrie, Munich

Sanides F (1964) The cyto-myeloarchitecture of the frontal lobe and its relation to phylogenetic differentiation of the cerebral cortex. J Hirnforsch 6:269–291

Sanides F (1971) Functional architecture of motor and sensory cortices in primates in the light of a new concept of neocortex development. In: Noback CR, Montana W (eds) Advances in primatology, vol I. Appleton-Century-Crofts, New York, pp 137–208

Sanides F (1972) Representations in the cerebral cortex and its areal lamination patterns. In: GF Bourne (ed) Structure and function of nervous tissue, vol 5. Academic, New York pp 329–453

Skinner JE (1978) A neurophysiological model for regulation of sensory input to cerebral cortex. In: Otto D (ed) Multidisciplinary perspectives in event-related brain potential research. U.S. Environmental Protection Agency, Washington DC, pp 616–625

Stamm JS, Kreder SV (1979) Minimal brain dysfunction: psychological and neurophysiological disorders in hyperkinetic children. In: Gazzaniga MS (ed) Neuropsychology. Plenum, New York, pp 119–150 (Handbook of behavioral neurobiology, vol 2)

Stuss DT, Benson DF (1986) The frontal lobes. Raven, New York

Stuss DT, Benson DF (1987) The frontal lobes and control of cognition and memory. In: Perecman E (ed) The frontal lobes revisited, IRBN Press, New York, pp 141–158

Tassin JP (1980) Approche du rôle fonctionnel du système méso-cortical dopaminergique. Psychologie Médicale 12:43–64

Unis AS, McMahon WM, Franz D (1986) A common neuropharmacological basis for the efficacy of tricyclic antidepressants, psychostimulants and clonidine in attention deficit disorder with hyperactivity. Scientific proceedings of the annual meeting of the Am Acad Child Adolesc Psychiatry 2:21

Weinberger DR (1988) Schizophrenia and the frontal lobe. Trends Neurosci 11:367–370

Yakovlev PI, Lecours AR (1967) The myelogenetic cycles of regional maturation in the brain. In: Minkowski A (ed) Regional development of the brain in early life. Blackwell, Oxford, pp 3–70

Zametkin A, Nordahl T, Gross M et al. (1986) Brain metabolism in hyperactive children. Scientific proceedings of the annual meeting of the Am Acad Child Adolesc Psychiatry 2:23

Zametkin AJ, Rapoport JL (1987) Neurobiology of attention deficit disorder with hyperactivity: where have we come in 50 years? J Amer Acad Child Adol Psychiatry 26:676–686

Monoaminergic Systems and Child Psychiatric Pathophysiology

B. Garreau, D. Lerminiaux, C. Barthélémy, J. P. Muh, and G. Lelord

Introduction

Since the demonstration in 1921 by Otto Loewi that a chemical substance was released after nervous stimulation, much neuroscientific research has been performed on neurotransmission and chemical synapse. Work in neuro- and psychopharmacology has permitted the comparison between notions of neuro-transmission and neuropsychiatric syndromes.

The antipsychotic effect of neuroleptics associated with an action of dopamine on synaptic transmission and the antidepressive effect of monoamine oxidase inhibitors associated with a change in serotonin transmission have both been at the origin of monoaminergic hypotheses for psychoses and depressions.

In children, it is by reference to adult pathology that monoaminergic hypotheses have been formulated to explain psychiatric syndromes. Even though an abnormality in the metabolism of dopamine, serotonin, or norepinephrine seems to be involved in certain behavioral deficits in children, it is rarely possible to know whether one is dealing with an abnormal synthesis of neurotransmitter due to an enzymatic deficiency (e.g. phenylketonuria), a decrease in the number of receptor sites, a block in the reuptake of neurotransmitter, a catabolic disorder affecting the degradation of the latter or, more generally, an imbalance in the interactions between different neurotransmitter systems.

Furthermore, studies on the development of the central nervous system during gestation and early childhood have indicated the vulnerability of certain developmental phases to endogenous or exogenous teratogenic agents. For this reason, we will commence with a brief presentation of ontogenetic data concerning the central nervous system. We will then proceed to a description of the major dysfunctions of the serotoninergic, dopaminergic, and noradrenergic systems, before summarizing our state of knowledge in this field, giving critical consideration to its limited extent.

Ontogenesis of Neurotransmission

Cerebral maturation corresponds to a series of anatomical and physiological events which can be summarized in six steps (Sarnat 1987):

1. Neurogenesis with production of neuroblasts
2. Neuronal migration
3. Cellular differentiation and axonal as well as dentritic growth
4. Synaptogenesis
5. Biosynthesis of neurotransmitters
6. Myelination of axons

These various stages overlap in time. They are sometimes almost simultaneous, such as the synaptogenesis immediately preceding the biosynthesis of neurotransmitters. Neuronal migration may also precede or follow synaptogenesis. However, for a given region of the central nervous system, all events are chronologically determined and similar from one individual to another.

The Major Stages of Neuronal Development

Neurogenesis and Neuronal Migration

The principal events responsible for the general brain morphology take place during the first 20 weeks of intrauterine life. At this time, the neurons originate and proliferate in the germinal zones surrounding cerebral cavities and then migrate to their final sites (Lyon et al. 1984).

In the brain stem, future neurons aggregate to form functional nuclei such as the locus coeruleus, the serotoninergic raphe nucleus, and the dopaminergic substantia nigra. At the telencephalic level, however, they migrate along radial glial fibers to the cortex. Considering laminar order in relation to time of generation, neurons that develop earlier are observed to occupy deeper layers than those that are generated later (Lyon et al. 1984; Coyle 1985).

Cellular Differentiation

The migration of the neuron to its final position plays an important role in the biochemical differentiation of the cell by virtue of which the synthesis of a specific neurotransmitter will be determined.

In animals, it seems in fact that the immature neuron can provisionally produce a neurotransmitter which does not correspond to that which the neuron will later manufacture after reaching its final destination (Coyle 1985). For instance, somatostatin and substance P were detected in fetal cerebellum of mammals at midgestation, but could no longer be found there at the end of maturation. Rather, glutamate and γ-aminobutyric acid (GABA) are the neurotransmitters found in granular and Purkinje cells of an adult cerebellum (Sarnat 1987).

The factors controlling the phenotypic expression of neurotransmitters are still obscure, but it is known that interactions between adjacent neurons as well as the action of neurotropic agents such as the nerve growth factor play a critical role at the level of the peripheral nervous system (Coyle 1985).

Synaptogenesis

Synaptogenesis is preceded by axonal growth. Then, at the end of the 3rd month of gestation, synaptogenesis starts taking place; it reaches its maximum at about 2 years after birth and continues throughout adulthood. This process consists in the cellular growth and the development of intercellular connections resulting in synaptic contact.

Relatively little is known about the factors controlling arborization of axons (whose length varies according to the GABAergic or monoaminergic neurotransmission under consideration). Nevertheless, recent studies have demonstrated the high precision of axonal contacts, in the case of serotoninergic and noradrenergic neurons, independently of the integrity of postsynaptic neurons (Coyle 1985).

Pathogenesis

Innate or acquired pathogenic factors exerting their influence during the first 20 weeks will give rise to macroscopic anomalies in the course of organogenesis; they will cause deficits in either the genesis or migration of neurons, or else a disturbance in the normal arrangement of neurons (i.e., a structural deficit in cerebral development). Pathogenic factors intervening later in development will have a more subtle effect on neuronal growth, synapse formation, and neuronal functions; such factors will be responsible for deficits that are more functional in nature (Evrard et al. 1984).

Multilevel Organization of Neurotransmission

The Reticular Formation

The cell bodies of the future neurons gather into different nuclei throughout the reticular formation of the brain stem and the midbrain. The appearance of these nuclei proceeds chronologically from the posterior to the more anterior regions of the reticular formation (Coyle 1985).

Neuronal Maturation in the Reticular Formation
Within the cell bodies of neurons forming the nuclei, the biochemical processes necessary for and specific to the synthesis and storage of neurotransmitter are elaborated, in a systematic manner, through a precise sequence of enzymatic processes during gestation. After birth, the enzymatic activity declines in the region of monoaminergic cell bodies and shifts toward the synaptic regions (e.g., the forebrain). In addition, the overall capacity for neurotransmitter uptake (V max) developed by the membrane during fetal life continues to increase gradually following birth, but the affinity of membrane reuptake mechanisms does not change (Young et al. 1984).

The Nuclei

The Noradrenergic Nucleus. The phylogenetically oldest nucleus of the reticular formation appears after approximately 6 weeks of gestation. It is the most posterior and results in the locus coeruleus located on the floor of the fourth ventricle. The noradrenergic axons sent by the few hundred neurons of the locus coeruleus are first to appear; their branching follows step by step the cerebral growth of the areas which they innervate. These axons account for up to 30% of the synapses of the immature cerebral cortex but contribute only a small percentage of the same synapses in adults. Early noradrenergic innervation is likely to play a critical role in primitive brain development. In fact, early lesions of the noradrenergic system seem to affect synaptic plasticity (Coyle 1985; Evrard et al. 1984).

The Serotoninergic Nucleus. Located in the midbrain, the serotoninergic raphe nucleus is second to appear. Despite its early development, the serotoninergic influence will start exerting itself more gradually. Like the noradrenergic neurons to which they may be antagonistic, the serotoninergic neurons send axons to every region of the central nervous system of the fetus. These projections are directed to, among other locations, the hypothalamus, the striatum, the limbic system, and the entire cerebral cortex. The early innervation of the primitive cortex by serotoninergic neurons seems to act as a modulator in neuronal development (influencing the rate of cellular division of the cerebral cortex in the fetus) (Coyle 1985).

The Dopaminergic Nuclei. The substantia nigra appears during a short period at the end of the first 3 months of pregnancy and constitutes the third nucleus of the reticular formation. These dopaminergic neurons project diffusely and abundantly to the striatum, for which they represent an important developmental factor. In fact, the pattern of dopaminergic innervation correlates with the activity of cell division found in intrinsic neurons of the striatum. After a gradual increase, the innervation becomes maximal at puberty (it is estimated that 15% of the nerve terminal synapses found in adults are dopaminergic in nature). The dopaminergic neurons located more medially and whose axons project to the limbic system and the frontal cortex have a more limited development although the latter goes on beyond childhood (Coyle 1985).

The Cholinergic Nuclei. The formation of the cholinergic complex requires a long time due to its size and its anteroposterior extension. It occurs, however, much earlier than the development of the structures it innervates, such as the cerebral cortex, the hippocampus, and the limbic system. As opposed to noradrenergic and serotoninergic fibres invading the cerebral cortex when it is still primitive, cholinergic fibres send their projections at a later stage, i.e., in the first years after birth. This late influence suggests a terminating or consolidating role of the cortical synaptogenesis (Coyle 1985).

The Cerebral Cortex

Intrinsic Neurons

The intrinsic neurons of the cerebral cortex play an important role in the process of local information (cell bodies and synaptic connections assigning it to circuits limited to the cortex only). Most of these neurons use GABA as neurotransmitter (it may be found in up to 30% of the cerebral synapses).

Neuronal Maturation in the Cerebral Cortex

The late maturation of GABAergic neurons coincides with the progressive differentiation of neurons in the cerebral cortex, and contrary to what happens at the level of the reticular formation, the processes of synthesis and inactivation, for one reason or another, do not take place in a coordinated manner. In fact, one finds in the immature brain a relatively high level of GABA compared with the weak activity of the enzyme responsible for its synthesis. The process of uptake inactivation is also characterized by a relatively high affinity in the immature brain as compared with values observed in adults. Moreover, the postsynaptic GABA receptors (associated with benzodiazepine receptors) show a gradual augmentation (Coyle 1985).

Neurotransmission and Behavioral Pattern

The major neurotransmitter systems become functional according to the following chronology: noradrenergic, serotoninergic, dopaminergic, cholinergic, and finally GABAergic. One can imagine the main phases of pre- and postnatal life as being marked by the respective development of these different neurotransmitter systems. Thus, the maturation of the noradrenergic system and the synaptic organization of the cerebral cortex parallel the maturation of vigilance states. The complex cognitive functions emerge at the end of the 1st year together with the cholinergic innervation of the cerebral cortex and the hippocampal formations. The development of functions such as attention, emotional behavior, and complex psychomotor performance corresponds to the establishment of an equilibrium state between the various neurotransmitter systems.

In this regard, it is known that the dopamine-to-serotonin concentration ratio is of importance for motor activity. A higher dopaminergic activity is associated with an increase as well as with a higher frequency of abnormal movements, whereas the serotoninergic activity inhibits them. The decrease in dopaminergic activity during childhood and adolescence, and the maintenance at constant levels of serotoninergic activity, leads up to predict a high motor activity during early childhood and a gradual decrease of this activity throughout development. Furthermore, Young et al. (1984) pointed out that monoaminergic systems are particularly active at transitional periods of development: the neonatal adaptation to the external environment or the maturational processes at puberty are associated with high levels of catecholamine turnover. Monoamines regulate the hypothalamic releasing factors which control pituitary functions and influence thereby the general homeostasis of the body.

The dopamine-to-serotonin concentration ratio is higher in boys: homovanillic acid (HVA)/5-hydroxyindoleacetic acid (5-HIAA) was found to be higher in the CSF (Young et al. 1984). This may explain the lower motor activity of girls and the higher prevalence among boys of neuropsychiatric disorders such as early infantile autism, Gilles de la Tourette's syndrome, or attentional deficit hyperactivity disorder (ADHD).

Next, we will attempt to describe the serotoninergic, dopaminergic, and noradrenergic systems by means of a physiopathological approach after having briefly discussed the metabolism of these monoamines as well as their topography.

Serotoninegic Systems

The role of serotonin in neurotransmission has been known for many years. Much more than to schizophrenia, however, this applies to the mechanisms involved in depression.

Metabolism of Serotonin

Serotonin is synthesized in the neuron from the precursor trytophan according to the following scheme:

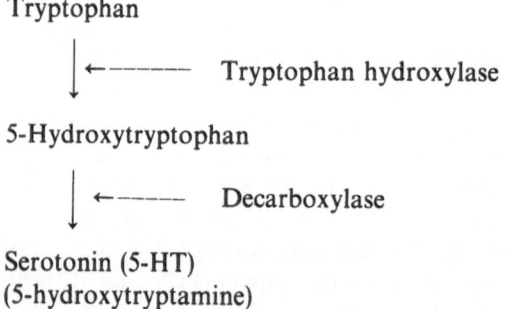

Tryptophan

    ↓ ←------- Tryptophan hydroxylase

5-Hydroxytryptophan

    ↓ ←------- Decarboxylase

Serotonin (5-HT)
(5-hydroxytryptamine)

The enzyme tryptophan hydroxylase (TH) is highly specific and is present exclusively in serotoninergic neurons. The activity of this enzyme increases along with the concentration of tryptophan free in the blood and thus in nerve endings. It increases also in the case of an elevated consumption of serotonin (e.g., electrical stimulation of serotoninergic neurons). Two important elements take part in the regulation of serotonin biosynthesis: on the one hand, the plasma concentration of free tryptophan, and on the other hand, the TH, the variations of which are related to the fluctuations in intrinsic nervous activity of the serotoninergic systems. Serotonin is stored in synaptic vesicles and released into the synaptic cleft upon arrival of an action potential at the nerve terminal. It then attaches itself to specific postsynaptic receptors and causes changes in the ionic conductance of the postsynaptic membrane. This leads to the appearance of a new nerve impulse in the postsynaptic neuron, or else to a blockage of the impulse, possibly mediated by another presynaptic neuron that is nonserotoninergic in nature (presynaptic inhibition) (Hamon 1981; Garnier et al. 1983). At the level of the synaptic cleft, the serotonin is either deactivated by monoamine oxidase (MAO), transformed into 5-HIAA which returns to the bloodstream, and then eliminated through urinary excretion, or else recaptured into the presynaptic process and stored in synaptic vesicles.

Topography

Fluorescence and autoradiographic studies in rats showed that serotoninergic neurons have their cell bodies essentially distributed in the medial area of the brain stem, i.e., in the raphe nuclei (Cohen et al. 1977; Hamon 1981). The medial and posterior groups of raphe nuclei send their fibers to the cerebellum and the spinal cord, respectively. As for the anterior raphe nuclei, they contain cell bodies with axons projecting to the anterior part of the brain and innervating all cerebral structures, i.e., the hippocampus and the cortex as well as the hypothalamus. The serotoninergic innervation is indeed extremely diffuse in the central nervous system and it can be said that the entire brain is provided with serotoninergic endings. The few maps obtained in humans, especially in fetuses, show that the spatial distribution of the different groups of serotoninergic cell bodies corresponds exactly to what is observed in rodents (Iversen 1984).

Functional Pole

The roles of the serotoninergic systems in physiology are numerous and involve sleep physiology, nociceptive control, behavioral regulation, and, as previously mentioned, phenomena of cerebral maturation.

Concerning the regulation of vigilance states, lesion studies as well as pharmacological experiments have pointed to the importance of serotonin in generating deep sleep. In fact, lesions in the rostral part of the raphe system containing the serotoninergic cell bodies result in abolition of slow wave sleep without any effect on paradoxical sleep. Blocking serotonin synthesis by p-chlorophenylalanine, which inhibits TH, also produces insomnia. In contrast, an intravenous injection of serotonin precursor induces a return to slow wave sleep (Jouvet 1969; Petitjean et al. 1978).

Following the observations of Reynolds (1969) describing the role of certain nuclei of the periaqueductal gray and of the medullary nucleus raphe magnus in some analgesia, Hosobuchi (1972) and Oliveras et al. (1978) suggested a contribution of serotonin in mediating analgesia. In fact, "analgesic sites" have been found in the raphe nuclei. Moreover, the analgesic effect of repeated central stimulations can be alleviated in man by administering serotonin precursors (Besson et al. 1987).

In the domain of behavioral regulation, blocking of serotoninergic transmission is generally accompanied, in animals, by a locomotor excitation, itself associated with a disinhibition inducing aggressive behavior. In contrast, facilitation of serotoninergic transmission sustains, in animals, the capacity to keep in check all motor action. In humans, Rydin et al. (1982) showed that there is an opposite relationship between anxiety and intensity of serotoninergic activity.

Blundell (1984) also described the role of serotonin in the regulation of feeding behavior. Some of the results obtained in the study show that serotonin inhibits the intake of food and contributes to the circadian control of feeding.

The serotoninergic systems are also involved in other functions such as regulation of body temperature, but these functions go beyond the frame of child psychiatric pathophysiology.

Pathophysiology

A deficit in the metabolism of serotonin seems indeed to be involved in certain behavioral disorders observed in children. However, it is on the role of this transmitter in depression that most of the extensive studies have been carried out.

The concept of a deficit in serotonin metabolism in depression is linked to the discovery of the mode of action of MAO inhibitors (MAOIs) and tricyclic antidepressants. Tricyclic antidepressants block at the synapse the active reuptake of serotonin into the storage vesicles of the presynaptic process. As for MAOIs, they inhibit serotonin degradation in the synaptic cleft, hence increasing the cerebral concentration of serotonin and prolonging the period over which this transmitter acts in the synaptic cleft. More recently, a certain amount of more specific serotonin molecules have proved their effectiveness (Willner 1985). These psychopharmacological data have prompted abundant work on depression and serotonin. Thus, many postmortem studies carried out on brains of depressed patients have shown a normal concentration of 5-HIAA and serotonin, but a decrease in the number of 3-imipramine binding sites (Stanley et al. 1982).

Mignot and Garcia (1986) report results of more than 15 CSF studies. Apart from a few methodological discrepancies, most of these studies report a constant diminution of about 30% in 5-HIAA measured in the CSF of depressed patients of all ages. However, the diverging results obtained from urine or plasma studies are not very convincing since they are contradictory. Mention should be made of the data gathered from serotonin metabolism at the level of blood platelets. In depressed patients Launay et al. (1983) observed a decrease in serotonin uptake by platelets and Langer et al. (1984) showed a decrease in the number of binding sites to tritiated imipramine.

Given the therapeutic action of antidepressants in eating disorders found in adolescents, such as anorexia nervosa or bulimia, a possible implication of serotonin producing these disorders was considered. In anorectic patients, Riederer et al. (1982) showed a decrease in 5-HIAA, a result similar to that found by Gerner et al. (1984), who performed their measurements in the spinal fluid. In contrast, Zemishlany et al. (1987) did not observe any change in serotonin reuptake in platelets of anorectic subjects.

Pharmacological approaches lead to less clear results. In fact, Hudson and Pope (1987) mention a positive effect of tricyclic antidepressants in bulimia, contrasting with the high variability of the results obtained in adolescents with anorexia nervosa.

A possible role for serotonin in specific sleep disorders in children and teenagers has been mooted in the discussion of sleep disorders associated with depression in children, although little information in respect of this is currently available.

The interest of studying serotonin in the context of research on infantile autism is discussed elsewhere in this book by Bruneau et al. One need only point out here that a potential role for serotonin in psychoses is suggested by the following findings:

1. Lysergic acid diethylamide, well known for its hallucinogenic properties, slows down serotonin turnover by inhibiting its formation from tryptophan and by blocking the specific serotoninergic receptors (Snyder 1976; Patay et al. 1980).
2. MAOIs regularly aggravate schizophrenic symptoms. MAOIs largely increase the content of serotonin in the brain by reducing the degradation of this amine by the enzyme MAO. This augmentation in serotonin concentration is all the more effective as the normal synthesis of serotonin itself remains unaffected.

Finally, it should be mentioned that maximal platelet imipramine binding sites correlated negatively with severity of aggression in conduct-disordered adolescents. These data were interpreted as reflecting a "serotonergic dyscontrol" syndrome in which decreased serotonergic activity is related to behavioral response disinhibition (Stoff et al. 1985).

Dopaminergic Sytems

Dopamine Metabolism

Dopamine is synthesized from an amino acid, i.e., tyrosine. Hydroxylation of tyrosine by TH results in 3,4-dihydroxyphenylalanine (Dopa). Dopa is then decarboxylated by Dopa-decarboxylase (DDC) to produce dopamine. In neurons where dopamine acts as neurotransmitter, i.e., in dopaminergic neurons, the synthesis stops at this stage. In noradrenergic neurons, however, dopamine is only a metabolic intermediate and is further hydroxylated by dopamine-β-hydroxylase (DBH) to form norepinephrine. The cerebral TH is concentrated in regions that are known to contain high concentrations of catecholamines, especially dopamine (Roth 1979). The DDC activity varies very much according to brain areas.

Regulation

The key to the regulation of the dopaminergic metabolism is the enzyme TH. Many observations indicate that TH activity accounts for the alterations in the level of physiological activity in catecholaminergic neurons. TH activity can be inhibited by catecholamines through negative feedback (Snyder 1976).

Since TH is a cytoplasmic enzyme, only dopamine which is free in the cytoplasm of dopaminergic neurons can have an inhibitory influence on it. On the other hand, it is a recognized fact that the presynaptic process contains receptors sensitive to the quantity of intrasynaptic dopamine and capable of regulating the release of dopamine as well as its synthesis. The activation of these receptors (inhibitory autoreceptors) by dopamine agonists decreases both the release and the synthesis of dopamine. A blockade of these receptors has the reverse effect.

Apart from these short-term regulatory mechanisms, involving quick adaptation to changes in physiological activity of dopaminergic neurons, there exists a long-term regulatory mechanism. The latter refers in fact to the enzymatic induction and quantitative increase of TH observed in response to prolonged neuronal activity. This has also been observed for DBH (Roth 1979).

Catabolism

Dopamine is catabolized into dihydroxyphenylacetic acid, HVA, and 3-methoxytyramine by means of three enzymes: MAO, catechol-*o*-methyltransferase, and ADH.

—Catechol-*o*-methyl transferase: At the CNS level, its distribution varies according to brain regions, and there is no correlation between this distribution and that of catecholamines.
—Monoamine oxidase: In the brain, MAO is found in catecholaminergic neurons. It is a mitochondrial enzyme.

Topography

The central dopaminergic systems have been mapped in rats by means of various methods: fluorescence microscopy employing fluorescent tracers, autoradiography using radiolabeled compounds, and histochemistry detecting dopamine and TH (Garnier et al. 1983). At the morphological level, two types of dopaminergic system are distinguished: long dopaminergic pathways (cells with long axons) and short dopaminergic pathways (cells with short axons; Nieoullon 1982).

Long Dopaminergic Pathways

Long dopaminergic pathways comprise ascending pathways projecting to the anterior part of the brain and originating from dopaminergic neurons that have their cell bodies located in the midbrain. This mesotelencephalic system is subdivided into three parts:

—The nigrostriatal system: Dopaminergic cell bodies are located in the nucleus A9 of the substantia nigra, with axons that project primarily to the putamen and the caudate nucleus.
—The mesolimbic system: Cell bodies are located in the nucleus A10 and their axonal projections are found in limbic structures, such as the olfactory tubercle, septum, nucleus accumbens, and amygdala.
—The mesocortical system: Cell bodies are also located in the mesencephalic nucleus A10. These nerve cells innervate various areas of the limbic and the prefrontal cortex, the ventral part of the frontal cortex, the cingulate cortex, and the entorhinal cortex. This means that cortical dopaminergic endings in rats project to areas which correspond in humans to prefrontal areas. The existence of such dopaminergic endings in prefrontal areas has indeed been verified in humans.

Short Dopaminergic Pathways

Short dopaminergic pathways contain periventricular, hypothalamic, and pituitary systems:

—The periventricular system: Cell bodies located in the periaqueductal gray area project their axons to the periaqueductal gray itself, the medial thalamus, and the hypothalamus.
—The incertohypothalamic system: Cell bodies located in the zona incerta send axons to various hypothalamic nuclei.
—The tuberoinfundibular system: Cell bodies in the periventricular nuclei of the thalamus have axons that innervate various nuclei of the hypothalamus and of the septum.
—The tuberohypophysial system: cell bodies located in the hypothalamus project to the median eminence and the pituitary gland.
—A system of very short interneurons within the olfactory lobe also belongs to the short dopaminergic pathways.

Functional Role

The functional role of the central dopaminergic systems remains fairly obscure, our knowledge being only fragmentary, except for the nigrostriatal system.

Nigrostriatal System

The study of the nigrostriatal system contributed to the discovery of a direct relationship between its dysfunction and Parkinson's disease. The blockade of dopaminergic receptors by neuroleptics induces secondary effects of parkinsonian type (Laverty 1978). Other signs such as hyperactivity and stereotypies are likely to involve different mechanisms (e.g., D2 vs D1 receptors). In fact, injecting different dopaminergic agonists into the striatum causes correspondingly different responses: apomorphine elicits stereotypies but no global hyperactivity, whereas amphetamine induces hyperactivity without stereotyped movements (Costall and Naylor 1979). It is possible that hyperkinetic states observed after treating parkinsonian patients with L-dopa are also due to hyperstimulation of dopaminergic receptors, possibly via receptor hypersensitivity.

Mesolimbic and Mesocortical Systems

One of the anatomical characteristics of the mesolimbic and mesocortical dopaminergic pathways is the divergence of their projections. Nevertheless, the position of their cell bodies situated in a circumscribed area of the midbrain renders their study delicate.

As pointed out above, their functions are not well known. Most of the work has been on the nucleus accumbens. In rats, this nucleus is in fact highly developed and receives a rich dopaminergic innervation from the midbrain. The main function of this meso-accumbens pathway is to control locomotor activity. Thus, it is common to observe, in animals, hyperexploration behavior after injecting dopamine or dopaminergic agonists into the nucleus accumbens. The injection of amphetamine reproduces this phenomenon. Furthermore, electrolytic destruction of the ventral tegmental area induces locomotor hyperactivity in association with hyperexploration behavior. Chronic injection of amphetamine then allows the animal to regain normal locomotor activity. It is worth mentioning that akinesia, as observed in parkinsonian patients, is related to the degeneration of dopaminergic nerve endings extending into the nucleus accumbens (Besson and Tassin 1985). Since the discovery of dopaminergic projections in frontal areas of the human brain, it has been suggested that functions of the aforementioned nerve endings could be closely linked to the mediating functions usually attributed to the prefrontal cortex, namely associative skills and emotional control.

Intrahypothalamic Dopaminergic Neurons

Among short pathways, the role of the tuberoinfundibular system is well known. Neurons of the infundibular nucleus secrete hormones which pass into the hypophysial portal system and act directly on the secretion from the anterior pituitary. The tuberoinfundibular dopaminergic neurons inhibit the secretion of thyrotropin releasing hormone and luteinizing hormone releasing hormone. Through the hypophysial portal system as an intermediate, the dopamine released by tuberoinfundibular neurons exerts its influence directly on prolactin secretion (Lichtensteiger 1979). Tuberohypophysial dopaminergic neurons act directly on pituitary secretions: they control the secretion of α-melanocyte stimulating hormone and growth hormone (Fuxe 1978).

Pathophysiology

We will not deal here with the possible implication of a dopaminergic dysfunction in autism, as this Pathophysiology aspect of dopaminergic systems is described elsewhere in this book (see chapter by Bruneau et al.). However, the hypothesis that schizophrenia results from hyperdopaminergia (Alpert and Friedhoff 1980) does deserve consideration. This hypothesis is based on solid pharmacoclinical evidence:

1. Neuroleptics blocking dopaminergic receptors are the only compounds able to produce a significant improvement in the behavior and in thought processes of schizophrenics.
2. Amphetamines which enhance the release of catecholamines into the synaptic cleft generally worsen schizophrenic symptoms. Furthermore, high amphetamine doses can induce an acute paranoid psychosis hardly distinguishable, at the clinical level, from true acute paranoid schizophrenia. This was observed in drug users injecting amphetamines intravenously. Angrist et al. (1974) found in the CSF of these subjects a considerable augmentation of accumulated HVA after probenecid, a result indicating that amphetamines increase dopaminergic turnover. Moreover, the storage of 3-Methoxy-4-hydroxyphenylglycol (MHPG), the principal catabolite of noradrenaline, is not modified.
3. L-dopa, when given to schizophrenic patients, aggravates their symptoms.
4. MAOIs, which slow down the breakdown of intracerebral dopamine, may also lead to a worsening of schizophrenic symptoms.

Nevertheless, studies on dopaminergic hyperactivity using biochemical assays to detect a possible augmentation of dopamine metabolites in CSF or in urine have proved unfruitful.

Bird et al. (1980) examined, postmortem, the brains of schizophrenics. They found an increase in dopamine concentration in the nucleus accumbens and in the olfactory tubercle area, but they could not conclude whether these results were related to the disease itself or to prolonged treatment with neuroleptics. MacKay et al. (1980) reported an increase in the number of dopaminergic receptors in brains of schizophrenics submitted to postmortem study. But, once more, this could be interpreted as the consequence of neuroleptic treatment. Meltzer postulated a decrease in the sensitivity of pre-synaptic autoreceptors which would lead to an increase in synthesis and use of dopamine (Meltzer 1980).

Thus, in the light of recent findings on the functioning of neuroleptics, a new concept of biological abnormalities in schizophrenia has developed. In this disease, one can distinguish two types of symptom which allow, according to their respective importance, the classification of pathologies into two groups:

—In one form of schizophrenia, negative symptoms predominate. Moreover, at the motor level one often finds hypertonia and akinesia, suggesting a dopaminergic deficit.

—In another form of schizophrenia, in which paranoid-hallucinatory elements predominate, stereotypies are more common, and probably indicate dopaminergic hyperfunction.

Thus, clinical and pharmacological data strongly suggest that there exists a dopaminergic dysfunction in schizophrenics, but the underlying pathophysiological bases are unknown. In fact a dysfunction in dopaminergic systems appears to be directly or indirectly involved in several childhood psychiatric disorders. Thus in Gilles de la Tourette's syndrome Riddle et al. (1988) found a reverse relationship between severity of symptoms and plasma levels of HVA. This inverse relationship concerns mostly motor activity and cognitive functions and not so much tics. Identical results were observed in the CSF by Cohen et al. (1978). Also, it is well known that dopamine receptor blocking agents like tiapride and haloperidol are helpful in many cases of Gilles de la Tourette's syndrome. Shaywitz et al. (1977) found low levels of HVA in the CSF of children with minimal brain dysfunction. Oades (1987), finally, described in an overview disturbances in dopamine and norepinephrine as well as in the balance between dopamine and norepinephrine metabolisms, and their implication in attentional disorders with hyperactivity.

Noradrenergic Systems

Metabolism of Norepinephrine

In noradrenergic neurons, dopamine is first synthesized in the cytoplasm and then taken up into synaptic vesicles. It is within these vesicles and during its uptake that dopamine is transformed into norepinephrine by DBH.

As with dopamine synthesis, the limiting step in norepinephrine synthesis is the enzyme TH. The norepinephrine which is produced exerts a negative feedback action to control TH activity.

The existence of such a feedback control of DBH activity has not yet been demonstrated. Intravesicular norepinephrine is released into synaptic space by peripheral and central noradrenergic neurons. The release of norepinephrine is accompanied by the release of the soluble envelope of storage granules via exocytosis. DBH is released at the same time as norepinephrine into synaptic space. For norepinephrine, there is a large reuptake by the presynaptic elements.

The catabolism of norepinephrine is mediated by four enzymes: MAO, catechol-o-methyl transferase, an aldehyde oxidase, and an aldehyde reductase. Catabolism results in formation of many catabolites, of which MHPG and VMA are the most important (Garnier et al. 1983). MHPG and VMA represent central (Garfinkel et al. 1977) and peripheral noradrenergic metabolism, respectively. For MHPG, two different fractions can be measured. The MHPG-glucuronide reflects mainly the peripheral and the MHPG-sulfate reflects the central noradrenergic metabolism (Filser et al. 1988).

Topography

Norepinephrine is widely distributed in the brain. Cell bodies of noradrenergic neurons are situated at the level of the brain stem, in structures that are anatomically ill-defined, outside the locus coeruleus.

Four ascending and three descending noradrenergic systems are distinguished:

—The ventral fasciculus issues from the pons and goes through the midbrain to end in the hypothalamus, the preoptic area, and the olfactory tubercle and bulb.
—The fasciculus originating from the locus coeruleus distributes itself within the raphe nuclei, the preoptic area, and some thalamic nuclei. It also gives off branches towards the limbic system.
—The dorsal periventricular fasciculus, the origin of which remains obscure, reaches the hypothalamus and the periventricular thalamic nuclei.
—The central periventricular fasciculus reaches the dorsomedial thalamus.
—The cerebellar fasciculus originating in the locus coeruleus projects to the entire cerebellar cortex.
—The descending noradrenergic fasciculus deriving from the locus coeruleus is a short bundle which ends at the level of the bulbar reticular formation.
—The bulbospinal fasciculus, also originating from the locus coeruleus, spreads throughout the spinal cord.

Functional Role

Noradrenergic systems are highly important for the fetus. Normal development of these systems is crucial for the maturation of other neurotransmitter systems. In fact, we have previously seen that noradrenergic nuclei are the first to develop. Noradrenergic systems influence neuronal migration and synaptogenesis as well as the regulation of other neurotransmitter systems, in particular dopaminergic systems (Holmes 1986). This role of norepinephrine in maturation appears to be widespread. The other roles seem secondary and, as stated by Tassin (1984), ascending noradrenergic systems are important for the autostimulating processes, especially learning capacities and organizational patterns of information.

Pathophysiology

The roles of noradrenergic dysfunction in generating sleep disorders and autism are described in this book by Garreau et al. and Bruneau et al., respectively. The noradrenergic hypothesis in respect of ADHD is discussed by Zametkin and Rapoport (1987).

A possible dysfunction of noradrenergic systems in depression was studied mainly by assays of MHPG in urine. This permitted two subgroups of patients to be distinguished: those who have low levels of MHPG and those who have high levels. In the first case, patients are sensitive to imipramine, and in the second, to amitryptiline (Schildkraut et al. 1973). However, no difference in psychopathology between the two subgroups has been reported. Nevertheless, most authors, such as Goodwin (1978), suggest that a dysfunction of noradrenergic systems is due to disturbances in other neurotransmitter systems, namely serotoninergic systems. On the other hand, Berger and Müller (1989) have presented evidence to support a cholinergic–noradrenergic imbalance hypothesis in depressed patients. Further, Rothenberger et al. (in press) investigated MHPG-sulfate, MHPG-glucuronide, and VMA in anorectic adolescents. They found that mainly the peripheral noradrenergic metabolism was altered (i.e., there was a peripheral metabolic shift from oxidative—VMA—to reductive—MHPG-glucuronide—noradrenergic pathways during weight gain in anorectic adolescents). Finally, unpublished results in children with tics (Rothenberger, personal communication) show that changes in the dopaminergic system in response to drug treatment with dopamine receptor blockers parallel changes in the noradrenergic system.

Conclusion

In the present work, we have described the possible implication of mono-aminergic disturbances in the pathogenesis of child psychiatric disorders. The essential element is the absence of nosological specificity. It is possible that this lack of nosological specificity found in biochemical markers is due to the paucity of our knowledge on the interactions between various aminergic systems and between all neurotransmitter systems, even though the current trend is to argue in terms of balance between different systems. We will mention a few examples of interaction between dopaminergic systems and other neurotransmitter systems.

In the striatum, nigrostriatal dopaminergic neurons synapse with acetylcholine interneurons and exert an inhibitory action on them. The cholinergic interneurons in turn act, in an excitatory manner, on striatal dopaminergic endings. This renders understandable the improvement of motor disorders observed in parkinsonian patients after administration of dopamin-ergic and anticholinergic agonists, and elucidates the potentiation of their effects.

The dopaminergic cell bodies of the substantia nigra also receive direct cholinergic afferent fibers of an excitatory nature.

In contrast to striatal dopaminergic neurons, mesolimbic dopaminergic neurons do not synapse with cholinergic nerve cells but rather with noradrenergic ones.

γ-Aminobutyric acid (GABA) is inhibitory to dopaminergic neurons and diminishes their turnover in dopamine. There is an important GABAergic pathway with cell bodies located in the striatum and pallidum projecting to the substantia nigra. The GABAergic neuronal endings synapse with dendrites and cell bodies of nigrostriatal dopaminergic neurons. Directly or indirectly via cholinergic interneurons, dopaminergic neurons exert their inhibitory action on striatal GABAergic neurons. GABA is also inhibitory to the activity of mesolimbic neurons.

Some serotoninergic neurons of the raphe project to the substantia nigra. It seems well established that serotoninergic action inhibits dopaminergic neurons. Limbic structures innervated by the mesolimbic dopaminergic system, especially by the nucleus accumbens, also receive a massive serotoninergic input. In addition, the existence of an influence of norepinephrine on dopaminergic activity in the brain became evident some time ago (Lloyds 1978).

More recent work performed by the group of J.P. Tassin demonstrated that dopaminergic neurons of the ventral tegmental area receive noradrenergic afferent neurons. These authors demonstrated that destruction of noradrenergic systems originating in the locus coeruleus and projecting, among other sites, to dopaminergic cell bodies, has two effects: on the one hand, a decrease in dopaminergic activity of the cortex, and on the other hand, a proliferation of dopaminergic nerve endings in the cortex. These systems, the activity of which decreases due to absence of norepinephrine, seem to some extent to compensate for this diminution by multiplying their branching.

The study of these interactions between neurotransmitter systems indicates the interest in approaching pathology by focusing on the biochemical "profile" rather than the metabolism of a given system.

Independence between dysfunction of monoaminergic systems and nosological classification (depression, attention deficit disorder, autism, tic disorder) may explain the lack of specificity previously mentioned. It is possible that certain behavioral symptoms are related to the functional state of one neurotransmitter system or the other. We will take as an example the work of Soubrie (1986), who studied the relationship between serotoninergic activity and behavior. In animals, destroying serotoninergic neurons results in a reduction of the cataleptogenic action of neuroleptics and favors locomotor excitation, be it in the form of focalized or stereotyped activities. In contrast, facilitating serotoninergic transmission sustains the capacity to curb motor action. Electrochemical lesions increase many forms of aggressive behavior. Thus, serotoninergic neurons seem to play an important role in behavioral control when the task is to produce a response requiring a long preparatory period or a high degree of inhibition (e.g., to avoid punishment). Soubrie considers the activation of serotoninergic systems as likely to be associated with tolerance to situational constraints. Lesions of serotoninergic afferents decrease the inhibition observed in animals facing a conflict situation. Thus, looking at animal as well as clinical data, the author favors a close relation between control of impulsiveness and serotoninergic activity in man.

Abnormally low levels of 5-HIAA were found in the CSF of subjects who had attempted to commit suicide (Lidberg et al. 1984), and even in anorectic patients with bulimia (Kaye et al. 1984).

Different approaches could explain the lack of specificity between nosological classification and central monoaminergic dysfunctions. For instance, very little is known about the role played by some false neurotransmitters, such as octopamine, which were thought to be involved in some pathological conditions. Gathering of more data in neurotransmitter and neuropeptide research is required, especially during child development.

Nevertheless, it seems that some biochemical abnormalities may be related to the intensity of certain clinical symptoms or syndromes. From a diagnostic point of view, these data show that a better clinical classification and more rigorous semiological approach should be considered.

Acknowledgments. We thank Mrs. L. Crespin and Mr. C. Chevet for their technical assistance and Prof. Dr. A. Rothenberger for valuable suggestions. This study was supported by grants from INSERM No 859014, Fondation pour la Recherche Médicale, Fondation H. Langlois, Sécurité Sociale, and Conseil Régional de la Région Centre. Consention d' animation de résedue INSERM "Etudes des functions neurophysiologiques et compréhension des compartiments autofigues" No 489001.

References

Alpert M, Friedhoff AJ (1980) An un-dopamine hypothesis of schizophrenia. Schizophr Bull 6:387–390

Angrist B, Sathananthan G, Wilk S, Gershon S (1974) Amphetamine psychosis: behavioral and biochemical aspects. J Psychiatr Res 11:13–23

Berger M, Müller W (1989) Psychologische Verlaufsuntersuchungen zur noradrenergen-cholinergen Imbalance—Hypothese affektiver Erkrankungen. Arbeits- und Ergebnisbericht des Sonderforschungsbereichs 258. University of Heidelberg, pp 1–62

Besson JM and Tassin JP (1985) Les systèmes dopaminergiques et leurs récepteurs. Neuropsy 1:21–29

Besson JM, Chaouch A, Chitour D (1987) Physiologie de la douleur. Encyl. Med. Chir. (Paris-France), Neurologie, 17003F10, 8P

Bird ED, Spokes EG, Iversen LL (1980) Dopamine and noradrenaline in post-mortem brain in Huntington's diesase and schizophrenic illness. Acta Psychiatr Scand Supplt 280, 61:63–73

Blundell JE (1984) Serotonin and appetite. Neuropharmacology 23:1537–1551

Bruneau N, Garreau B, Barthélémy C, Martineau J, Muh J, Lelord G (1990) Clinical, electrophysiological, and biochemical markers and monoaminergic hypotheses in autism. This volume, pp 217–234

Cohen DJ, Caparullo BK, Shaywitz BA and Bowers MB (1977) Dopamine and serotonin metabolism in neuropsychiatrically disturbed children. Arch Gen Psychiatry 34:534–537

Cohen DJ, Schaywitz BA, Caparulo B, Young JG and Bowers MB (1988) Chronic, multiple tics of Gilles de la Tourette's disease: CSF acid monoamine metabolites after probenecid administration. Arch Gen Psychiatry 35:245–250

Costall B, Naylor RJ (1979) Behavioral aspects of dopamine agonists and antagonists. In Horn AS, Korf J, Westerink BHC (eds) Neurobiology of dopamine. Academic Press Publ 31: 555–576

Coyle JT (1985) Development of brain neurotransmitters and child psychiatry. In RF Hales and AJ Frances (eds) Annual review of the American Academy of Child Psychiatry, Vol. 4, Washington D.C., American Psychiatry Press

Evrard Ph, Lyon G and Gadisseux JF (1984) Le développement prénatal du système nerveux et ses perturbations: les mécanismes généraux. In: Progrès en néonatologie. Karger, Basel, pp 63–69

Filser JG, Spira J, Fischer M, Gattaz WF, Müller WE (1988) The evaluation of MHPG-sulfate as a possible marker of the central norepinephrine turnover. Studies in healthy volunteers and depressed patients. J Psychiatr Res 22:171–181

Fuxe K (1978) Place de la dopamine dans les amines biogènes à fonction de neurotransmetteur. Triangle. J Sandoz Sci Méd 28:125–136

Garfinkel PE, Warsh JJ, Stranger HC, Godse DD (1977) CNS monoamine metabolism in bipolar affective disorder. Evaluation using a peripheral decarboxylase inhibitor. Arch Gen Psychiatry 34:735–739

Garnier C, Barthélémy C, Garreau B, Jouve J, Muh JP and Lelord G (1983) Les anomalies des monoamines et de leurs enzymes dans l'autisme de l'enfant. L'encéphale, IX 3:201–261

Garreau B, Barthélémy C, Bruneau N, Martineau J (1990) Sleep disturbances in children: from the physiological to the clinical. This volume, pp 317–342

Gerner RH, Cohen DJ, Fairbanks L, Anderson GM, Young JG, Scheirin M, Linnoila M, Shaywitz BA, Hare TA (1984) CSF neurochemistry of women with anorexia nervosa and normal women. Am J Psychiatr 141:1441–1444

Hamon M (1981) Les systèmes sérotoninergiques corticaux. Séminaire de psychiatrie biologique, Hôpital Ste Anne 1:107–126

Holmes GL (1986) Morphological and physiological maturation of the brain in the neonate and young children. J Clin Neurophysiol 3, 3:209–238

Hosobuchi Y (1972) Tryptophan reversal of tolerance to analgesia induced by central gray stimulation. Lancet 2:47

Hudson JI and Pope HG (1987) Newer antidepressants in the treatment of bulimia nervosa. Psych Pharmacology Bull 23:52–57

Iversen SD (1984) 5-HT and anxiety. Neuropharmacology 23:1553–1560

Jouvet M (1969) Biogenic amines and the states of sleep. Science 163:32–41

Kaye WH, Ebert MH, Gwirtman HE, Weiss SR (1984) Differences in brain serotoninergic metabolism between nonbulimic and bulimic patients with anorexia nervosa. Am J Psychiatry 141:1598–1601

Langer SZ, Raisman R, Sechter D, Gay C, Loo H, Zarifian E (1984) ^{3}H-Imipramine and ^{3}H-desipramine binding sites in depression. In: Frontiers in biochemical and pharmacological research in depression. E Ushdin et al. (eds). Raven, New York, pp 113–125

Launay JM, Pasques D, Dreux C (1983) Données nouvelles sur le métabolisme de la sérotonine dans les plaquettes sanguines: modèles de neurones. Act Phar Biol Clin 2:155–159

Laverty R (1978) Catecholamines: role in health and disease. Drugs 16:418–440

Lichtensteiger W (1979) The neuroendocrinology of dopamine systems In: Horn AS, Korf J, Westerink BHC (eds). Neurobiology of dopamine. Academic, London New York San Francisco, pp 491–521

Lidberg L, Asberg M, Sunquist-Stensman UB (1984) 5-Hydroxyindolacetic acid levels in attempted suicides who have killed their children. Lancet 2:928

Linnoila M, Virkkunen M, Scheinin M, Nuutila A, Rimon R, Goodwin FK (1983) Low cerebrospinal fluid 5-hydroxyindolcetic acid concentration differentiates impulsive from nonimpulsive violent behavior. Life Science 33:2609–2614

Lloyds KG (1978) Neurotransmitter interactions related to central dopamine neurons. Essays Neurochem Neuropharmacol 3:131–207

Lyon G, Evrard Ph and Gadisseux JF (1984) Les anomalies du dévelopement du télencéphale human pendant la période cytogénèse-histogénèse. Progrès néonatalogie: 70–84

Mac Kay AvP, Bird ED, Iversen LL, Spokes EC, Creese I, Snyder SH (1980) Dopaminergic abnormalities in post-mortem schizophrenic brain. Adv Biochem Psychopharmacol 24:325–333

Meltzer HY (1980) Relevance of dopamine autoreceptors for psychiatry: preclinical and clinical studies. Schizophr Bull 6:456–475

Mignot E and Garcia A (1986) Sérotonine et dépression. Neuropsy 12:353–362

Nieoullon A (1982) La transmission chimique de l'influx nerveux. Encycl Med Chir. Paris, Neurologie 17003, A10

Oades RD (1987) Attention deficit disorder with hyperactivity: the contribution of catecholaminergic activity. Progress in Neurobiology 29:365–391

Oliveras JL, Hosobuchi Y, Guilbaud G, Besson JR (1978) Analgesic electrical stimulation of the feline nucleus raphe magnus. Development of tolerance and its reversal by 5-HTP. Brain Research 146:404–409

Patay M, Scharbach H and Levron M (1980) Glycine, acétylcholine et sérotonine: neuro-transmédiateurs ou neuromodulateurs? Psycho Med 12:25–34

Petitjean F, Buda C, Janin M, Sakai K, Jouvet M (1978) Patterns of sleep alteration following selective raphe nuclei lesions. Sleep Research 7:40

Reynolds DV (1969) Surgery in the rat during electrical analgesia induced by focal brain stimulation. Science 104:444–445

Riddle MA, Leckman JF, Anderson GM, Ort SI, Hardin MT, Stevenson J, Cohen DJ (1988) Tourette's syndrome: clinical and neurochemical correlates. J Am Acad Child Psychiatry 27:409–412

Riederer P, Toifl K and Kruzik P (1982) Excretion of biogenic amine metabolites in anorexia nervosa. Clin Chim Acta 123:27–32

Roth RH (1979) Tyrosine hydroxylase. In: Horn AS, Korf J, Westerink BHC (eds) The neurobiology of dopamine. Academic, New York, pp 101–122

Rothenberger A, Müller HU, Müller W (1990) Central versus peripheral disturbances in noradrenergic metabolism of adolescents with anorexia nervosa. In: Remschmidt H, Schmidt MH (eds). Child and Youth Psychiatry–European Perspectives, vol. 1, Huber, Houston Toronto Berlin, in press

Rydin E, Schalling D, Asberg M (1982) Rorschach ratings in depressed and suicidal patients with low levels of 5-HIAA in CSF. Psychiatr Res 7:229–243

Sarnat HB (1987) Troubles de la migration neuronale terminale au cours de la période prénatale. Am J Dis Child 6, 59:380–390

Schildkraut JJ, Keeler BA, Grab LL (1973) MHPG excretion and clinical classification in depressive disorders. Lancet 1:1251–1252

Shaywitz BA, Cohen DJ and Bowers MB (1977) CSF monoamine metabolites in children with minimal brain dysfunction: evidence for alteration of brain dopamine. Pediatrics 90:67–71

Synder SH (1976) Catecholamines, serotonin and histamine. In: Siegel GJ, Alberts RW, Katzman R, Agranoff BW (eds) Basic neurochemistry. Little Brown, Boston, pp 203–217

Soubrie P (1986) Neurones sérotoninergiques et comportement. J Pharmacol 11–2:107–112

Stanley N, Vingilio J, Gershon S (1982) triated imipramine binding sites are decreased in the frontal cortox of suicide. Science 216:1337–1338

Soubrie P (1986) Neurones sérotoninergiques et comportement. J Pharmacol 11–2:107–112

Stoff DM, Pollock L, Bridger WH (1985) Platelet imipramine binding sites correlate with aggression in adolescents. In: Shagass C, Josiassen RC, Bridger WH, Weiss KJ, Stoff DM, Simpson G (eds) Biological psychiatry. Elsevier, New York, pp 180–183

Tassin JP, Les Systèmes noradrénergiques centraux. Séminaire de psychiatrie biologique, Hôpital Ste Anne 1:81–96

Willner P (1985) Antidepressants and serotoninergic neurotransmission: an integrative review. Psychopharmacology 85:307–404

Young JG, Cohen DJ, Anderson GM and Shaywitz BA (1984) Neurotransmitter. In: Greenhill L and Shopsin B (eds). Ontogeny and perspective for studies of child development and pathology of childhood. Spectrum, New York, pp 51–83

Zemishlamy Z, Modai I, Apter A, jerushalmy Z, Samuel E and Tyano S (1987) Serotonin (5-HT) uptake by blood platelets in anorexia nervosa. Acta Psychiatr Scand 75:127–130

Zametkin AJ, Rapoport JL (1987) Neurobiology of attention deficit disorder with hyperactivity: where have we come in 50 years? J Amer Acad Child Adol Psychiatry 26:676–686

Genetic Aspects of Brain Maturation and Behavior

J. Hebebrand and P. Propping

Introduction

The application of molecular biology to the neurosciences offers a promising approach to uncover some underlying mechanisms relevant for the maturation of the central nervous system. Our knowledge of the involved genes and their products is increasing at a dramatic rate. However, in the light of the fact that approximately 30% of human genes are expressed in the brain, we are just beginning to evaluate these 15–30 000 genes. At the DNA level every 100th base pair is polymorphic and the average heterozygosity rate is about 1:270. At the gene product level the average heterozygosity ranges between 2.2% and 15% depending on the tissue examined. If these figures are applied to the central nervous system, it becomes evident that an appreciable genetic variation probably exists in the brain that should also be relevant for brain maturation and function. The genetic variation can ultimately lead to variable neurophysiological systems that interact differentially with environmental factors (Propping 1987a).

According to Volpe (1987), brain maturation can be subdivided into six major developmental events (Table 1). Both exogenous and genetic factors can induce aberrant development at all stages. Disorders usually result in rather uniform clinical symptoms encompassing mental retardation, neurological abnormalities, and/or seizures. However, normal intelligence is compatible with some disorders independent of the developmental stage at which the aberration occurred (for detailed clinical descriptions see Warkany et al. 1981). At birth the total incidence of congenital malformations is approximately 3.5%. Up to age 14 slightly more than 4% of children will have suffered from cerebral palsy, epilepsy, and/or mental retardation (von Wendt and Rantakallio 1986).

In genetic terms monogenic and polygenic as well as chromosomal disorders may lead to aberrant brain maturation. Not infrequently, different genetic disorders can result in the same delineated phenotype such as micrencephaly, holoprosencephaly, or agyria. These genetic disorders offer an approach to define genetic influences on brain maturation and subsequent behavior. We will attempt to show the complex influence of genetic factors on brain maturation by selectively describing some relatively frequent disorders. Special emphasis will be placed on the *genetic mechanisms* involved and on the *neuropathological substrate* of the respective disease and its *neuropsychological consequences*. We will use the

Table 1. Normal and aberrant development of the central nervous system, including examples for causative genetic disorders (essentially according to Volpe 1987). Exogenous and genetic factors can induce aberrant development at each major maturational event. Among genetic disorders monogenic, polygenic (examples not included in table), and chromosome anomalies must be considered

Major events in maturation of central nervous system	Pathoanatomical disorders resulting from aberrant development at stage in question	Single examples of specific genetic disorders which induce aberrant development at stage(s) in question, including chromosome anomalies and monogenic diseases
Dorsal induction a) Primary neurulation, 3–4 weeks of gestation b) Caudal neural tube formation (secondary neurulation), 4 weeks of gestation to after birth	Craniorachischisis Anencephaly Myeloschisis Encephalocele Myelomeningocele	Trisomy 18 Meckel's syndrome (a.r.)
Ventral induction 5–6 weeks of gestation	Holoprosencephaly Faciotelencephalic malformations	Trisomies 13 and 18; 13q⁻; 18p⁻ Meckel's syndrome (a.r.) Alobar holoprosencephaly (a.r.)
Neuronal proliferation 2–4 months of gestation	Micrencephaly Macrencephaly	Trisomies 13, 18, 21, 22; 4p⁻; 5p⁻ X-linked recessive micrencephaly Achondroplasia (a.d.) Neurofibromatosis (a.d.) Tuberous sclerosis (a.d.) Tay-Sachs disease (a.r.)
Migration 3–5 months of gestation	Schizencephaly Agyria Macrogyria Polymicrogyria Neuronal heterotopias Agenesis of corpus callosum	Trisomies 13 and 21 Miller-Dieker syndrome (17p⁻ and a.r.) Neurofibromatosis (a.d.) Tuberous sclerosis (a.d.) Zellweger syndrome (a.r.) Aicardi's syndrome (X.d.) Incontinentia pigmenti (X.d.)
Organization 6 months of gestation—years postnatal	Abnormal development of dendritic branching and spines Disturbance of cytoskeletal structures	Trisomies 13, 15, 21 Phenylketonuria (a.r.)
Myelination birth—years postnatal	Cerebral white matter hypoplasia Myelination disorders	Maple syrup urine disease (a.r.) Phenylketonuria (a.r.) Homocystinuria (a.r.) Pelizaeus-Merzbacher disease (X.r.)

Abbreviations: a.d., autosomal dominant; a.r., autosomal recessive; X.d. and X.r., X-linked dominant and recessive

selected disorders only to illustrate this approach, which reflects upon the interdependency of these variables. At the same time it will become quite clear that this analytical approach leads to more unanswered than solved questions. Despite the fact that some of the selected disorders have been thoroughly described from a geneticist's, a pathologist's, and a psychiatrist's viewpoint, the interdependency of the underlying genetic mechanisms and the resulting symptoms has presently been understood at a rather superficial level only.

In the second section we will discuss some investigations concerning the genetic influence on normal development and behavior. The limited methodology presently available to investigate the genetic influence on normal maturation and behavior makes this an even more formidable task than the evaluation of genetic mechanisms that lead to aberrant development and/or behavior. Finally, we propose and outline a strategy involving research at the molecular level of brain function.

Sex Chromosome Abnormalities

Sex chromosome abnormalities make it possible to assess the influence of the sex chromosome complement on specific aspects of personality and functional development. Upon the discovery of X chromosomal aneuploidies in 1959 (Jacobs and Strong 1959; Ford et al. 1959) many cases were subsequently ascertained in mental and penal institutions or because of phenotypic abnormalities. For example, a criminobiological study of patients with the XYY syndrome and Klinefelter syndrome revealed that the incidence of sex chromosome aberrations among institutionalized males was 2.5% (12/480). The criminal offenses consisted mainly of petty thieving, indecency, sexual offenses, arson, and less violent crimes (Tsuboi 1970). However, prospective longitudinal studies have subsequently revealed that in contrast to disorders of the autosomes, sex chromosome aneuploidy is not consistently associated with a readily recognizable phenotype. We discuss these abnormalities in greater detail, because the respective genotypes are clearly defined. Additionally, many neuropsychological studies have been performed in individuals affected with a disorder of the sex chromosomes. Unfortunately, neuropathological observations are scanty. We focus on disorders presenting with a supernumerary or lacking X chromosome to allow an evaluation of the influence of quantitative imbalances in the number of X chromosomes on cognitive development and behavior. Despite the clearly defined genotypes the effect of an additional or missing X chromosome on behavior and intelligence does not follow any recognizable rule.

Neuropsychological Studies

Prospective longitudinal studies were initiated in Denmark in the mid-1960s in an attempt to obtain unbiased information on the physical and mental development of affected individuals. In a subsequent investigation Witkin et al.

(1976) were able to show that XYY men have a slightly reduced intelligence but no particular propensity for aggressive behavior as believed before. Instead the authors concluded that the higher rate of criminality in XYY men could partially be explained by their reduced IQ. In addition, personality traits were held responsible. On the other hand, in XXY men the increase in criminality could be completely explained by the IQ deficit.

The cytogenetic surveys on newborns and the consecutive longitudinal studies carried out in Europe and North America have tremendously helped to delineate the specific consequences of the sex chromosome aneuploidies. In the Danish study conducted by Nielsen et al. (recently updated in 1986) 58 children with sex chromosome abnormalities were ascertained by screening 20 222 newborns. The incidences of the major sex chromosome abnormalities 47,XXY, 47,XYY, 47,XXX, and 45,XO were 1.5, 1.1, 1.3, and 0.7/1000 of the respective sex. Thus, sex chromosome abnormalities are not rare events. It was estimated that in the general population only a total of 15% of all sex chromosome abnormalities are likely to be diagnosed throughout life.

Klinefelter Syndrome (47,XXY)

In all prospective longitudinal studies boys with a supernumerary X chromosome manifested delayed speech and language development. Accordingly, verbal IQs were significantly lower than performance IQs in most of the studies (Walzer et al. 1986; Ratcliffe et al. 1986; Stewart et al. 1986; Evans et al. 1986). This deficit in the verbal IQ (order of magnitude, 10 IQ points) was not only evident in childhood, but continued to be a problem in prepubertal boys (Walzer et al. 1986). The language abilities were characterized by a pattern of deficits that included problems in rate and order processing of auditory stimuli, problems in understanding of complex grammatical constructions, and difficulties with oral language production whereas receptive language scores were largely found to be age appropriate (Graham et al. 1988). Not surprisingly, the majority of boys had persistent difficulties in reading and spelling, often requiring speech therapy and remedial teaching (40%–80%). Furthermore, 12 out of 14 boys (Robinson et al. 1986) and 11/13 boys (Walzer et al. 1986) were identified by school personnel as having learning difficulties. In the first study, it was specifically mentioned that the teachers had been unaware of the karyotype. Two of the 19 boys in the second study had done consistently well in all areas of academic achievement. Multiple regression analysis, including variables of social class, mother's educational level, and handedness, revealed that the WISC scores for 12 XXY boys and 23 control boys (ages 6–8 years) were significantly different in verbal ($P < 0.01$), mean performance ($P < 0.05$), and mean full-scale ($P < 0.01$) scores (Ratcliffe et al. 1986). The latter ranged from 67 to 130, with a mean of 100.8 (115.7 in control sample). Four XXY boys had significantly lower verbal than performance scores (in each case $P < 0.01$), and one boy had a significant difference in the opposite direction ($P < 0.05$).

Neuromaturational lags encompassing reduced gross and fine motor skills, coordination, speed, dexterity, and strength were observed in the XXY boys.

Slightly more than 50% showed diminished self-esteem, anxiety, and immature behavior which are probably secondary to school problems (Robinson et al. 1986; Salbenblatt et al. 1987). XXY boys had significantly lower scores than controls for the categories of activity and intensity of responsiveness. They showed a significant tendency to withdraw from novel situations and experiences, and their pliancy scores were high (Walzer et al. 1986). Stewart et al. (1986) found the XXY boys to be less social, assertive, and active than XY boys.

Triple X Syndrome (47,XXX)

Of the major sex chromosome abnormalities the triple X syndrome is consistently associated with the lowest IQ values, which in the Edinburgh study ranged from 57 to 94 (at ages 7–10). The mean of 75.9 was significantly ($P < 0.001$) lower than in controls (105.8). Verbal scores were more depressed than performance scores. The cognitive ability was found to be at the lower end of the normal range (Ratcliffe et al. 1986). Stewart et al. (1986) also observed greater deficits of the XXX girls in verbal as compared to performance scores. The scores for arithmetic and reading were significantly lower than in sibs. The investigators concluded that XXX girls were impaired verbally and to a lesser extent spatially.

Robinson et al. (1986) give an impression of the psychosocial abilities of XXX girls: Two of 11 girls functioned normally; eight exhibited various degrees of developmental difficulties but were not perceived as conspicuously different from other family members. One girl contrasted sharply from her family. Five girls had a below average intellectual ability, and six revealed severe language impairment. Eight of the 11 girls had a motor dysfunction, and three were unable to achieve a satisfactory psychosocial adjustment. Major stressful events with long-lasting consequences occurred in 8 of the 11 families (e.g., divorce of parents, alcoholic father, allegation of child abuse). Except in two of the families (those of the most successful girls) the affected children had suffered from ambiguous relationships at home. Thus, the psychosocial capabilities of the XXX girls in this study possibly cannot be considered as representative of all girls with this sex chromosome abnormality. Instead, the study suggests that the adaptation of an XXX girl is dependent on the efforts of her family to provide supportive encouragement. The XXX and XXY syndromes are possibly associated with an increased frequency of psychoses diagnosed as schizophrenia (Propping and Friedl 1988).

Turner Syndrome (45,XO)

In contrast to the two disorders with supernumerary X chromosomes, girls with the 45,XO complement have verbal scores in the normal range. Mean performance and full-scale IQ scores for nine girls with XO and partial X monosomy were significantly below average; the data also suggest a lower mean verbal IQ, but the difference was not significant. Deficits were observed in perceptual skills, in contrast to the presence of adequate skills in receptive and expressive language. Five of the nine girls showed poor concentration and were easily distractible. Six had learning difficulties which required special education (Robinson et al. 1986). The girls with Turner syndrome revealed a sensory–motor integration dysfunction, leading to a decreased perceptual awareness of the body in space.

The physical limitations associated with the syndrome, especially short stature, have a pervasive influence on psychosocial adaptation. Some girls are impulsive. Turner girls have been described as being conscientious and diligent. They tend to have a good relationship with teachers and schoolmates. Their school performance levels were found to be quite similar to those of their normal sisters. Only 7% were in special groups and 22% had received extra teaching in one or more subjects (figures correspond to average). Forty percent had special difficulties in arithmetic (cf. 20% of sisters). Of the girls 17 years old and older, 26% were in college (Nielsen and Sillesen 1981). Twenty-seven percent of the girls had minor mental problems which in no case led to an admission to a psychiatric hospital or treatment for mental illness. This rate was not considered as above the expected frequency in the general population. An extensive study concerning the psychopathology and social functioning in women with Turner syndrome, in comparison with matched women of constitutionally short stature, revealed that women with Turner syndrome display less mental illness according to positive symptom-oriented criteria, but also less mental health, when day-to-day functioning is considered (Downey et al. 1989). Psychological characteristics of girls with Turner syndrome include a low responsiveness to emotional stimuli, poor emotional expressions, passive attitude, passive personal relations, and lack of aggressiveness.

Conclusions

The longitudinal studies have helped tremendously to define the variability of mental and psychological development of persons with sex chromosome abnormalities. Some affected individuals can function perfectly normally, and their intelligence levels can even be above average. Accordingly, in these subjects the chromosomal aberration would surely not have been suspected on the basis of mental or behavioral problems. On the other hand the majority of affected individuals show learning disabilities, as clearly demonstrated by the high percentage who were thus identified by their teachers and who required remedial teaching. Additionally, psychological problems exist to a significantly higher extent than in controls. In general terms, the encountered symptoms are similar to those attributed to the minimal brain dysfunction syndrome (Theilgaard 1984). In XXY males and to a greater extent in XXX females verbal test scores are depressed. Performance scores are seemingly not reduced to a similar extent: they are either (low) average (XXY) or clearly below average (XXX). In contrast, Turner syndrome is associated with a more or less average verbal test score and deficits in the performance scores.

Obviously, genetic factors other than the specific chromosome aberrations also influence the extent of the deviation from normal intellectual skills and behavior. Furthermore, environmental factors, such as compensatory training, certainly improve the ability of a specific individual to cope with limitations posed by the chromosome aberration.

Genetic Mechanisms

The mechanisms by which the sex chromosomal abnormalities lead to the above-mentioned behavioral phenomena are unclear. A clear-cut separation of genetically from environmentally induced problems cannot be accomplished. Nevertheless, it seems reasonable to assume that the underlying genetic mechanisms are responsible for a large portion of the behavioral phenomena by acting indirectly (via endocrinological mechanisms) and/or directly on central neurons.

Regarding the sex chromosome aneuploidies involving a supernumerary or lacking X chromosome, it has to be kept in mind that in normal 46,XX women one of the two X chromosomes is inactivated presumably slightly prior to the 16th day of embryonic development, at which time the X chromatin is formed rather suddenly in the entire embryo. Inactivation of one or two X chromosomes, respectively, is also apparent in the XXY and XXX syndromes. However, inactivation is not complete: genes localized in the distal part of the short arm of the X chromosome (Xp) escape inactivation. The term pseudoautosomal was introduced to describe the genetic behavior of these sequences: This region pairs with a homologous section on the distal short arm of the Y chromosome in male meiosis, so that recombination can occur between the pseudoautosomal regions of the X and Y chromosomes. Thus, these sequences fail to show classic sex linkage and simulate autosomal behavior (Goodfellow et al. 1987).

Accordingly, the biological substrate of the physical and behavioral manifestations of the X0,XXY, and XXX syndromes may be caused by the numerical aberration present during early ontogenesis until X-inactivation occurs. However, the early inactivation makes it appear unlikely that the neuropsychological manifestations are solely sequelae of gene dosis effects induced at this developmental stage. Alternatively, the behavioral phenomena result from the chromosome imbalance that either leads to a deficient (Turner syndrome) or a triplicate (XXX, XXY) dosis of genes that are located on distal Xp. Furthermore, single other X chromosomal genes outside the pseudoautosomal region are possibly not inactivated and thus contribute to gene dosis effects on brain function. Another conceivable mechanism is that the numerical X chromosome aberration per se negatively influences cell maturation and division rates.

Since hormonal disorders are a major feature of sex chromosome abnormalities, it is well conceivable that they cause or contribute to the slight mental deficits observed (Nyborg 1984). A comparison of the cognitive style of 47,XXY males with 46,XY hypogonadal males revealed that the former were characterized by relative weakness in verbalization, arithmetic, and reading abilities. Furthermore, the males with Klinefelter's syndrome had less developed body concepts, less freedom of distractibility, and less psychological maturity and differentiation. These findings seem to suggest that the hormone disorders responsible for hypogonadism do not by themselves explain the cognitive style of XXY males. Triple X females are fertile; their hormone levels deviate only slightly from those of XX women. Nevertheless, their mental abilities are reduced more than in the XXY and X0 syndromes. In addition, mental retardation is a common feature of the rare syndromes in which further X chromosomes exist (e.g., 48,XXXY or 48,XXXX). It thus appears probable that the supernumerary X chromosomes influence cognition not only via abnormal hormone levels.

Neurophysiological and Neuropathological Observations

Neuropathological studies of the sex chromosome abnormalities are unfortunately scarce. Reske-Nielsen et al. (1982) performed a neuropathological study of two women with Turner syndrome and found pronounced neuropathological aberrations in the right temporoparietal occipital area in one of them. Nielsen and Tsuboi (1974) mentioned that 4 of 11 XXY men revealed a dilatation of the ventricular system. However, the representativeness of these studies must be questioned.

In contrast to the scanty neuropathological studies performed in adults, numerous investigations have shown a high rate of the karyotype 45,X0 among spontaneous abortions. More than 50% of spontaneous abortions are chromosomally abnormal. About 97% of autosomally abnormal fetuses are spontaneously aborted, but this is the fate of only 83% of these with sex chromosome anomalies. The latter group is almost entirely accounted for by the 45,X0 karyotype. Fewer than 1 in 300 45,X0 fetuses survive to term. The widest variability in phenotypic manifestation has been encountered among X0 zygotes that at the same time constitute the most frequent single abnormal karyotype among all observed abortions (Vogel and Motulsky 1986). It has been suggested that the 45,X0 constitution is virtually always lethal prior to birth and that all liveborn 45,X0 individuals are, in actuality, mosaics in whom the second cell line is not demonstrable (Hook and Warburton 1983).

An early effect of additional X chromosome material possibly leads to slower brain growth, as illustrated by smaller head circumferences in XXY males and XXX females upon birth (Ratcliffe et al. 1986). While Stewart et al. (1986) also found the head circumferences in XXY and XXX subjects to be in the lower range, differences from controls were not significant. Furthermore, there was a tendency to approach the 50th percentile at later age.

EEG studies have been performed in persons with a sex chromosome aberration. Due to ascertainment of institutionalized individuals the earlier studies did not lead to unbiased results. Epilepsy is not increased in the XYY and XXY males. However, XYY males recruited from the Danish longitudinal study showed slower EEGs than controls, and a similar tendency was apparent in the XXY males. Theta and slow wave alpha activity were greater in the XYY group;

XXY males again showed a similar tendency but were within the normal range. The strewing of alpha and an excess of theta activity, which were interpreted as maturational deficits of cerebral organization, have been reported to occur frequently in personality disorders (Volavka et al. 1977b). In further studies using computer analysis Volavka et al. (1977a, 1979) found a significant reduction in the average alpha frequency in XYY males. Possibly, these findings represent developmental lags. No significant differences were apparent between the XXY group and controls.

In an examination of 64 Turner girls, dominant EEG rhythm was more rapid and the aging effect more pronounced (Tsuboi and Nielsen 1985). Alpha activity was more rapid, of a lower amplitude, and of a lower amount. In contrast, beta activity was of a slower frequency, higher amount, and higher amplitude. Theta activity was also increased. A marked or moderate asymmetry was confined to the occipital and parietal regions. The investigators concluded that a functional brain disorder exists more often in Turner girls than in controls. The findings were more pronounced in complete X0 girls than in those with mosaics. Since a selective deficit in the control functions of the right hemisphere was apparent, the authors point out that this might be related to the degree of dyscalculia and lower perceptual organization present in Turner syndrome.

Several findings suggest a lack or an inhibition of *lateralization* in persons with sex chromosome aneuploidies. Among 16 individuals with a supernumerary X or Y chromosome, only six had a strong right-hand dominance (Nielsen et al. 1981). Dichotic listening revealed that 57% of Turner girls failed to show the typical right ear (left hemisphere) advantage (Stewart et al. 1986). It appears that X0 females do not use their left hemisphere in processing verbal stimuli to the same degree as XX females. Studies in XXY males also suggested lower degrees of left hemisphere specialization for language and greater than normal degrees of right hemisphere specialization for nonverbal processing (Netley and Rovet 1984). It was postulated that the slower fetal growth rates of the extra-X group might contribute to their atypical hemispheric specialization and to the failure of their left hemisphere to gain dominance over their right in language processing.

Autosomal Chromosome Aberrations

Imbalances of the autosomes practically always negatively influence brain function. In general, the respective phenotypes, including the malformations caused by the specific chromosomal aberration, result from aneuploidy. This state of chromosomal imbalance is due either to a loss or an extra dose of euchromatic chromosome material (total or partial monosomy vs total or partial trisomy). The aberration is visible upon either routine or high resolution chromosome analysis. Molecular genetics has revealed the existence of even smaller deletions or duplications than those encompassing approximately 3–6 million base pairs, which represent the resolution limit of chromosome analysis. Accordingly, the resulting disorders are not classified as chromosomal disorders.

The more euchromatic chromosome material is involved in aneuploidies, the more genes are present in an imbalanced state. In trisomic states, the phenotype is presumably in some way related to a 1.5-fold expression of the genes involved. However, as Epstein (1986) points out, the consequence of the 50% increase in gene expression depends on the nature of the gene products. Whereas a 50% increase in the synthesis rate of an enzyme is not likely to have an adverse effect on the phenotype, the reverse is presumably true for other proteins such as receptors, growth factors, and morphogens.

Down Syndrome

Down syndrome (DS) is the most prevalent chromosomal disorder in man. Depending on the age structure of mothers and the frequency of prenatal diagnosis, its incidence at birth is about 1–2/1000. The characteristic phenotype and mental retardation, which is practically always associated with the disorder, usually make a clinical diagnosis easy. Only chromosome analysis, however, can differentiate translocation trisomies from free trisomies. The former imply a high recurrence risk for subsequent children of the parents and other relatives, if they are balanced translocation carriers.

Neurophysiological and Neuropsychological Studies

In DS individuals IQ scores range between 20 and 70 (reviewed by Lott 1986). In exceptional cases higher values are achieved. The interpretation of IQ studies has been hampered by the fact that intelligence in DS can decrease with age. Furthermore, many investigations have been performed on institutionalized persons and are thus not representative for the syndrome. Specific deficits in auditory and visual sequential memory appear to be associated with the syndrome. DS children lag behind in their language development as compared with nonretarded children of the same developmental stage, presumably due to deficits in vocal imitation skills. Within the Stanford-Binet intelligence scale, DS individuals seemed to score higher than IQ-matched controls on items involving figural content and visual motor abilities. They scored lower on semantic content, general comprehension, judgment, and reasoning. Their ability to communicate understandably is reduced.

It appears that certain personality traits are associated with DS. The children have been found to be outgoing and affectionate, with unusual skills in social competence and a paucity of maladaptive behavior.

The neurological features of DS show considerable interindividual variability. Hypotonia is a neurological symptom in DS infants. Possibly, a reduced kinesthetic, tactile, and cutaneous judgment is the basis for the inert listlessness and for the muscular and tendon hypotonia in affected infants. Disturbances in visual scanning strategies and intrinsic oculomotor programming support the hypothesis that a lack of perceptual integration is related to motor delay.

Apparently, age-related changes in the EEG occur in DS individuals, which are paralleled by an increasing incidence of seizures. Paroxysmal seizures have been found to occur in 20% of DS patients under age 20. When the EEG was repeated in these individuals over age 45, deterioration was observed in about one-third of them. However, if DS individuals are matched with mentally retarded subjects with respect to age, sex, length of hospitalization, and IQ, a dramatic reduction in the relative incidence of epilepsy and the frequency of seizures in the DS group is readily apparent. EEG recordings in DS subjects and an age-matched group of undifferentially retarded patients revealed that 77% of the DS subjects had normal EEGs, in contrast to only 4% of the matched group. An excess of theta activity and other EEG findings have been regarded as evidence for immaturity of cerebral development. As Galbraith (1986) concludes: The nervous system of DS subjects is hypoexcitable for neuropathological manifestations.

Evoked response testing has revealed that the amplitude of late wave components of visual, auditory, and somatosensory evoked responses is larger among DS subjects than controls. In addition, cortical evoked potentials fail to habituate. These results were interpreted as indicating a deficient inhibitory capacity in DS individuals, which possibly leads to an inability of the brain to adapt to constantly changing demands of the environment. It has also been considered that a hippocampal dysfunction leads to abnormal evoked potentials.

Neuropathology

Of the trisomies that have been studied during infancy, both trisomy 13 and 18 display more severe gross structural abnormalities of the brain than trisomy 21. Whereas major malformations of the brain are occasionally also encountered in trisomy 21, the macroscopic abnormalities in this disorder are predominantly characterized by a lower weight of the brain and altered convolutional patterns (reviewed in Coyle et al. 1986). A study of 100 DS brains of various ages showed a 20%–50% reduction in brain weight after midinfancy. Whereas no significant differences from age-matched controls were apparent in infancy and early childhood, a reduction was evident at 3–5 years, suggesting postnatal brain growth arrest (Wisniewski et al. 1986). The altered convolutional pattern is the most distinctive gross neuropathological feature of a DS brain. The "embryonic simplicity" (Davidoff 1928) of the convolutional pattern of the cerebral hemispheres (narrow superior temporal gyrus, shallow primary sulci, decreased number of secondary sulci) may reflect a marked diminution in cortical surface area and thickness. A substantial number of DS individuals have calcifications in their basal ganglia.

A number of histological changes, none of which are pathognomic, however, have been observed in DS brains. Many but not all studies showed a reduced neuronal density in several brain areas. Attempts have been made to correlate these reductions with specific mental deficits apparent in DS individuals (Wisniewski et al. 1986). The disturbance in the development of neurons probably occurs in the early prenatal and to a lesser degree peri- and postnatal periods of brain maturation. The cause of the possible arrest of neurogenesis could be secondary to a defect in production or recognition of trophic substances. Beside a reduced neuronal cell density, other histological changes include significantly altered dendrites and dendritic spines. No statistical differences in cortical synaptic density in DS and controls were apparent at ages 9, 13, 18, and 27 years. In newborns with DS the synaptic density is apparently slightly higher than in controls. In contrast to the similar synaptic density, the surface area of synaptic contact is 20%–35% lower in DS cases when compared to the control group.

Neurochemical Findings

Abnormalities in the pharmacological response of the autonomic nervous system in DS subjects have been reported upon administration of atropine, which produced a significantly greater cardioacceleratory response than did comparable doses in age-matched normal or mentally retarded controls (reviewed in Coyle et al. 1986). Studies of peripheral noradrenergic markers in DS revealed that some were decreased, such as serum dopamine-β-hydroxylase, serum norepinephrine, and platelet monoamine oxidase, whereas others were increased, e.g., β-receptor sensitivity in fibroblasts and α-receptor sensitivity in platelets. A disturbed interrelationship between adrenergic and cholinergic autonomic systems in DS has been proposed.

Several groups have observed a significant reduction in the levels of serotonin in whole blood and in the platelet fraction of DS individuals. These results point to an abnormality in transport and storage of serotonin within platelets, perhaps due to changes in the cell membrane and its fluidity.

The composition of cell membranes in DS individuals is altered (e.g., reduction in polyunsaturated fatty acids). It has been suggested that genes encoded on chromosome 21, when present in three copies, alter the composition of cell membranes, change proportions of cytoskeletal elements, and alter membrane fluidity, effectively rendering the cell membrane more rigid (Coyle et al. 1986). This may adversely affect the function of membrane-bound enzymes or receptors. Indeed, electrophysiological studies have detected abnormal excitability of DS neurons in culture.

The fact that brains of virtually all DS individuals 40 years old and older exhibit neurofibrillary tangles and amyloid plaques is of considerable interest, since these same pathological changes are the hallmark of Alzheimer's disease. Additionally, both disorders have a reduction in cholinergic and noradrenergic markers in common. It was hoped that the same pathogenetic mechanism could be elucidated which leads to dementia of the Alzheimer type and which is encountered in aging individuals with DS. However, the proportion of DS subjects who develop dementia is unclear. Whereas some studies have suggested that dementia develops in nearly all DS individuals, others have questioned this high rate by pointing out that dementia is possibly the result of institutionalization. When it became apparent that the gene for the amyloid β precursor protein, a main constituent of the amyloid plaques, is localized on chromosome 21, a common basis for dementia in both disorders seemed almost certain. However, recent studies have excluded a linkage between the locus for the amyloid β precursor protein and both sporadic and familial Alzheimer's disease. These research efforts have verified the chromosomal localization of the gene for the amyloid β precursor protein and have thus led to a probable explanation for accumulation of its degration product in DS brains. It seems likely that the precursor protein is expressed 50% more in trisomy 21 than normal. Undoubtedly, its role in the pathogenesis of the neurological abnormalities in DS warrants further attention (Glenner 1988).

Genetic Mechanisms

In order to understand the pathogenesis of the abnormalities in trisomy 21, the relevant genes on chromosome 21 need to be identified. It is estimated that chromosome 21 encompasses 1.6% of total haploid genome size, equivalent to approximately 320–2400 active genes. Observations based on some rare cases with reciprocal translocations suggest that the distal part of the long arm of chromosome 21 and especially band 21 q22 are responsible for the DS phenotype (Vogel and Motulsky 1986). This roughly reduces by one-half the number of genes which possibly account for the DS phenotype. Furthermore, if only those genes coding for molecules of special developmental, structural, or functional significance to the nervous system are counted, anywhere from between 50 and 1000 genes are potentially involved in the neurological abnormalities of DS (Epstein 1986). Presumably, only an elucidation of these genes and their protein products will eventually enable us to understand how trisomy 21 adversely affects the nervous system. Presently, approximately 20 gene loci have been localized to this small chromosome. So far a 50% increase in all the respective gene products that have been quantified in DS has been demonstrated. It seems highly likely that this 1.5-fold increase in the respective gene products is involved in the pathogenesis of the DS phenotype.

It is not known whether the 50% increase in the concentration of gene products so far found in trisomies will apply to all products synthesized from genes in the trisomic state. At present, a posttranslational regulation of gene product concentration cannot be ruled out. Such a mechanism might apply to those gene products that are more tightly regulated.

Thus, two different mechanisms are conceivable: (a) The quantitative gene imbalances directly affect the phenotype. Accordingly, the nature of the genes (and their products) that are present in the trisomic state directly influences the phenotype. (b) Alternatively, the nature of the involved genes may not be so important as the quantitative gene imbalance that is caused. Thus, the extra chromosome in autosomal trisomies might merely act as a "diffuse disturbance" on the genome. Such a disturbance may manifest itself in the overproduction of particular proteins, the uncontrolled regulation of certain genes, causing them to be permanently switched on and expressed, or the expression of tissue-specific proteins in all tissues (discussed in Stefani et al. 1988).

In favor of the latter hypothesis are results obtained at the molecular level: Quantitative differences in the expression of anonymous chromosome 21-specific RNA sequences revealed no consistent patterns when DS tissues were compared with age- and sex-matched normal tissues. These results were not compatible with the 3:2 ratio of expression that is known from some gene product determinations (Stefani et al. 1988). However, since it is not known whether the employed sequences are translated, the investigation cannot be viewed as proof for the second hypothesis. Further research is required to elucidate the underlying genetic mechanisms responsible for the phenotype in autosomal trisomies.

Monogenic Disorders

General Considerations

Almost all monogenic disorders are rare; however, their total number is quite impressive: McKusick (1986) lists 3907 loci which either certainly or presumably follow monogenic inheritance. Monogenic diseases account for approximately three-quarters of the listings. Whereas most psychiatric disorders have a multifactorial basis, it is essential for psychiatrists to be able to filter out correctly the monogenic diseases that may present as mental disorders. The undifferentiated phenotype subsequently falls apart in distinct disease entities. Such a diagnosis sets the foundation for an evaluation of the prognosis. If possible, specific therapeutic steps can be initiated. Furthermore, parents of the affected child can receive genetic counseling accordingly.

In more abstract terms, monogenic disorders also provide us with an opportunity to study the effects of a single (dominant) or two (recessive) disease alleles on the phenotype. A large number of monogenic disorders can lead to a psychiatric symptomatology as exemplified for schizophrenia (Propping and Friedl 1988). Among psychiatric disorders all modes of mendelian inheritance— autosomal dominant and recessive, and X-linked—are represented. At present, linkage studies are helping us to define the chromosomal localization of a gene locus and enable us to verify or exclude heterogeneity of a given disorder. In future, the application of "reversed genetics" will eventually allow the character-ization of proteins, defects of which lead to monogenic disorders. In the following, tuberous sclerosis is used as a good example of a monogenic disorder that may impair normal development.

Tuberous Sclerosis

Tuberous sclerosis (TS) in its classic form encompassing mental retardation, epilepsy, and adenoma sebaceum is a well-known autosomal dominant disease. Whereas its incidence has been presumed to range between 1:100000 and

1:170 000, these estimates represent minimal figures, since cases with subtle or no clinical manifestations are easily overlooked. An incidence of 1:10–20 000 seems more correct (Gomez 1979; Hunt and Lindenbaum 1984).

Genetics

Linkage studies have revealed that the locus for the TS disease allele is presumably on the long arm of chromosome 9. Initially, a linkage, which has subsequently been confirmed using restriction fragment length polymorphisms, was established to the AB0 blood group locus (Fryer et al. 1987; Connor et al. 1987). However, the linkage to the AB0 blood group locus has not been confirmed in all families. At present, heterogeneity of TS seems a distinct possibility. From a genetic viewpoint TS is an interesting disorder because of the variability of its manifestations. While most autosomal dominant disorders tend to show a relatively large variability in age of onset and symptoms, TS is extreme in that it is compatible with a normal life but can also lead to severe mental retardation.

Seizures and Mental Retardation

Slightly more than one-third of TS individuals are of average intelligence. Gomez (1979) has pointed out that apparently only those individuals with TS who develop seizures are at risk of becoming mentally retarded. Of his 160 patients, 88% suffered from seizures. Of these epileptic patients slightly more than 70% were of subnormal intelligence; the remaining patients with seizures exhibited an average intelligence. The age of seizure onset and the presence and the severity of mental subnormality are directly related. This can be clearly illustrated when only children with an onset of seizures in the first 2 years of life are considered: of these almost all have a subnormal intelligence. At this age especially infantile spasms are a common manifestation of TS. Depending on the patient group, more than 25% of children with infantile spasms can have TS (Pampiglione and Pugh 1975).

Seizures in TS patients that begin later in childhood have a better prognosis concerning mental developmental. Gomez (1979) discusses that a pronounced pathology of the central nervous system can cause seizures and that the seriousness of the cerebral pathology also precludes a normal development. Alternatively, the early onset of seizures in patients with TS could directly interfere with the functional organization of the central nervous system. Perhaps the early onset of seizures is facilitated by another factor(s), which lowers the seizure threshhold. Indeed, it has been suggested that modifying genes genetically predispose to seizures in TS.

Neuropathology

Approximately 85% of TS subjects have mostly multiple subependymal nodules (Gomez 1979). The vast majority are mineralized (calcium, iron). Mineralization may be present during infancy and can be progressive. Other neuropathological features of TS include cortical tubers, dilatation of lateral ventricles, and subependymal giant cell astrocytomas. Both megalocephaly and microcephaly have been observed on rare occasions. Gross heterotopias can sometimes be seen. In addition, there are often scattered heterotopic neurons, which are felt to represent migration arrest between the periventricular germinal matrix and the cortex.

The primary lesions of TS are generally considered to be hamartomas. They result from focally disordered organogenesis. The tumorlike lesions are composed of cellular elements (parenchymal and supporting cells) native to the tissue in which they arise, but abnormal in their numbers, location within the organ, organization, and cellular morphology.

Psychiatric Symptoms

According to Hunt and Dennis (1987) TS can be associated with psychiatric disorders, particularly with severely impaired behavior. In their study of 89 children and one adult with TS they were able to show that infantile spasms commenced in 69 patients prior to 17 months. This high rate possibly represents an ascertainment bias, since it is clearly above the expected rate. Nevertheless, the investigation offers an evaluation of the psychiatric problems that affect children with a severe manifestation of TS.

Of the TS children with infantile spasms, 58% revealed autistic behavior at 5 years. This rate increases to 70% if those children with severe mental retardation who were unable to walk and talk are excluded. This incidence of autistic behavior is much higher than that which would be expected simply as a sequela of infantile spasms. It is also above the 13% manifestation rate of autistic behavior in severely retarded mobile children. TS is thus a genetic disease that is associated with autistic behavior.

As was already mentioned for schizophrenia, a psychiatric disease as defined by clinical criteria cannot be expected to represent an etiological entity. It is a general experience of medical genetics that "common" diseases as a rule consist of many etiological entities to which genetic factors contribute to a varying extent. Here autism may be used as an example relevant for child psychiatry (Table 2).

In the TS study all of the children with autistic behavior were mentally retarded, but as the authors point out, it is not known to what extent their development had been limited by their avoidance of social contact. Almost half of the 90 TS patients showed hyperkinetic behavior. Two-thirds of the patients had severe behavioral problems which impaired social interaction. The behavior of the majority of children seriously disrupted family life.

What Is the Neuropathological Substrate for the Development of the Neurological and Psychiatric Symptomatology in TS Patients?

Hunt and Dennis (1987) ask whether the calcified lesions in the periventricular zone, which belongs to the limbic system, are responsible. However, the correlation of intracranial calcification with the clinical syndrome is poor: neither

Table 2. Genetic disorders that can lead to autistic behavior (Propping 1989)

Disorder	Genetic basis
Tuberous sclerorsis	Autosomal dominant
Neurofibromatosis	Autosomal dominant
Laurence-Moon-Biedl syndrome	Autosomal recessive
Phenylketonuria (untreated)	Autosomal recessive
Albinism	Autosomal recessive
Histidinemia	Autosomal recessive
Hurler's disease	Autosomal recessive
Williams-Beuren syndrome	Autosomal dominant
Rett syndrome	Heterogeneous?
	X-linked dominant?
de Lange's syndrome	Heterogeneous?
Martin-Bell syndrome	Fragile site on Xq27
Down syndrome	Trisomy 21
XYY syndrome	Gonosomal aberration
XXX syndrome	Gonosomal aberration

seizures nor mental retardation appear to be directly related to the presence or absence of calcifications.

A radiological study of 110 cases showed that patients with clinically established TS were more likely to be retarded by a factor of at least 3, whereas those whose only evidence of TS was radiological were nearly twice as likely to have normal mental development (Kingsley et al. 1986). No consistent relationship was found between the presence of cerebral lesions, the number of periventricular nodules, and the intellectual level of the child, apart from a trend towards poorer performance in those with the largest number of lesions, both periventricular and cerebral. Further research is necessary to explain why some children with TS develop infantile seizures and subsequently mental retardation, while others show similar radiological features of the disease without developing neurological symptoms.

Since both severe and mild manifestations of the disease can occur within members of the same family, the effect of the same disease allele is dependent on other genetic and possibly exogenous factors. Families in which the disease follows a mild course in most or all members have been observed. Do different disease alleles contribute to the interfamilial clinical variability? Undoubtedly, molecular genetics will answer this question in the near future. Once the gene has been elucidated, it will be possible to deduce the amino acid sequence of the involved protein and perhaps its structure and function.

Genetic Aspects of Normal Behavior

Whereas genetic diseases enable us to study the consequences of a more or less specifically defined genetic aberration for brain maturation and behavior, the evaluation of genetic factors contributing to the variation in normal behavior is extremely difficult. The range of normal behavior is so vast that only single aspects can be compared at a time, thus resulting in a reductionistic approach. Social factors strongly influence perception of normal behavior, and history has shown that the definition of normal behavior can even change within a society. This points to the fact that our genome allows us to adapt to different cultures which each have distinct perceptions of normal behavior.

Interaction of Environment and Genome. In evolutionary terms, environmental changes lead to selection; thus the environment in the long run directly influences the genetic composition. Additionally, the power of the brain to innovate and learn leads to culturally driven evolution. The rise of agriculture, for instance, imposed new selection pressures that led to genetic changes in human populations (Wilson 1985). To complicate matters even further, modern human beings are changing the natural environment at a pace previously unknown. The impact of molecular biology on modern society and its practical implications, including the distinct possibility that mankind can influence its own genome, cannot be adequately assessed at the present time. In conclusion, the diverse interactions of our natural environment, our brain in itself, which makes cultural evolution possible, and our genome evidently make every attempt at an evaluation of genetic factors involved in normal behavior a formidable task.

Twin Studies

Twin research has been greatly stimulated by the hope of being able to solve the nature–nurture question. An evaluation of the influence of genetic factors on behavior and intelligence has been one of the major preoccupations of the investigators of twin studies. The social consequences are obvious; nevertheless, research has often not fulfilled the high standards required for the study of such a delicate issue. Often twin data have been abused. Before behavioral studies of monozygotic (MZ) as compared to dizygotic (DZ) twins or singletons can be interpreted, it is essential to know how twinning in itself affects maturation and intelligence.

Inherent Biological and Methodological Limitations

Biological Aspects
The *mortality* of twins in the fetal and neonatal period is higher than that of singletons. These deaths are almost entirely accounted for by male pairs (Myrianthopoulos 1975). Second-born twins show a somewhat higher fetal mortality than those born first.

Malformations occur more often in twins than in singletons. The difference between twins and singletons (18.33% vs 15.56%) is solely accounted for by an increase in malformations in MZ twins. Among malformations of the central nervous system both macrocephaly and encephalocele are significantly increased in twins. MZ twins can be discordant for a specific malformation. It is conceivable that disadvantages of cell number and possibly of biochemical timing, which are brought about by the twinning process, are in themselves sufficiently severe to increase significantly the malformation frequency in MZ twins. Alternatively and possibly more likely, is a synergistic action of a disadvantaged embryo and the exposure to environmental factors that would be nonteratogenic or only minimally teratogenic in DZ twins or singletons (Myrianthopoulos and Melnick 1977).

At birth, MZ twins are less concordant for both *body weight* and *length* than DZ twins (Wilson 1976). The lighter and smaller MZ twin, however, usually shows a catch up effect, so that at the age of 1 year MZ twins are significantly more alike than DZ twins. Correlation coefficients of MZ twins increase even further with age, whereas those of DZ twins decline. Approximately 70% of MZ twins have monochorionic placentas and are thus subject to varying degrees of vascular anastomoses. A reexamination of ten MZ twin sets with large birth weight differences (> 750 g) revealed that whereas the smaller twin still remained significantly lighter and marginally shorter, the IQ was not significantly lower. Wilson (1979) concludes that *intelligence* is apparently buffered against effects of nutritional deficit in the prenatal period.

The average IQ of twins is 4–10 points lower than that of singletons (Myrianthopoulos et al. 1976). As the investigators were able to show in their prospective study of more than 50 000 pregnancies, the mean IQ of twins showed a slight increase from 4 to 7 years, whereas the IQ in singletons declined slightly. This supports the contention that twins tend to catch up with time. Both pre- and postnatal factors have been held responsible for the lower IQ in twins. Myrianthopoulos et al. were able to follow up 44 twins whose co-twins had died or had been separated at birth or soon thereafter. The mean 4-year and 7-year IQs were practically identical with the IQs of twins raised together. These findings suggest that lower intellectual performance is an attribute of twinning which must be due to factors associated with the prenatal environment.

Almost six times as many twins as singletons show mental retardation at age 8 months, and over seven times as many motor retardation. At 4 and 7 years this difference drops to threefold. Concordance rates for undifferentiated mental retardation are significantly higher in MZ than in DZ twins; its presence in both twins of a pair did not seem to be related to any antecedent events in the family (Myrianthopoulos et al. 1972).

Methodological Aspects

Many methodological problems limit the value of twin studies. Instead of systematically pointing them out, we will summarize the limitations which confront investigators who strive to separate genetic from environmental factors by evaluating MZ twins reared apart.

Separated MZ twins can be regarded as nature's unique controlled laboratory experiment on the interplay between heredity and environment. However, reality shows that this is an extremely rare experiment. Ideally, the separation should have occurred soon after birth, the separated twins should have been placed beyond the boundaries of their biological and social milieu, and physical contact (and knowledge of twinning) should not have occurred. Optimally, both premature and complicated births should be excluded. Zygosity determination must fulfill modern standards. A bias towards ascertainment of more alike MZ twins should be avoided. According to Farber (1981), of the 121 cases reported in the last 50 years, only three are "twins reared apart" in the classic sense.

Results

As Farber (1981) points out in her reevaluation of 121 sets of separated MZ twins, the data forcefully speak for the influence of both heredity and environment. Nevertheless, the diverse studies have uncovered high concordance rates for a number of *behavioral aspects* (Table 3), many of which are hard to define and evaluate systematically. This, in addition to the small numbers involved, makes the interpretation subject to criticism. However, a picture (p. 87) included by Farber in her reevaluation study showing a group of MZ twin pairs, of which both twins of a single pair have unconsciously put their hands in a characteristic position, strikingly documents the concordance for gestures, and raises the question of whether written information can convey the astonishing degree of similarity apparent for some behavioral aspects.

The changing influence of genetic factors on temperamental individuality was studied longitudinally in a group of 44 same-sexed twin pairs at four different ages from infancy to puberty (Torgersen 1987). Eight different temperamental categories, such as mood, activity, adaptability, and intensity, were investigated. At age 15 the within-pair similarities in temperament were even higher in MZ twins than at earlier ages. The within-pair differences were smaller for all categories in MZ than in DZ twins. Activity was a category where the low within-pair differences in MZ and high differences in DZ twins were still stable at all ages from 9 months on.

The setting of most previous twin studies has only allowed an estimation of the relative contribution of genetic and environmental variance at a given age for a particular point of time. The more important question, however, as to what extent developmental change can be accounted for as a function of environmental and preprogrammed maturational factors, has hardly been addressed adequately (Hay and O'Brien 1983). Thus, behavior genetics should address changes that are consistent across all children and not the idiosyncratic spurts and lags of each family. For example, considerable evidence of greater MZ than DZ similarity in the age changes in IQ as well as the general level of performance has accumulated (Wilson 1979). However, the results do not allow an evaluation of the entire course of development. Attempts have hardly been made to examine systematic patterns of change across the entire period of childhood and across all children.

It is especially important to keep the limitations of twin studies in mind when investigating the genetic influence on *intelligence*. Vogel and Motulsky (1986) conclude that these studies have proven very little regarding genetic variability of

Table 3. Concordance of behavioral aspects in twins reared apart (according to Farber 1981). Farber reevaluated the literature on 121 monozygotic twins brought up apart. Unsystematic observations of behavioral aspects are summarized in the table. It has to be kept in mind that only three twin pairs fulfill the requirements in order to be classified as twins reared apart in the classic sense

Gifts and talents	Almost always similar; same or similar modes of expression, similar in degree of interest and ability 5/33 dissimilar concerning interest or ability in music, drawing, dance, athletics, or drama
Sexual functioning	Single pairs concordant for frigidity, unsatisfactory sexual adjustments, high libido, low libido, lack of satisfaction, dislike of sex, lack of interest in the opposite sex, one pair discordant for homosexuality
Drinking	Predominantly concordant, even to the degree of alcohol consumption. All cases of heavy drinking were concordant
Smoking	Predominantly concordant, including a similarity in the amount of tobacco consumed
Gestures, movement, voice, habits, nervous mannerisms	A high degree of similarity is consistently reported in almost all twins: e.g., high-pitched voice, limp or firm handshake, nervous laugh, rubbing of nose 7/11 sets concordant for mail biting
Learning disorders	One set concordant for dyslexia (diagnostic term used only once); two additional sets concordant for learning disabilities that appear to be dyslexic in nature
Criminality	No substantive evidence to support genetic determination; potent environmental factors discernible
Mood, affective liability	Majority of sets similar in characteristic mood, tone, and affective style. Sets reporting depressive affects or sets reporting symptoms diagnosed as depressive were predominantly discordant. Suicide attempts discordant Strong environmental contribution to reactive, neurotic, or characterological depressive traits
Fearfulness, pattern of anxiety	Pattern of experience is strikingly similar, e.g. similarity in symptoms and in description of state
Enuresis	4/11 concordant sets
Encopresis	2/2 discordant sets
Sleep disorders	3/9 concordant sets for nightmares 3/6 concordant sets for sleepwalking/sleeptalking 5/6 concordant sets for insomnia 8/9 concordant sets for very light sleep The concordance rates for the later two disorders suggest high heritability
Menstrual symptoms	Symptoms like irregular menstruation, dysmenorrhea, menostasis, headache, emotionality/mood swings associated with menses, vomiting, mittelschmerz are strikingly similar in the reared apart twins Not only did the twins have the same menstrual symptoms, but they also tended to describe them in identical ways and they tended to have them at about the same ages or periods of life

intelligence in the normal range. Indeed, opposite interpretations of the same data have been forwarded, ranging from very low to very high estimates of heritability.

General Genetic Considerations. If interindividual variation in intelligence and behavior is influenced by genetic factors, it should be the result of the combined effect of an interplay of single genes, which are each inherited in a mendelian fashion, and their interaction with the environment. Thus, theoretically, a specific set of genes might positively influence intelligence in a defined environment, but have an opposite effect in a different environment. Recently, a full cross-fostering study dealing with IQ was presented providing evidence for the independent contribution of both genetic and environmental factors. The study included children whose natural and adoptive parents were of the most highly contrasting socioeconomic status. The data showed that the background of the natural parents had an effect on the children's IQ independent of the socioeconomic status of the adoptive parents. An independent effect of the postnatal environment was also readily discernible (Capron and Duyme 1989).

Presently very little is known about single genes that influence intelligence in the normal range. One of the few known single gene effects is the influence of heterozygosity for metabolic disorders. These heterozygotes, who were ascertained via their diseased children, have a slightly but significantly lower IQ than controls (Propping 1982, 1987b). Thus, heterozygosity for certain genes can contribute to the genetic variability of intelligence in the normal range. Studies of this kind offer an alternative to twin research and allow an investigation of the influence of a single gene on the variability of intelligence.

Neurophysiological Studies

In the light of the inaccessibility of the brain for genetic studies, modern diagnostic methods offer a possibility to investigate genetic variation in brain functioning. Mainly the *electroencephalogram* (EEG) has been employed as a noninvasive method for this purpose. The EEG shows a characteristic development from infancy to adult stage, which is highly genetically determined. Premature infants will show the same age-corrected EEG patterns as infants born at term, which in itself is evidence of a genetic "blueprint" and of a maturation largely independent of environmental factors (Volpe 1987).

The resting EEGs of MZ twins are extremely alike, so that the slight differences are not larger than those observed upon a reexamination of a single proband. This still holds true when quantitative EEG analysis is used (Stassen et al. 1987). The similarity persists into adulthood; focal changes and vascular problems account for a discordance in some older MZ twins. Thus, the time rate of maturation in MZ twins is practically identical, whereas in DZ twins differences are apparent (Vogel 1981).

EEG studies of MZ twins reared apart have excluded similar environmental factors as the reason for the virtually identical EEGs in MZ twins (Juel-Nielsen and Harrald 1958). Whereas the physiology that leads to the respective EEG patterns shows a polygenic determination, family studies have revealed that the so-called low-voltage EEG and possibly some other EEG variants are inherited in an autosomal dominant fashion (Vogel 1970). This points to the strong influence of a single alde on the EEG, which has a prevalence of approximately 5% in Caucasians. Since individual differences in neurophysiological parameters can be regarded as the prerequisite of individual behavior, the question arises as to whether genetically determined EEG patterns are associated with specific personalities. Vogel and Schalt (1979) were indeed able to delineate certain personality types based on their EEG patterns. Whereas considerable overlap exists, probands with low-voltage EEGs tended to be group orientated and extraverted. Spatial orientation was found to be especially good.

Outlook

We have attempted to evaluate the genetic influence on brain maturation and behavior. Using single genetic disorders as examples, we have tried to illustrate the genetic factors, the neuropathological substrate, and its impact on behavior. In principle, this approach can be applied to every genetically determined psychiatric disease. The mechanisms by which the genetic disorders lead to neuropathological changes and altered behavior are not understood in any of our examples, despite their relative frequency and numerous attempts to uncover the underlying pathology.

We believe that the elucidation of the underlying genetic mechanisms will provide us with a powerful instrument with which we can determine how the neuropathological changes in a certain disease came about. This approach requires research at the molecular level. Interdisciplinary collaboration is required to elucidate further the cause of behavioral phenomena observed in psychiatric disorders.

Receptor Research

We would like to conclude by briefly discussing the promising field of receptor research, which is progressing at a tremendous pace. Since receptors are essential for the functioning of the central nervous system, the elucidation of their structure, function, and regulation will undoubtedly contribute to our understanding of the brain. The c-DNAs for single or all of the subunits of several central nervous system receptors have been cloned, including the nicotinic and muscarinic acetylcholine, the GABA/benzodiazepine, the glycine, and the β-adrenergic receptors (reviewed in Hebebrand et al. 1990). This allows a deduction of their amino acid sequences and their presumable transmembrane topologies. For several subunits of specific receptors isoforms exist, which when assembled to a functioning receptor can be regarded as isoreceptors (Hebebrand et al. 1988a).

Does genetic variability of the genes encoding subunits of central receptors and their regulators exist? What are the functional implications? Do the isoforms exhibit polymorphisms (Hebebrand and Friedl 1987)? Are the distributions of these isoforms differentially regulated? In many molecular genetic studies of psychiatric diseases genes for receptor subunits are regarded as "candidate genes." They are used to establish whether psychiatric diseases are associated or linked with a specific polymorphism of a given c-DNA probe for a receptor subunit.

The advent of the molecular approach to the elucidation of brain function has begun. The new methods will allow us to increase our basic knowledge. For example, the ontogeny of most central receptors in humans is not known. Nevertheless, knowledge of this sort is of importance to determine when the neurochemical premise for an interaction of neurons is first established. The muscarinic receptor is already detectable at the 10th gestational week (Ravikumar and Sastry 1985). We have been investigating the ontogeny of the GABA/benzodiazepine receptor, which is also already detectable at a very early stage, in the 7th gestational week (Hebebrand et al. 1988b). It seems possible that receptors in this early developmental stage are not located at synapses. They might be directly involved in the morphogenesis of neurons (Wolff et al. 1987). It thus seems conceivable that genetic variation in these structures can influence brain maturation and function at several stages. In conclusion, the aspect of genetic variation of central receptors warrants further attention in order to find out whether aberrations can lead to genetically determined psychiatric or neurological disease.

References

Capron C, Duyme M (1989) Assessment of effects of socio-economic status on IQ in a full cross-fostering study. Nature 340:552–554

Connor JM, Pirrit LA, Yates JRW, Fryer AE, Ferguson-Smith MA (1987) Linkage of the tuberous sclerosis locus to an DNA polymorphism detected by v-abl. J Med Genet 24:544–546

Coyle JT, Oster-Granite ML, Gearhart JD (1986) The neurobiologic consequences of Down syndrome. Brain Res Bull 16:773–787

Davidoff LM (1928) The brain in mongolian idiocy. Arch Neurol Psychiatry 20:1229–1257

Downey J, Ehrhardt AA, Gruen R, Bell JJ, Morishima A (1989) Psychopathology and social functioning in women with Turner syndrome. J Nerv Ment Dis 177:191–201

Epstein CJ (1986) Trisomy 21 and the nervous system: from cause to cure. In: Epstein CJ (ed) The neurobiology of Down syndrome. Raven, New York, pp 1–15

Evans JA, Flindt de von R, Greenberg C, Ramsay S, Hamerton JL (1986) Physical and psychologic parameters in children with sex chromosome anomalies: further follow-up from the Winnipeg cytogenetic study of 14,069 newborn infants. In: Ratcliffe SG, Paul N (eds) Prospective studies on children with sex chromosome aneuploidy. Liss, New York, pp 183–207

Farber SL (1981) Identical twins reared apart. Basic Books, New York

Ford CE, Miller OJ, Polani PE, Almeida JC, Briggs JH (1959) A sex-chromosome anomaly in a case of gonadal dysgenesis (Turner's syndrome). Lancet 1:711–713

Fryer AE, Chalmers A, Connor JM, Fraser I, Povey S, Yates AD, Yates JRW, Osborne JP (1987) Evidence that the gene for tuberous sclerosis is on chromosome 9. Lancet 1:659–667

Galbraith GC (1986) Unique EEG and evoked response patterns in Down syndrome individuals. In: Epstein CJ (ed) The neurobiology of Down syndrome. Raven, New York, pp 109–119

Glenner GG (1988) Alzheimer's disease: its proteins and genes. Cell 52:307–308

Goodfellow PJ, Darling SM, Pritchard C, Goodfellow PN (1987) Homologies between the sex chromosomes of man. In: Vogel F, Sperling K (eds) Human genetics. Proceedings of the 7th international congress, Berlin, 1986. Springer, Berlin Heidelberg New York, pp 145–151

Gomez MR (1979) Tuberous sclerosis. Raven, New York

Graham JM, Bashir AS, Stark RE, Sibert A, Walzer S (1988) Oral and written language abilities of XXY boys: implications for anticipatory guidance. Pediatrics 81:795–806

Hay DA, O'Brien PJ (1983) A genetic approach to the structure and development of cognition in twin children. Child Dev 54:317–330

Hebebrand J, Friedl W (1987) Phylogenetic receptor research: implications in studying psychiatric and neurological disease. J Psychiatr Res 21:531–537

Hebebrand J, Friedl W, Propping P (1988a) The concept of isoreceptors: application to the nicotinic acetylcholine receptor and the gamma-aminobutyric acid$_A$/benzodiazepine receptor complex. J Neural Transm 71:1–9

Hebebrand J, Hofmann D, Reichelt R, Schnarr S, Knapp M, Propping P, Födisch HJ (1988b) Early ontogeny of the centra benzodiazepine receptor in human embryos and fetuses. Life Sci 43:2127–2136

Hebebrand J, Reichelt R, Körner J (1990) Receptor heterogeneity at the molecular level: implications for neuropsychiatric research. J Psychiatr Gen (in press)

Hook EB, Warburton (1983) The distribution of chromosomal genotypes associated with Turner's syndrome: livebirth prevalence rates and evidence for diminished fetal mortality and severity in genotypes associated with structural X abnormalities or mosaicism. Hum Genet 64:24–27

Hunt A, Dennis J (1987) Psychiatric disorder among children with tuberous sclerosis. Dev Med Child Neurol 29:190–198

Hunt A, Lindenbaum RH (1984) Tuberous sclerosis: a new estimate of prevalence within the Oxford region. J Med Genet 21:272–277

Jacobs PA, Strong JA (1959) A case of human intersexuality having a possible XXY sex-determining mechanism. Nature 183:302–303

Juel-Nielsen N, Harrald B (1958) The electroencephalogram in uniovular twins brought up apart. Acta Genet Stat Med 8:57–64

Kingsley DPE, Kendall BE, Fitz CR (1986) Tuberous sclerosis: a clinicoradiological evaluation of 110 cases with particular reference to atypical presentation. Neuroradiology 28:38–46

Lott IT (1986) The neurology of Down syndrome. In: Epstein CJ (ed) The neurobiology of Down syndrome. Raven, New York, pp 17–28

McKusick VA (1986) Mendelian inheritance in man. Johns Hopkins University Press, Baltimore

Myrianthopoulos NC (1975) Congenital malformations in twins: epidemiologic survey. Birth Defects 11 (8):1–39

Myrianthopoulos NC, Melnick M (1977) Malformations in monozygotic twins: a possible example of environmental influence on the developmental genetic clock. In: Inouye E, Nishimura H (eds) Gene-environment interaction in common diseases. University of Tokyo Press, Tokyo, pp 206–220

Myrianthopoulos NC, Broman SH, Nichols PL, Anderson VE (1972) Intellectual development of a prospectively studied population of twins and comparison with singletons. In: DeGrouchy J, Ebling FJG, Henderson IW (eds) Human genetics, Elsevier, Amsterdam, pp 244–257

Myrianthopoulos NC, Nichols PL, Broman SH (1976) Intellectual development of twins — comparison with singletons. Acta Genet Med Gemellol (Roma) 25:376–380

Netley C, Rovet J (1984) Hemispheric lateralization in 47,XXY Klinefelter's syndrome boys. Brain Cogn 3:10–18

Nielsen J, Sillesen I (1981) Turner's syndrome in 115 Danish girls born between 1955 and 1966. Acta Jutlandica 56, Medicine series 22, Aarhus

Nielsen J, Tsuboi T (1974) Electroencephalographic examination in the XXY syndrome and in Klinefelter's syndrome. Br J Psychiatry 125:236–237

Nielsen J, Nielsen T, Sorensen AM, Sorensen K (1981) Mental development of unselected children with sex chromosome abnormalities. Hum Genet 59:324–332

Nielsen J, Wohlert M, Faaborg-Andersen J, Eriksen G, Hansen KB, Hvidman L, Krag-Olsen B, Videbech P (1986) Chromosome examination of 20,222 newborn children: results from a 7.5-year study in Aarhus, Denmark. In: Ratcliffe SG, Paul N (eds) Prospective studies on children with sex chromosome aneuploidy. Liss, New York, pp 209–219

Nyborg H (1984) Performance and intelligence in hormonally different groups. In: Devries GJ, Debruin JPC, Uylings HBM (eds) Sex differences in the brain. Prog Brain Res 61:491–508.

Pampiglione G, Pugh E (1975) Infantile spasms and subsequent appearance of tuberous sclerosis syndrome. Lancet 2:1046

Propping P (1982) Genetik und Intelligenz. Z Kinder Jugendpsychiatr 10:110–130

Propping P (1987a) Introduction to the symposium on psychobiological genetics. In: Vogel F, Sperling K (eds) Human genetics. Proceedings of the 7th International Congress, Berlin 1986. Springer, Berlin Heidelberg, pp 450–451

Propping P (1987b) Single gene effects in psychiatric disorders. In: Vogel F, Sperling K (eds) Human genetics. Proceedings of the 7th international congress, Berlin, 1986. Springer, Berlin Heidelberg New York, pp 452–457

Propping P (1989) Psychiatrische Genetik. Springer, Berlin Heidelberg New York Tokyo

Propping P, Friedl W (1988) Genetic studies of biochemical, pathophysiological and pharmacological factors in schizophrenia. In: Tsuang MT, Simpson JC (eds) Handbook of schizophrenia, vol 3. Elsevier, Amsterdam, pp 579–608

Ratcliffe SG, Murray L, Teague P (1986) Edinburgh study of growth and development abnormalities. III. In: Ratcliffe SG, Paul N (eds) Prospective studies on children with sex chromosome aneuploidy. Liss, New York, pp 73–118

Ravikumar BV, Sastry PS (1985) Muscarinic cholinergic receptors in human foetal brain: chracterization and ontogeny of (^{3}H) Quinuclidinyl benzilate binding sites in frontal cortex. J Neurochem 44:240–246

Reske-Nielsen E, Christensen A-L, Nielsen J (1982) A neuropathological and neuropsychological study of Turner's syndrome. Cortex 18:181–190

Robinson A, Bender BG, Borelli JB, Puck MH, Salbenblatt JA, Winter JSD (1986) Sex chromosomal aneuploidy: prospective and longitudinal studies. In: Ratcliffe SG, Paul N (eds) Prospective studies on children with sex chromosome aneuploidy. Liss, New York, pp 23–71

Salbenblatt JA, Meyers DC, Bender BG, Linden MG, Robinson A (1987) Gross and fine motor development in 47,XXY and 47,XYY males. Pediatrics 80:240–244

Stassen HH, Bomben G, Propping P (1987) Genetic aspects of the EEG: an investigation into the within-pair similarity of monozygotic and dizygotic twins with a new method of analysis. Electroencephalogr Clin Neurophysiol 66:489–501

Stefani L, Galt J, Palmer A, Affara N, Ferguson-Smith M, Nevin C (1988) Expression of chromosome 21 specific sequences in normal and Down's syndrome tissues. Nucleic Acids Res 16:2885–2896

Stewart DA, Bailey JD, Netley CT, Rovet J, Park E (1986) Growth and development from early to midadolescence of children with X and Y chromosome aneuploidy: the Toronto study. In: Ratcliffe SG, Paul N (eds) Prospective studies on children with sex chromosome aneuploidy. Liss, New York, pp 119–182

Theilgaard A (1984) A psychological study of the personalities of XYY- and XXY-men. Acta Psychiatr Scand [Suppl] 315:69:1–131

Torgersen AM (1987) Longitudinal research on temperament in twins. Acta Genet Med Gemellol (Roma) 36:145–154

Tsuboi T (1970) Crimino-biologic study of patients with the XYY syndrome and Klinefelter's syndrome. Humangenetik 10:68–84

Tsuboi T, Nielsen J (1985) Electroencephalographic examination of 64 Danish Turner girls. Acta Neurol Scand 72:590–601

Vogel F (1970) The genetic basis of the normal human electroencephalogram (EEG). Humangenetik 10:91–114

Vogel F (1981) Genetisch bedingte Variabilität in der geistig-seelischen Entwicklung. Klin Wochenschr 59:1009–1018

Vogel F, Motulsky AG (1986) Human genetics. Problems and approaches. Springer, Berlin Heidelberg New York Tokyo

Vogel F, Schalt E (1979) The electroencephalogram (EEG) as a research tool in human behavior genetics: psychological examinations in healthy males with various inherited EEG variants. III. Interpretation of the results. Hum Genet 47:81–111

Volavka J, Mednick SA, Rasmussen L, Sergeant J (1977a) EEG spectra in XYY and XXY men. Electroencephalalogr Clin Neurophysiol 43:798–801

Volavka J, Mednick SA, Sergeant J, Rasmussen L (1977b) Electroencephalograms of XYY and XXY men. Br J Psychiatry 130:43–47

Volavka J, Dednick SA, Rasmussen L, Teasdale T (1979) EEG response to sine wave modulated light in XYY, XXY and XY men. Acta Psychiatr Scand 59:509–516

Volpe JJ (1987) Neurology of the newborn. Saunders, Philadelphia

von Wendt L,. Rantakallio P (1986) Congenital malformations of the central nervous system in a 1-year birth cohort followed to the age of 14 years. Childs Nerv Syst 2:80–82

Walzer S, Bashir AS, Graham JM, Silbert AR, Lange NT, DeNapoli MF, Richmond JB (1986) Behavioral development of boys with X chromosome aneuploidy: impact of reactive style on the educational intervention for learning deficits. In: Ratcliffe SG, Paul N (eds) Prospective studies on children with sex chromosome aneuploidy. Liss, New York, pp 1–21

Warkany J, Lemire RJ, Cohen MM (1981) Mental retardation and congenital malformations of the central nervous system. Year Book Medical Publishers, Chicago

Wilson AC (1985) The molecular basis of evolution. Sci Am Oct: 148–157

Wilson RS (1976) Concordance in physical growth for monozygotic and dizygotic twins. Ann Hum Biol 3:1–10

Wilson RS (1979) Twin growth: initial deficit, recovery, and trends in concordance from birth to nine years. Ann Hum Biol 6:205–220

Wisniewski KE, Laure-Kamionowska M, Connell F, Wen GY (1986) Neuronal density and synaptogenesis in the postnatal stage of brain maturation in Down syndrome. In: Epstein CJ (ed) The neurobiology of Down syndrome. Raven, New York, pp 29–44

Witkin HA, Mednick SA, Schulsinger F, Bakkestrom E, Christiansen KO, Goodenough DR, Hirschhorn K, Lundsteen C, Owen DR, Philip J, Rubin DB, Stocking M (1976) Criminality in XYY and XXY men. Science 193:547–555

Wolff JR, Joo F, Kasa P (1987) Synaptic, metabolic and morphogenetic effects of GABA in the superior cervical ganglion of rats. In: Neurotrophic activity of GABA during development. Liss, New York, pp 221–252

The Importance of Genetic Factors in the Etiology of Child Psychiatric Disorders

P. Lombroso, D. L. Pauls, and J. F. Leckman

Introduction

The genetic study of childhood psychiatric disorders has advanced considerably over the past 20 years. All methodologies previously employed to investigate adult psychiatric disturbances are useful in the study of children. This chapter will review the approaches that have been utilized in the study of the genetic and environmental factors which predispose children to a number of psychiatric disturbances. Selected data from twin, aggregation, and adoption studies which exemplify these techniques will be presented. In addition, recent advances in molecular biology will be reviewed. We believe that these techniques hold the greatest promise for elucidating the fundamental biological processes responsible for childhood psychiatric disorders which have a genetic basis.

Traditional Approaches

Twin Studies

Twin studies grew out of an attempt to partition the variance of behavioral disturbances into their genetic and environmental components. The theoretical underpinnings for this methodology are straightforward. Monozygotic (MZ) twins share identical genes and in most cases a common environment. Dizygotic (DZ) twins share, on average, one-half of their genes and in most cases a common environment. If one assumes that the environmental influences are of the same magnitude, then differences found in the correlations between MZ and DZ twins for a particular trait shoud be due to the differences in shared genotypes. For a trait completely determined by heredity, the expected correlations are 1.0 for MZ twins and 0.5 for DZ twins. If the pattern of correlations were 0.75 and 0.50 for MZ and DZ twins, respectively, inherited factors would be estimated to explain half of the phenotype variance of the trait.

Although an extensive critique of the twin method is beyond the scope of this chapter, a few of its complexities should be mentioned. For example, the assumption that the DZ pairs only share half their genes in common would be erroneous if there were assortative mating for the trait in question. Conversely, if the trait were determined by nonadditive genetic factors, which would increase the variability between individuals, the observed DZ twin correlation would be lower. In addition, the estimate of the genetic contribution to the inheritance of a trait would be incorrect if the similarity of the MZ twins' experience was greater than that between DZ twins.

A number of studies have examined twins to estimate the contribution of genetic factors in various child psychiatric disturbances. Studies of autism offer a good example of how a twin study paradigm can be applied to a child psychiatric disorder. Early work in childhood autism generally dismissed the importance of genetic influences. This argument was based largely on the low recurrence risk of autism among at-risk relatives. However, this low recurrence rate is misleading unless it is compared to the general population rate. Specifically, Rutter (1967) found that 2% of autistic children had siblings with the same disturbance. This number was more than 50 times greater than the frequency found in the general population. These findings led to a detailed study of MZ and DZ twins in an effort to clarify what role, if any, genetic factors have in the transmission of autism.

In 1977, Folstein and Rutter published a systematic study of 21 same-sexed twin pairs (11 MZ and 10 DZ) in which at least one twin showed the disturbance of infantile autism. Their results indicated a 36% pair-wise concordance rate for autism among the MZ twins, while the DZ twins had a 0% concordance rate. The concordance rate for cognitive abnormalities was 82% among the MZ pairs and 10% for the dizygotic twins. If their cognitive abnormalities are truly "lesser variants" of autism, the findings from this study support an etiological role for genetic factors in this disorder. However, the low recurrence rate for the full syndrome among DZ twins is problematic.

One of the advantages of the twin methodology is that it is possible to study the contribution of genetic factors to traits that may not present a stable phenotype throughout the course of development. For example, demonstration of a hereditary pattern of transmission of attention deficit with hyperactivity disorder (ADHD) has been obtained using a variety of research strategies, including familial aggregation studies (Morrison and Stuart 1971; Cantwell 1972; Biederman et al. 1984; Pauls et al. 1983) and adoptee and cross-fostering studies (Safer 1973; Cantwell 1975a, b) as well as twin studies (Vandenberg 1962; Scarr 1966; Willerman 1973; Graham and Stevenson 1987).

A major complication of the familial aggregation and adoptee studies is the determination of the "affected status" of adults within the family units under study. Although various attempts have been made both to establish methods to assess ADHD in adulthood and to develop methods to make retrospective diagnosis of childhood ADHD, major questions remain concerning their reliability and validity. The twin studies, however, readily surmount these difficulties by selecting twin pairs that fall within the appropriate development epoch so that identical assessment protocols can be employed.

The greatest weakness pointed to by critics of the twin studies is the assumption that the environments for MZ and DZ twins are comparable. The assumption has been examined (Cohen et al. 1975) and no relevant differences have been found. In addition, as Rowe (1983) has indicated, it is not enough for critics to show that DZ twins differ from MZ twins or that singletons differ from twins and to use this to condemn twin studies in general. One must demonstrate that such differences in biology or the environment are etiologically important for the particular disturbance under investigation.

As has been pointed out (Pauls 1985), one of the most interesting applications of the twin study paradigm will be the examination of discordant MZ twin pairs. Assuming a genetic etiology for a behavioral disturbance, MZ twins should show full concordance rates. The fact that they do not implies that environmental factors are also of etiological importance. A closer examination of discordant twins should help to identify important nonheritable factors that mediate the expression of the underlying genetic vulnerability.

Familial Aggregation Studies

Two types of family study have been commonly used to investigate genetic factors involved in child psychiatric disturbances. The first, termed "family aggregation" studies, looks for an increased frequency of a disorder among biological at-risk family members. The second, adoption or "cross-fostering" studies, investigates biologically related individuals who have been reared apart and biologically unrelated individuals who have been reared in similar environments.

The paradigm for aggregation studies has been used for many years. If an illness has a genetic component then an increased number of cases should be found within the family. Such studies do indeed yield data suggesting genetic involvement. However, if environmental factors are etiologically involved, they can confound these data. In fact, certain environmental factors, for example cultural ones, could be transmitted from one generation to the next in a pattern that is indistinguishable from genetic transmission. Even so, a number of studies have been conducted which have yielded suggestive results.

One example concerns the familial aggregation of Tourette's syndrome (TS) and other chronic tic disorders. Gilles de la Tourette in his initial report in 1885 suggested that TS and chronic tics (CT) were likely to be etiologically related to one another and to be transmitted by hereditary factors. Over the past decade, several familial aggregation studies have been published which have found that the recurrence risk of TS among first-degree relatives is 6%–12% and the risk for chronic tics is 15%–20% (Comings and Comings 1984; Comings et al. 1984; Pauls et al. 1981, 1984; Shapiro et al. 1978).

Aggregation studies have also been used to suggest that obsessive–compulsive disorder (OCD) is associated with TS. This hypothesis is based on the finding of (a) an increased number of OC symptoms among patients with TS (Fernando 1967; Nee et al. 1982; Yaryura-Tobias et al. 1981) and (b) an increased rate of OC symptoms and frank OCD among family members of TS probands regardless of whether the proband carried a diagnosis of OCD (Pauls et al. 1986).

Adoption Studies

The fundamental problem in family aggregation studies is that resemblance among relatives could be due to shared heredity or shared environment. Adoption or cross-fostering studies powerfully separate these two source of variance. If a trait is genetic, then adopted children should resemble their biological parents and siblings more than their adoptive ones. Conversely, if environmental effects predominate, then adoptees should resemble their new family more than their biological family.

A number of criticisms have been directed against this methodology. First, adoption may not occur until the child is several years old, when early and perhaps critical development has already occurred in the environment of the biological family. Second, adoptees are often placed into families which resemble their families of origin (termed selective placement), confounding the assumption of differing environments. Finally, there is often difficulty in obtaining information regarding biological relatives, especially fathers. Despite these limitations, adoption studies were seminal in the earlier investigations of the genetic contribution in adult disorders, such as schizophrenia (Kety 1983), antisocial personality (Hutching and Mednick 1975), and alcoholism (Cloninger et al. 1981).

Adoption studies have been used to study alcoholism and antisocial personality. The past decade has seen the publication of separate adoption studies conducted in Denmark, Sweden, and the United States (Goodwin 1979, 1985; Cloninger et al. 1981; Cadoret et al. 1985, 1986). The results have been remarkably similar. The sons of alcoholics were three to four times as likely to become alcoholics whether they were raised by their alcoholic biological parents or by their nonalcoholic adoptive parents. Sons of alcoholics were no more susceptible to nonalcoholic psychiatric disturbances (i.e., depression or sociopathy) than were sons of nonalcoholics when both groups were raised by nonalcoholic adoptive parents.

It is noteworthy that few adoption studies for childhood disorders have been done. One of the reasons for this is that it is essential to obtain reliable assessments of the parents as children. It is often impossible to obtain the necessary information to make a retrospective diagnosis of a childhood disorder in the parents (both biological and adoptive) of the children being studied. One way to overcome this limitation is through the use of prospective longitudinal designs, such as the Colorado Adoption study (Plomin and DeFries 1985; Plomin et al. 1988). This paradigm would be particularly useful for personality factors and conditions of high prevalence.

Genetic Models

Genetic models have been developed to test specific hypotheses that relate to various patterns of genetic transmission. The general approach is to collect data on a homogeneous group and ask whether or not a particular model will provide an adequate fit to the observed data. If the best fit for the model yields expected prevalence rates that are significantly different than the observed rates then the model can be rejected. On the other hand, if the expected and observed values are in close agreement, then the model cannot be rejected. Such models are of greatest value when they clearly discriminate between competing models. Before reviewing various models some key concepts should be defined.

For a single major locus model, the genotype is the actual genetic constitution of an individual at a particular locus on a chromosome. A specific locus may have two or more alleles, the product of which are alternative protein variants. For example, different hemoglobin molecules have been described, each with a small variation in amino acid sequence. All are termed alleles of the same hemoglobin locus. Of course, an allele may encode for an abnormal protein, resulting in disease states (e.g., sickle cell anemia).

If a particular locus has two alleles, A and a, then an individual may have one of three different genotypes: AA, Aa, and aa. The phenotypic expression will vary, however, depending on the gene product. For example, in phenylketonuria two alleles are found in the population: the product of P is normal whereas the product of p is an abnormal enzyme. Individuals with genotype PP (homozygotes for the normal gene) will be phenotypically normal, that is, able to metabolize phenylalanine in the diet. Individuals with genotype pp (homozygotes for the abnormal gene) will be unable to tolerate phenylalanine. The third genotype, Pp, is the heterozygote. In the past, such carriers were phenotypically indistinguishable from the homozygote, PP. Today, however, an intermediate level of enzymatic activity can be detected after a loading dose of the amino acid. Thus, three different genotypes that appeared in the past as two phenotypes can today be broken down into three due to advances in protein chemistry.

Phenylketonuria is an example of a disorder that is due to a single major locus. In this simple model, an illness is attributed entirely to a single locus with two, or more, alternative alleles. The mathematical formulation for this model was developed in the 1970s (Morton et al. 1971; Elandt-Johnson 1971). It is quite general and includes classic mendelian autosomal dominant and autosomal recessive models as subhypotheses. It is attractive because it postulates a single major locus and hence offers the possibility of eventually obtaining a more precise understanding of gene action in the etiology of the disorder. Although many have abandoned the search for a major locus associated with psychiatric disorders, genetic linkage studies have shown that some complex human disorders long considered genetically intractable are determined primarily by one locus, for example, bipolar disease (Egeland et al. 1987; Baron et al. 1987) (see below).

When many loci produce an illness, the inheritance is polygenic. When, in addition, environmental factors are involved, the inheritance is said to be multifactorial. The mutlifactorial-polygenic model has been developed to account for these additional factors. The application of multifactorial-polygenic models to psychiatric disorders has been most thoroughly developed by the research group at Washington University at St. Louis (Rice et al. 1979).

Linkage Studies

The detection of genetic linkage is one of the strongest clinical techniques available to test the genetic basis for an illness. The following section will describe the underlying theory and methodology of genetic linkage studies. In addition, recent advances in molecular biology will be presented which make this technique even more powerful.

Linkage analysis is designed to estimate the distance between genetic loci. Two loci on separate chromosomes will be transmitted independently from parent to child. This phenomenon is encapsulated in Mendel's law of independent assortment. Genes which lie at opposite ends of a single chromosome will also be transmitted independently of one another since the distance between them is so large that the probability of a crossover event occurring (and hence the allelic configuration of the parent recombining) is equal to the probability that a crossover event will not occur. When genetic loci are closer together the probability that a crossover will occur between them decreases, and hence the probability that they will be transmitted independently decreases. Therefore, if a sufficient number of matings is observable, it is possible to estimate the relative distance between genetic loci by tabulating the number of recombination events which occurred.

Before the development of sophisticated enzymological techniques for studying proteins, linkage analysis was based solely on phenotype. It was noted that certain diseases were found in individuals with certain traits. For example, certain bleeding disorders were often seen together with some forms of color blindness, and the gene responsible for one type of hemophilia was correctly thought to be closely linked to genes involved in color vision.

As protein chemistry advanced, it became possible to do similar linkage studies with serum proteins that were found to be polymorphic based on their electrophoretic mobilities. Researchers could then ask whether a particular protein was transmitted in close linkage to an illness.

For example, suppose allele A^1 at locus A is responsible for a specific trait and allele A^2 is the allele responsible for the normal phenotype. At another locus, allele B^1 codes for a particular polymorphic variant of a protein encoded by locus B and B^2 is the allele responsible for another variant. If the two genetic loci are closely linked, then the allelic combination present in the parents at the two loci will be transmitted to the offspring. That is, suppose one of the parents has the trait in question and has genotype $B^1 B^2$ at locus B. Suppose the other parent is unaffected and has genotype $B^1 B^1$ at locus B. If the couple has two affected and two unaffected children and the genotype at locus B is $B^1 B^2$ for the two affected and $B^1 B^1$ for the two unaffected, then it is likely that locus B is linked to the trait locus A, particularly if a similar pattern of association is seen in a number of families. Locus B is then called a market locus for trait A and can be used to identify the most likely genotype at locus A.

However, the vast majority of diseases could not be shown to be linked to any of the known protein marker genes. The reason for this was trivial: there were not enough sufficiently polymorphic markers spread throughout the human genome to provide adequate linkage data. If linkage analysis was to advance, a new methodology was needed which would somehow saturate the genome with market loci.

One application of linkage analysis has been in the area of learning disabilities. Learning disorders are one of the most common handicapping conditions of childhood. They are a heterogeneous group of disorders, both in their etiologies and in their presentation. Specific learning disabilities lie within this broad group of cognitive disturbances and are defined as an unexpected difficulty in learning how to read and spell—unexpected in the sense that the child has no known neurological disorder, is of normal intelligence, and has no particular emotional or socioeconomic handicap. Originally described by Kerr (1897) and Morgan (1896), it was given the name of "congenital word blindness."

Specific reading disability (RD), or dyslexia, is thought to be a heterogeneous disorder with a number of etiologies, some of which are genetic. Family and twin data suggest that some forms of RD are genetic but the mode(s) of transmission have not been clearly elucidated. Some reports suggest polygenic inheritance, others autosomal dominant inheritance, and still others autosomal recessive inheritance in families of female probands (DeFries and Decker 1981; Lewitter et al. 1980). These results suggest that RD may be genetically heterogeneous. If that is true, then in order to identify one of the subtypes, it is important to select groups of patients who are likely to have the same underlying genetic etiology. Unrelated individuals are the least likely to be genetically homogeneous whereas individuals in the same family are the most likely to have a common genetic etiology. For a linkage study it is most advantageous to select very large, multigenerational families which are large enough to reach statistical significance on their own. Several large families will allow examination of the hypothesis of genetic heterogeneity.

Pennington and Smith (1983) selected large multigenerational families with several affected individuals for their linkage studies of reading disability. Only families which met criteria consistent with an autosomal mode of transmission were included. All members of these families were given a battery of cognitive tests as well as interviews relevant to eliciting a history of dyslexia.

Linkage analyses were performed using both genotyping markers (i.e., proteins) and chromosomal banding markers (which show specific banding patterns under the microscope and are passed genetically). Only one marker (a banding marker for chromosome 15) gave significant ($P < 0.001$) evidence for linkage. These results suggest that the genetic locus for at least one form of RD is tightly linked to a site on the short arm of chromosome 15. However, the findings must be seen as tentative as the same group was unable to confirm the results using additional chromosome 15 markers (Kimberling et al. 1985).

Advances in Molecular Biology

Restriction Enzymes

As was mentioned above, the vast majority of genetic illnesses could not be shown to be linked to known genes due to a paucity of sufficiently polymorphic marker loci. The human genome as encoded in the DNA molecules is enormous, and the number of genes that were known to encode specific proteins was too small to be of significant use in genetic linkage studies.

It was soon realized that the DNA molecule itself was a potential source for markers. For example, suppose gene A has two protein variants, A1 and A2, which have equivalent enzymatic activity but vary slightly in their electrophoretic mobility. The difference in their mobilities is due to variations in their amino acid sequences, variations which do not cause noticeable enzymatic differences.

These two proteins are called polymorphic. The DNA sequences which encode for A1 and A2 will necessarily be polymorphic as well. That is, if one could cut out exactly that portion of the DNA that encoded for allele A1 and compare it nucleotide by nucleotide with the DNA sequence for allele A2, there would be nucleotide differences at exactly those areas encoding for the different amino acids. An individual with protein A1 would have DNA sequence A1, which was different from an individual with protein A2 and DNA sequence A2.

The strategy employed to increase vastly the number of polymorphic markers can now be described. This technique relies on enzymes which cut the DNA molecule at precise points as determined by specific nucleotide sequences. These enzymes, called restriction enzymes (REs), will cut a DNA molecule wherever this exact sequence of nucleotides occurs (Fig. 1). A second RE will recognize another specific nucleotide sequence at which it will cleave the DNA molecule. DNA cut with a particular RE will produce many millions of DNA fragments, the exact number being determined by the amount of the input DNA, the number of restriction sites for the RE, and the precision with which the RE cuts.

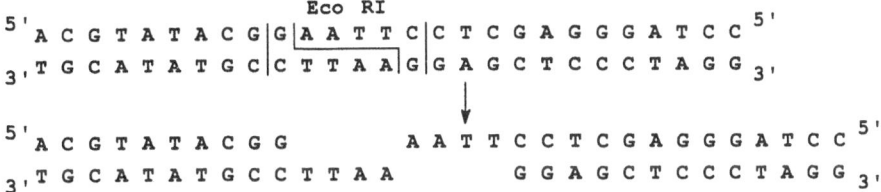

Fig. 1. Restriction enzymes recognize specific nucleotide sequences and cleave DNA molecules at those sites. In this example, Eco RI recognizes the sequence "GAATTC" and makes a staggered, symmetrical cut in the DNA and releases two fragments

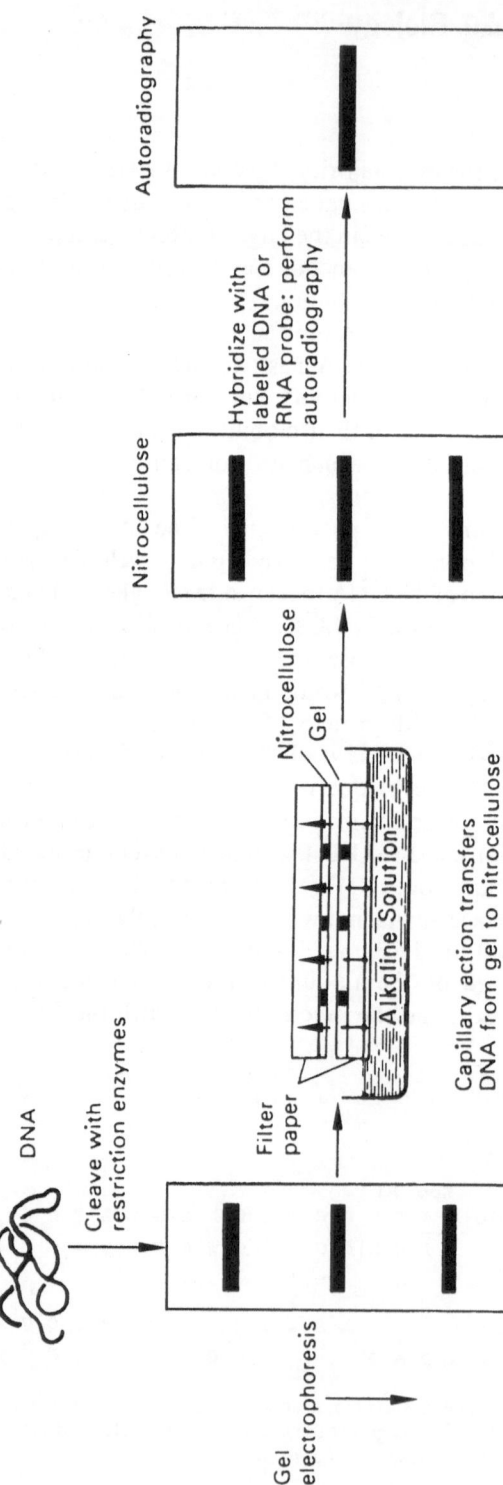

Fig. 2. Steps involved in a Southern blot include cutting DNA with one or several restriction enzymes. The DNA fragments are separated by size using agarose gel electrophoresis. The gel is then laid on a piece of nitrocellulose and a flow of buffer through the gel transfers the fragments to the filter. The filter can be hybridized to a suitably labeled probe, and specific DNA fragments that hybridize to the probe will give a signal following autoradiography

Southern Blots

Specific DNA fragments can be visualized using a procedure termed a Southern blot (Fig. 2). After digestion, the DNA fragments can be separated using agarose gel electrophoresis. The fragments will migrate through the gel according to their size, with the smaller fragments traveling more easily and, thus, farther than the larger fragments. After several hours a smear of DNA fragments will have migrated through the gel. Thus separated, the cut DNA fragments are transferred to a filter paper to which they are tightly bonded.

The DNA can be seen if a probe is added and autoradiography performed. A probe is a specific fragment of DNA which has been isolated and made radioactive, and which has the property of tightly binding (hybridizing) to a complementary sequence of nucleotides. If the fragments of DNA resulting from digestion with a unique RE are bound to filter paper and the filter is now washed in the presence of a radioactive probe, the probe will anneal to the precise fragment(s) of DNA which are its complement. It is then a simple matter to lie the filter paper on photographic paper and expose the paper over several days. One or more bands will become visible in the smear of DNA fragments. If the process were repeated with a different RE, whose restriction site was a different sequence of nucleotides, another series of bands might emerge.

Restriction Fragment Length Polymorphisms

The DNA from two individuals can be isolated easily from their cells (usually from lymphocytes in the bloodstream), and a Southern blot can be made comparing their DNA fragments. Their DNA will have many nucleotide differences (unless they are monozygotic twins). It is probable that one or more of these base differences will define a new site for the RE being used, or remove ones that were present before. This in turn will change the fragment lengths around the RE sites and change their mobility within the gel. The net result will be a possible change in the banding pattern on a Southern blot. It may require several different REs or probes, but eventually new bands will emerge or old ones will disappear. This difference in banding pattern reflects the individual variation in DNA sequences. Their discovery has given birth to a new generation of markers termed restriction fragment length polymorphisms (RFLPs).

There are two stages in conducting linkage analyses with REs and probes. The first stage is defining a series of RFLPs between various samples of DNA. The second is to then ask whether a particular RFLP segregates with the trait of interest, in a similar fashion as was done earlier with polymorphic proteins. If there is no linkage, one moves to the next REP and repeats the procedure. Figure 3 shows a hypothetical family with an illness which is transmitted in a dominant fashion. An RE was used to cut the DNAs from various family members. When the DNA was then mixed with a probe and exposed on radiographic film, a pattern of RFLPs was noted. In this case, an RFLP was

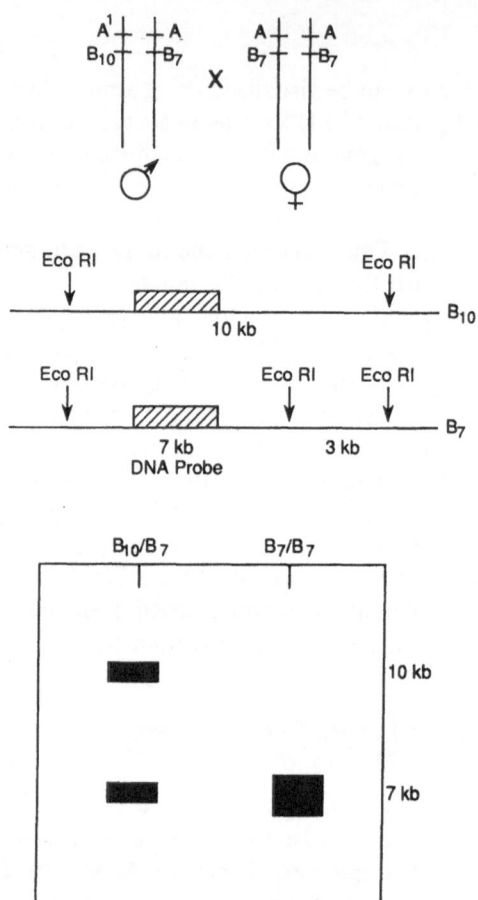

a.

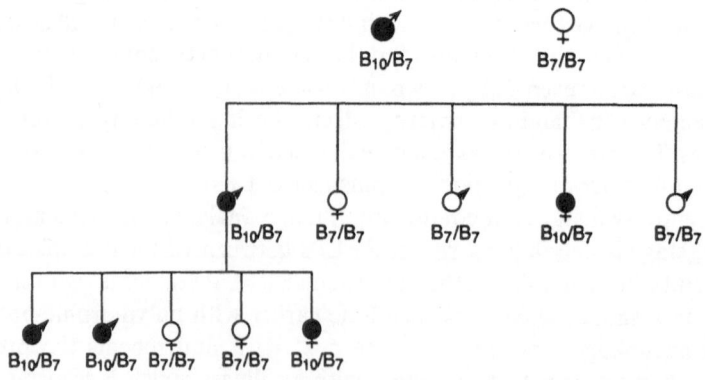

b

found closely linked to the abnormal gene since whoever had the illness also carried the particular RFLP banding pattern when their DNA was cut with Eco RI and hybridized with a probe.

Large pedigrees are the most useful families in which to perform linkage studies of the type just described. In the study of a rare disease, large families minimize the risk that another phenotypically similar but genotypically different gene is involved. An excellent example of the usefulness of large families is the Venezuela study on Huntington's chorea (Gusella et al. 1983) in which over 300 family members had been studied. Currently, over 2000 family members have been screened. In this case an RFLP was found closely linked to the Huntington's gene and located on the tip of the short arm of chromosome 4. Once this linkage had been identified, other families were studied to confirm the linkage relationship.

In the study of a more common illness, such as affective disorder, large pedigrees are necessary but may not be sufficient. An affected individual with a genetically distinct but phenotypically similar disease might marry into a large family in which affective illness is already present. This would make accurate linkage studies considerably more difficult.

To avoid this problem, it is useful to use large pedigrees which have isolated themselves, either culturally or geographically. The Old Order Amish population meets these conditions. Bipolar illness was found to be prevalent among this group (Egeland and Hostetter 1983), and their cultural background not only isolated them from the surrounding community but also minimized confounding variables, such as alcoholism, drug abuse, and antisocial behavior. An RFLP was found closely linked to a putative gene for bipolar illness on chromosome 11 (Egeland et al. 1987). Several other research groups published their linkage findings from separate pedigrees with bipolar disease. Baron et al. (1987) showed linkage on the X chromosome, and Hodgkinson et al. (1987) demonstrated that bipolar disease among their Icelandic families was not linked to chromosome 11. These disparate findings are in fact quite compatible with what is currently understood about the etiology of bipolar illness. Several distinct genes may produce an illness which clinicians identify as bipolar illness. That is, several genotypes exist which are expressed as a single phenotype. It is probable that similar results will be found in other illnesses which on clinical grounds appear homogeneous.

Fig. 3a, b. Analysis of large pedigrees is particularly useful to demonstrate linkage between genes of interest and marker loci. **a** In this example, the father has a gene (A^1) producing an abnormal protein on one chromosome, and the normal variant (A) on the other. The mother has the normal variant on both of her chromosomes. Eco RI digestion reveals a tightly linked RFLP. An Eco RI site has disappeared on one of the father's chromosomes' resulting in a larger (10-kb) fragment appearing on a Southern blot. Digestion of the mother's DNA with Eco RI produces 7-kb fragments with a signal of approximately doubled intensity. **b** Demonstration of how a large pedigree can be used to follow the transmission of an abnormal gene and restriction fragments. Any child with A^1 (*dark symbol*) shows the B10/B7 banding pattern, while a child with the normal variants A will show a B7/B7 banding pattern

Linkage Map of the Human Genome

In 1980, Botstein predicted that 150 evenly spaced RFLPs would be sufficient to saturate the human genome with marker loci (Botstein et al. 1980). At that point, any genetically transmitted illness would be found to be tightly linked to at least one RFLP and its chromosomal position could be accurately defined. Such a map has recently been published by Helms et al. (1988) using 545 RFLPs. This work will permit the eventual isolation of abnormal genes by recombinant DNA techniques and the functional analysis of their gene products. It may also be useful to develop maps of the human genome that are based solely on those regions which code for protein synthesis, so-called "exon" or "cDNA" maps (McKusick and Ruddle 1987).

At that point, we will be in a better position to return once again to the question of environmental effects on psychiatric disturbances. For example, let us assume that a large pedigree with bipolar illness is identified. The specific variety of bipolar illness is established through linkage mapping with RFLPs. The children can be examined and some will be identified as carrying the abnormal gene. These children may be studied longitudinally. Some will show disturbances and others will not, and hopefully environmental stressors can then be identified. Obviously, this type of research can be done with any illness for which a definitive genetic basis has been found.

Chromosomal Walking

Genetic linkage studies can identify a region of DNA in which a suspected gene lies. This region, one to several million bases, is quite large and may encode for many separate gene products. To isolate the particular gene of interest involves recombinant DNA techniques, the details of which are beyond the scope of this chapter (see Watson et al. 1983 for an excellent review). There are several different methods to "walk along" the chromosome for the purpose of isolating a specific gene in preparation for cloning the region.

Functional studies of the putative gene product can then be undertaken. Determination of whether the abnormal protein is involved in an enzyme pathway, is a membrane protein such as a neurotransmitter receptor, or is a transport protein are important steps in understanding the pathophysiology of the illness. Such functional analyses will not only lead to an understanding of the underlying molecular basis for the disease but may lead to more rational therapeutic interventions.

Complicating Factors

Genetic studies of childhood behavioral disorders have been hampered in the past by the complexity and changeable character of many of the phenotypes of interest as well as the imprecision and frequently changing diagnostic criteria.

Infantile autism is now believed to be the final common expression of several different etiological processes. Any study which hopes to clarify basic mechanisms in autism must begin by studying homogeneous populations. Those children whose autistic symptoms are a consequence of maternal rubella infection may not differ phenotypically from those children whose autistic diathesis is secondary to a specific genetic abnormality, but the latter will be difficult to find in linkage studies using a mixed population which includes both groups. The work on bipolar affective disorder demonstrates that large pedigrees are helpful in studying common heterogeneous disorders since they assure relatively homogeneous samples of patients within one kindred.

In child psychiatry, several factors have worked against the collection of such homogeneous groups. Diagnostic instruments which have been developed are rather imprecise in their ability to produce accurate subtypes. Parent–child agreement on these tests has been found to be low and the test–retest reliability has been shown to be poor (Chambers et al. 1985).

The actual collection of information has also proven problematic. The most dependable methodologies rely on face-to-face interviews with all available relatives. This has often been impossible to do. In addition, in the study of child psychiatric disorders, one necessarily must rely on the ability of adults who are being interviewed to remember accurately their childhood symptomatology. The underreporting of such symptoms has been well documented (Andreasen et al. 1977). Finally, the confounding variable of assortative matings must be taken into account in any study which attempts to address genetic issues.

Genetic heterogeneity is likely to exist for most childhood (as well as adult) behavioral disorders. Therefore, much effort must be made to obtain a homogeneous group of subjects. The pedigree approach has been discussed above. It is true that the population being studied is very restricted and the results obtained may not be generalized to the population at large. This should not be a deterrent as the insights gleaned from such a study may provide useful clinical differences that could then be applied to other groups.

Although the number of psychiatric disorders which lend themselves to linkage study and recombinant DNA technology is unknown and may indeed be small, it is important to attempt to pursue this avenue of research. As has been mentioned, these techniques are currently useful for illnesses in which only a single major locus accounts for a sizeable portion of the total phenotypic variance. In addition to pursuing these single locus disorders, new strategies need to be developed to isolate the various genes contributing to polygenic disturbances.

Future Directions

Genetic linkage is a powerful technique to demonstrate that a genetic factor is etiologically involved in the transmission of a particular disorder. As has been suggested in previous sections, the current revolution in molecular biology is being applied in human genetics, with the specific goal of isolating and characterizing the genes involved in the expression of various disorders. Similar work in psychiatry and, in particular, child psychiatry needs to be supported in the years to come.

Once the basic constitutional events that predispose an individual to a disorder are known, it will be possible to ascertain which individuals and their offspring are at risk. Earlier diagnosis will be possible and intervention programs may be instituted. At the same time, the design of appropriate longitudinal studies will make it feasible to investigate which environmental factors modulate the inherent genetic basis; that is, which are protective in nature and which are not.

Only recently, the methodology of recombinant DNA techniques had limited applicability to human disorders, because of the paucity of polymorphic markers. At the present time, RFLPs have been located throughout the human genome and it is theoretically possible to identify the chromosomal region to which a gene for any particular illness maps.

The question arises as to how many complex behavioral disorders will lend themselves to this type of analysis. The answer remains unclear. However, the linkage studies described earlier for manic-depressive disorder suggest that at least some complex illnesses may be the result of single mutations. Moreover, the fact that a genetic illness is polygenic in nature does not exclude the possibility that future molecular techniques will be able to tease apart the genetic mechanisms involved.

The number of psychiatric disorders for which a single major locus accounts for a sizeable fraction of the total phenotypic variance is unknown. However, the understanding gleaned from any successful project would certainly be exciting if the end result were a furthering of our understanding of how the mind and body function.

Finally, child psychiatric research should not only concentrate on understanding the molecular basis for specific childhood disorders. It should also apply itself to the more basic questions of the developing CNS. The isolation and characterization of genes involved in CNS development and maturation will be a long and arduous field of endeavor, and should be an area pursued by future researchers in child psychiatry.

References

Andreasen N, Endicott J, Spitzer R, Winokur G (1977) The family history method of diagnostic criteria. Arch Gen Psychiatry 34:1229–1235

Baron M, Risch N, Hamburger R, Mandel B et al. (1987) Genetic linkage between X-chromosome markers and bipolar affective illness. Nature 326:289–292

Biederman J, Munir K, Knee D (1984) Attention deficit disorder with and without conduct and oppositional disorder: a controlled family study. Annual meeting of the American Academy of Child and Adolescent Psychiatry, Toronto

Botstein D, White R, Skolnick M, Davis R (1980) Construction of a genetic linkage map in man using restriction fragment length polymorphisms. Am J Hum Genet 32:(344)312–331

Cadoret R, O'Gorman T, Troughton E, Heywood E (1985) Alcoholism and antisocial personality. Arch Gen Psychiatry 42:161–167

Cadoret R, Troughton E, O'Gorman T, Heywood E (1986) An adoption study of genetic and environmental factors in drug abuse. Arch Gen Psychiatry 43:1131–1136

Cantwell D (1972) Psychiatric illness in the families of hyperactive children. Arch Gen Psychiatry 27:414–417

Cantwell D (1975a) Genetics of hyperactivity. J Child Psychol Psychiatry 16:261–264

Cantwell D (1975b) Genetic studies on hyperactive children. Psychiatric illness in biologic and adoptive parents. In: Fieve R, Rosenthal D, Brill H (eds) Genetic research in psychiatry. Johns Hopkins University Press, Baltimore, pp 275–280

Chambers W, Puig-Antich J, Hirsh M, Paez P et al. (1985) The assessment of affective disorders in children and adolescents by semi-structured interview: test-retest reliability of the K-SADS. Arch Gen Psychiatry 42:696–702

Cloninger C, Bohman M, Sigvardsson S (1981) Inheritance of alcohol abuse: cross-fostering analysis of adopted men. Arch Gen Psychiatry 38:861–868

Cohen D, Dibble E, Grawe J et al. (1975) Reliably separating identical from fraternal twins. Arch Gen Psychiatry 32:1371–1378

Comings D, Comings B (1984) Tourette's syndrome and attention deficit disorder with hyperactivity: are they due to the same gene? J Am Acad Child Psychiatry 23:138–146

Comings D, Comings B, Devor E, Cloninger C (1984) Detection of a major gene for Gilles de la Tourette syndrome. Am J Hum Genet 36:586–600

DeFries J, Decker S (1981) Genetic aspects of reading disability: the Colorado family reading study. In: Aaron P, Malatesha M (eds) Neuropsychological and neuropsycholinguistic aspects of reading disability. Academic, New York

Egeland J, Hostetter A (1983) Amish study I: affective disorders among the Amish. Am J Psychiatry 140:56–61

Egeland J, Gerhard D, Pauls D et al. (1987) Bipolar affective disorders linked to DNA markers on chromosome 11. Nature 325:783–787

Elandt-Johnson R (1971) Probability models and statistical methods in genetics. Wiley, New York

Fernando S (1967) Gilles de la Tourette's syndrome. Br J Psychiatry 113:607–617

Folstein S, Rutter M (1977) Infantile autism: genetic study of twin pairs. J Child Psychol Psychiatry 18:297–321

Goodwin D (1979) Alcoholism and heredity: a review and hypothesis. Arch Gen Psychiatry 36:57–61

Goodwin D (1985) Alcoholism and genetics. Arch Gen Psychiatry 42:171–174

Graham P, Stevenson J (1987) Temperament and psychiatric disorder: the genetic contribution to behavior in childhood. Aust NZ J Psychiatry 21:267–274

Gusella J, Wexler N, Conneally P, Naylor S et al. (1983) A polymorphic marker genetically linked to Huntington's disease. Nature 306:234–238

Helms C, Green P, Weiffenbach B, Bowden D et al. (1988) A human genetic linkage map with 545 genetic loci. Am J Hum Genet (Suppl) 43:A147

Hodgkinson S, Sherrington R, Gurling H, Marchbanks R et al. (1987) Molecular genetic evidence for heterogeneity in manic depression. Nature 325:805–806

Hutchings B, Mednick S (1975) Registered criminality in the adoptive and biological parents of registered male criminal adoptees. Proc Annu Meet Am Psychopathol Assoc 63:105–116

Kerr J (1897) School hygiene, in its mental, moral and physical aspects. Howard medical prize essay. J R Stat Soc 60:613–680

Kety S (1983) Observations on genetic and environmental influences in the etiology of mental disorder from studies on adoptees and their relatives. In: Kety S, Rowland L, Sidman R, Matthysse S (eds) Genetics of neurological and psychiatric disorders. Raven, New York, pp 105–114

Kimberling W, Fain P, Ing P, Smith S, Pennington B (1985) Genetic linkage studies of reading disability with chromosome 15 markers (Abstract). The 15th annual meeting of Behavior Genetics Association, Pennsylvania State University

Lewitter F, DeFries J, Elston R (1980) Genetic models of reading disabilities. Behav Genet 10:9–30

McKusick V, Ruddle F (1987) Toward a complete map of the human genome. Genomics 1:103–106

Morgan W (1896) A case of congenital word-blindness. Br Med J 2:1543–1544

Morrison J, Stewart M (1971) A family study of the hyperactive child syndrome. Biol Psychiatry 3:189–195

Morton N, Yee S, Lew R (1971) Complex segregation analysis. Am J Hum Genet 23:602–610

Nee L, Polinsky R, Ebert M (1982) Tourette syndrome: clinical and family studies. In: Friedhoff A, Chase T (eds) Gilles de la Tourette syndrome. Raven, New York, pp 291–295 (Advances in Neurology, vol. 35)

Pauls D (1985) Strategies for the genetic study of child psychiatric disorders. In: Michels R, Cavenar J, Brodie H et al. (eds) Psychiatry vol 2. Lippincott, Philadelphia

Pauls D, Cohen D, Heimbuch R, Dettor J, Kidd K (1981) The familial pattern and transmission of Tourette syndrome and multiple tics. Arch Gen Psychiatry 32:1091–1093

Pauls D, Shaywitz S, Kramer P, Shaywitz B, Cohen D (1983) Demonstration of vertical transmission of attention deficit disorder. Ann Neurol 14:363

Pauls D, Kruger S, Leckman J, Cohen D, Kidd K (1984) The risk of Tourette's syndrome and chronic multiple tics among relatives of Tourette's syndrome patients obtained by direct interview. J Am Acad Child Psychiatry 23:134–137

Pauls D, Towbin K, Leckman J et al. (1986) Gilles de la Tourette syndrome and obsessive compulsive disorder: evidence supporting an etiological relationship. Arch Gen Psychiatry 43:1180–1182

Pennington B, Smith S (1983) Genetic influences on learning disabilities and speech and language disorders. Child Dev 54:369–387

Plomin R, DeFries JC (1985) Origins of individual differences in infancy: the Colorado adoption project. Academic, New York

Plomin R, DeFries JC, Fulker DW (1988) Nature and nurture during infancy and early childhood. Cambridge University Press, New York

Rice J, Cloninger CR, Reich T (1979) Multifactorial inheritance with cultural transmission and assertative mating. I. Description and basic properties of the unitary models. Am J Hum Genet 30:618–627

Rowe D (1983) Biometrical genetic models of self-reported delinquent behavior: a twin study. Behav Genet 13:473–489

Rutter M (1967) Psychotic disorders in early childhood. In: Copper A, Walk A (eds) Recent developments in schizophrenia (Special publication). Br J Psychiatry 1:1–14

Safer D (1973) A familial factor in minimal brain dysfunction. Behav Genet 3:175–186

Scarr S (1966) Genetic factors in activity motivation. Child Dev 37:663–673

Shapiro A, Shapiro E, Bruun R, Sweet R (1978) Gilles de la Tourette syndrome. Raven, New York

Vandenberg S (1962) The hereditary abilities study: hereditary components in a psychological test battery. Am J Hum Genet 14:220–237

Watson J, Tooze J, Kurtz D (1983) Recombinant DNA: a short course. Freeman, New York

Willerman L (1973) Activity level and hyperactivity in twins. Child Dev 37:663–673

Yaryura-Tobias J, Neziroglu F, Howard S, Fuller B (1981) Clinical aspects of Gilles de la Tourette syndrome. Orthopsychiatry 10:263–268

Cognition

Cognitive Correlates of Abnormal EEG Waveforms in Children

J. Martinius

Cognitive processes are central to mental and psychic development and well-being in childhood, in particular during school age. Prerequisites for cognition and, in a broader sense, for learning are composite brain functions, which in children who suffer from brain dysfunction are disturbed in one way or another. Abnormal EEG waveforms are one of the many functional expressions of brain pathology. If they occur in the surface EEG with a particular configuration, distribution, and time course, such waveforms may be the correlate of epilepsy or, in the absence of overt clinical signs of seizures, an indication of an increased tendency to have epileptic seizures. In children with proven epilepsy and in some without manifest seizures, abnormal waveforms of the hypersynchronous type do occur, although paroxysmally and frequently at times of apparently intact consciousness. This observation has aroused interest in its significance and stimulated research studying and analyzing in detail cognitive functioning in relation to the "interictal" occurrence of abnormal waveforms in the EEG.

Behavior and Learning Disorders in Children with Epilepsy

In epileptic children learning problems are quite frequent. Despite the fact that the majority of epileptic children react favorably to treatment by stopping having fits, in excess of 30% of those attending ordinary schools and presumably having normal intellectual capacities exhibit problems which interfere with adequate academic attainment. These problems may affect any component of the learning process. The latter involves perceptual, cognitive, and effector functions such as recognition, discrimination, understanding, memory, and producing a response, all of which are task specific. In addition, the individual is enabled to engage in learning by a certain state of consciousness, i.e., arousal and attention, the latter subserving selectivity of cognition. Learning problems are constituted and need to be disentangled individually, just as a seizure problem demands individual clarification. Learning problems may consist of a pure and primary attention deficit. They may, on the other hand, consist of specific problems of reading or arithmetic and may ultimately be linked to a memory impairment, affecting either recognition memory or associative memory. To complicate things even further, there may not be an attention deficit or a specific cognitive problem but an emotional or conduct disorder which, in a classroom situation, keeps the child from learning. When investigating cognitive correlates of abnormal EEG waveforms one therefore has to distinguish the specific aspects of cognition and learning (Baird et al. 1980).

Most of these processes have been studied in epileptic children of school age. Over the course of some 15 years Stores et al. (Stores 1973, 1981, 1982; Stores and Hart 1976; Stores et al. 1978, Stores and Lewin 1981) produced comprehensive evidence of behavior and learning disorders and of their relation to type of epilepsy, drug treatment, gender, emotional dependence, and situational context. Epileptic boys with a persistent left temporal spike focus were found to run the greatest risk of reading retardation, inattentiveness, emotional dependence, and other types of disturbed behavior. Previously, specific patterns of cognitive impairment had been observed in children by others (Fedio and Mirsky 1969) in the presence of right hemisphere foci as well as generalized forms of epilepsy, the findings differing somewhat from those of Stores et al. (1978).

Transitory Cognitive Impairment

Obviously, cognitive impairment and learning problems in children with epilepsy have various sources which interact and are not easily separable. Still, a strong suggestion remains that one of the sources of cognitive malfunctioning despite adequate seizure control may be paroxysmal disturbances of a more delicate epileptiform nature. In contrast to observable and typical changes of behavior occurring in strict temporal relation to partial or generalized hypersynchronous activity in the EEG, we are concerned here with subtle changes in the central processing of information which display a strict temporal relation to epileptiform activity in the EEG, formerly called "epileptic equivalents" and "subclinical seizures," and more recently named "transitory cognitive impairment" (TCI), a term proposed by Aarts et al. (1984).

Previous Studies on TCI

The number of studies devoted to this question is not impressive if only those are taken into consideration which have been conducted with sufficient reliability. As long ago as 1939 Schwab discovered that reaction time to light stimuli is delayed during brief bursts of generalized 3/s spike-wave. He also found that light and sound stimuli themselves have an effect upon the seizure by occasionally terminating it. Prechtl et al. (1961) demonstrated a decrease in performance during interictal episodes of flattening (suppression) of EEG activity. Kooi and Hovey (1957) found disturbances of higher mental processes by administering subtests of the Wechsler Intelligence Scale and recording responses in relation to burst activity in the EEG. Cognitive difficulties were found not only during bilateral synchronous 3/s spike-wave paroxysms but also with irregular sharp and slow wave activity. More precise answers came from an investigation by Mirksy and Van Buren (1965). Their results indicated that well-organized bilateral synchronous bursts produce greater deficits of performance than irregular bursts. Tests requiring higher attentive functions were more impaired than pure motor responding. Attention deficits were already noticeable prior to the beginning of a burst. But in some patients there was no change in performance even in the presence of well-organized symmetrical bursts, meaning that the relationship between burst activity and deficits of performance is not one-to-one. Penry (1973) pointed out that spike-wave paroxysms had to last longer than 3s to cause errors in a simple pursuit reaction test, and it became clear that in children the same holds true (Hutt et al. 1976, 1977). These studies confirmed that subclinical paroxysms of spike-wave reduce the information-handling capacity of patients with epilepsy. However, it was also found that the effect of spike-wave is not all-or-none in nature, but is proportional to the information-processing demands of a task. Hutt et al. argued that it is not so much signal detection which is impaired as the child's assessment of signal criteria.

In a further study on memory functions during photically evoked spike-wave in children, Hutt and Gilbert (1980) found the impairment to be greatest if paroxysms fell between the end of presentation and the beginning of recall. Recall was weakest for the last part of digit span presentation.

A most pertinent study on cognitive impairment during epileptiform EEG activity was done by Aarts et al. (1984). This study marks the present state of knowledge and shall therefore be reviewed in greater detail. The authors studied 46 persons who had been referred for intensive EEG investigation concerning epilepsy. Among them was an unspecified number of children, the youngest being 10 years old. The subjects underwent computer-generated testing with display presentation, not unlike popular computer games, in which short-term memory (verbal and topographical) was the tested variable. The procedure included simultaneous EEG and video recording. The occurrence of errors in relation to paroxysmal activity was analyzed. TCI was demonstrable in half of all patients studied, in 50% of those with symmetrical and in 40% of those with asymmetrical generalized discharges, and in 58% of those with focal epileptiform activity. The effects of focal discharges were specific to the task lateralized to the appropriate hemisphere, i.e., left-sided epileptiform activity caused higher error rates in the verbal tasks, while right-sided activity caused higher error rates in the topographical task. The error rate associated with discharges during the stimulus was strikingly elevated over the basal level and was significantly higher than that when discharges occurred during response. The results clearly show that TCI is not necessarily a consequence of a general impairment of attention or arousal but rather is related to specific cognitive functions located in the region or regions where the epileptiform discharges arise.

Epileptiform EEG Activity in Natural Situations

Although the study by Aarts et al. (1984) relied upon laboratory tests simulating "natural" situations (computer games), the results still may not be transferable to classroom learning or to other conditions of everyday life which tend to be much more open to emotional influences. Epileptic fits are known to vary with the influence of emotions. The occurrence and strength of TCI may be even more subject to this influence. In a case study Zegans et al. (1964) reported a marked decrement of burst activity and a simultaneous increase in GSR (galvanic skin response) and heart rate during a psychiatric interview in a disturbed child with petit mal epilepsy. Ounsted and Hutt (1964) noticed that if task difficulty is increased beyond a certain point the number of spike-wave paroxysms rises, following a prior decrease. Guey et al. (1969) used EEG telemetry to study children under more or less natural conditions and were able to demonstrate clearly strong psychological influences upon the rate of absence seizures, both clinical and subclinical. The frequency of paroxysms varied significantly, being highest during inactivity and during school exercises and lowest during a psychiatric interview. They concluded that "the situation in which the epileptic patient finds himself determines the number of seizures (or discharges respectively) he may suffer. In particular, the significance of a situation for the patient seems to have importance for triggering or inhibiting discharges."

This means that if we want to know more about cognitive correlates of abnormal EEG waveforms we have to examine children in natural situations and collect as much information as possible about the emotional significance of a given situation to the individual child. Although such studies have not been systematically carried out, considerable progress has been made in that direction. Well-founded knowledge exists about triggering mechanisms evoking reflex seizures by sensory stimulation (Gastaut and Tassinary 1966). Stimuli may be visual, auditory, proprio- and exteroceptive, or vegetative. They may be provoked by visual patterns (Wilkins et al. 1979) or by visual exploration, particularly by reading. Patients with this type of epilepsy are rare. Yet it may be worthwhile finding out whether such reflex-provoking mechanisms are operant on the "subclinical" level. Children with epilepsy frequently have reading problems, which in part could be due to TCI.

Situational correlates for both precipitation and inhibition of burst activity have to be looked at separately. They were discovered soon after the introduction of ambulatory cassette recording devices which allowed for continuous recording of the EEG along with behavioral observation over periods of 24 h or longer. By now evidence has accumulated that emotional stress may trigger epileptiform bursts but that pleasant emotional excitement may have the same effect (Sato et al. 1976). Again, the work of Stores (1981, 1982) and others with the ambulatory recording system has shed much light on the occurrence of epileptiform EEG activity in real-life situations. It has become quite clear that factors of an environmental and particularly of an emotional nature, such as anxiety, boredom, or stress, precipitate seizure activity, while concentration, being interested in something, and physical activity have inhibiting effects upon discharges. However, there appear to be thresholds below and above which the inhibiting effect, for instance of concentration, can be reversed to facilitation. According to Ricci et al. (1972) this model holds true also for focal discharges.

Electro-Clinical Correlations

One must, of course, be careful not to overinterpret the pathogenic influence of TCI as the cognitive correlate of abnormal EEG waveforms. In recognizing TCI as a possibility one should still remain hesitant in giving anticonvulsant medication to children who show interictal discharges in their EEG. Cognitive impairment related to such discharges needs to be proven, requiring intensive diagnostic monitoring in situations relevant to cognition and learning, taking into account emotional concomitants. In this context the question may arise of how important it is to quantify the effects of single bursts of epileptiform activity on cognition. The answer would certainly be of theoretical importance. In practical terms, a major gain would be achieved simply by identifying critical situations and states of consciousness or even mood in individual children and trying to have them do their exercises and classroom learning under conditions which counteract an abundance of potentially harmful brain dysfunction. And since drug action via arousal and emotion can indirectly affect the frequency of interictal discharges, their recording would in turn help to guide drug treatment.

Conclusion

A correlation between abnormal, epileptiform EEG waves and disturbances of cognitive functions and learning is well established in adults and children alike. Such disturbances express themselves in a subtle manner, affecting cognitive processes such as signal detection, recognition, and in particular short-term memory. Effector functions are inhibited to a lesser degree. The relationship is not of an all-or-none type. In a controlled experimental situation about 50% of epileptic subjects show selective cognitive impairment in strict temporal relation to "interictal" epileptiform EEG activity. The relation appears to become more positive with increasing information-processing demands. The emotional impact of a given situation appears to exert another important influence, operating in relation to thresholds. If interest is the prevailing drive, epileptiform discharges tend to be suppressed, whereas boredom and stress increase the occurrence of discharges. Further research will have to take these intervening variables into consideration. In order to obtain better insight, diagnostic monitoring incorporating ambulatory recording in natural situations will have to be performed.

References

Aarts JHP, Binnie CD, Smit AM, Wilkins AJ (1984) Selective cognitive impairment during focal and generalised epileptiform EEG activity. Brain 107:293–308

Baird ER, John ER, Ahn H, Maisel E (1980) Neurometric evaluation of epileptic children who do well and poorly in school. Electroencephalogr Clin Neurophysiol 48:683–693

Fedio P, Mirsky AF (1969) Selective intellectual deficits in children with temporal lobe or centrencephalic epilepsy. Neuropsychologia 7:287–300

Gastaut H, Tassinari CA (1966) Triggering mechanisms in epilepsy. The electroclinical point of view. Epilepsia 7:85–138

Guey J, Dureau M, Dravet C, Roger J (1969) A study of the rhythm of petit mal absences in children in relation to prevailing situations. The use of EEG telemetry during psychological examination, school exercises and periods of inactivity. Epilepsia 10:441–541

Hutt SJ, Gilbert S (1980) Effects of evoked spike wave discharges upon short term memory in patients with epilepsy. Cortex 16:445–457

Hutt SJ, Denner S, Newton J (1976) Auditory thresholds during evoked spike-wave activity in epileptic patients. Cortex 12:249–257

Hutt SJ, Newton J, Fairweather H (1977) Choice reaction time and EEG activity in children with epilepsy. Neuropsychologia 15:257–267

Kooi KA, Hovey HB (1957) Alterations in mental function and paroxysmal cerebral activity. AMA Arch Neurol Psychiatry 78:264–271

Mirsky AF, Van Buren JM (1965) On the nature of the "absence" in centrecephalic epilepsy: a study of some behavioral, electroencephalographic and autonomic factors. Electroencephalogr Clin Neurophysiol 18:334–348

Ounsted C, Hutt SJ (1964) The effect of attentive factors on bioelectrical paroxysms in epileptic children. Proc R Soc Med 57:1178

Penry JK (1973) Behavioral correlates of generalized spike-wave discharge in the electroencephalogram. In: Brazier M (ed) Epilepsy. Its phenomena in man. Academic, New York, pp 171–188

Prechtl HFR, Boeke PE, Schut T (1961) The electroencephalogram and performance in epileptic patients. Neurology 11:296–302

Ricci G, Berti G, Cherubini E (1972) Changes in interictal focal activity and spike-wave paroxysms during motor and mental activity. Epilepsia 13:785–794

Sato S, Penry JK, Dreifuss FE (1976) Electroencephalographic monitoring of generalized spike-wave paroxysms in the hospital and at home. In: Kellaway P, Petersen I (eds) Quantitative analytic studies in epilepsy. Raven, New York, pp 237–251

Schwab RS (1939) Method of measuring consciousness in attacks of petit mal epilepsy. AMA Arch Neurol Psychiatry 41:215–217

Stores G (1973) Studies of attention and seizure disorders. Dev Med Child Neurol 15:376–382

Stores G, Lewin R (1981) A study of factors associated with the occurrence of generalized seizure discharge in children with epilepsy using the Oxford Medilog System for ambulatory monitoring. In: Dam M, Gram L, Penry JK (eds) Advances in epileptology: XIIth Epilepsy International Symposium. Raven, New York, pp 421–422

Stores G (1981) Memory impairment in children with epilepsy. In: Melin KA (ed) Second workshop on memory functions. Acta Neurol Scand [Suppl] 89, pp 21–29

Stores G (1982) Psychological factors and seizure occurrence in children. In: Apley J, Ounsted C (eds) One child. Clinics in developmental medicine, no 80. Heinemann Medical, London, pp 75–83

Stores G, Hart J (1976) Reading skills in children with generalized or focal epilepsy. Dev Med Child Neurol 18:705–716

Stores G, Hart J, Piran N (1978) Inattentiveness in school children with epilepsy. Epilepsia 19:169–175

Wilkins AJ, Darby CE, Binnie CD (1979) Neurophysiological aspects of pattern-sensitive epilepsy. Brain 102:1–25

Zegans LS, Kool KA, Waggoner RW, Kemph JP (1964) Effect of psychiatric interview upon paroxysmal cerebral activity and autonomic measures in a disturbed child with petit mal epilepsy. Psychosomat Med 26:151–161

A Neuropsychophysiological Approach to Specific Developmental Learning Disabilities

G. A. Chiarenza

Introduction

Learning to read and write is one of the aspects of scholastic success on which the expectations of the family, the teacher, and the child itself weigh most heavily. The child's ability to satisfy these expectations is frequently a crucial factor in the child's current and future adaptation to life.

Various hypotheses have been advanced to explain reading and writing learning difficulties based on function models and knowledge of the central nervous system. Thus, from time to time, the key to interpretation was thought to have been found in an alteration of a given cognitive process. Further, since the dysfunction manifests itself at school age, educational specialists have seen the problem in terms of their own area of competence and highlighted sometimes its psychological, sometimes its neurological or pedagogic aspects. This variety of interpretation has not helped a fluid or immediately fruitful exchange of knowledge because of the different jargons involved, so an organic and overall formulation of the problem has only been arrived at with difficulty.

The definitions of dyslexia proposed up to now have been either (a) pre-dominantly clinical, concentrating on a profile of reading and writing errors and associated disturbances, such as difficulties in visual-spatial abilities, temporal analysis of rhythm, motor coordination, and mixed cerebral domi-nance, or (b) limited to describing the disturbance as the alteration of specific cognitive processes at certain stages of information processing. Furthermore, since the children examined were subjected to a variety of tests, many of the results are not comparable owing to the use of both different definitions and methodologies.

A further difficulty lies in the fact that reading and writing learning difficulties appear when the child first goes to school, without premonitory specific symptoms, without obvious neurological or personality disturbance, and when there has already been development of quite complex linguistic ability.

It is known that learning reading and writing requires the separate and integrated processing of auditory and visual information. A hypothesis developed in this respect based on neuropsychological research by Luria (1973) sees the learning of reading and writing as a two-stage process: encoding and comprehension (Aaron and Bakker 1982). The encoding stage has in its turn two components: simultaneous processing of the word and sequential processing of certain units within the word or sentence. These processes need not be mutually exclusive and can occur simultaneously. The inability to read could be due to incorrect use of the simultaneous or sequential strategy or a defective comprehension process (Bateman 1968; Ingram et al. 1970; Bannatyne 1966; Kinsbourne and Warrington 1966; Myklebust 1965). A similar approach was used by E. Boder (1973). Boder also maintains that reading and spelling are closely connected interdependent functions and that the diagnosis of dyslexia can be seen by looking at the reading and writing performance only as a whole and not just at individual errors. *Reading* requires visual perception and discrimination, visual memory sequencing, and directional orientional processes (Benton 1962; Birch 1962); it also requires the integration of different sensory modalities and the translation of visual symbols into meaningful auditory equivalents (Ingram 1963; Birch and Belmont 1964; Rabinovitch 1968). *Speech* requires the conversion of sounds into their visual symbol equivalents and depends on the auditory perception and discrimination processes, on auditory memory sequencing, and on recall (Wepman 1962; Bannatyne 1966; Bakker 1970). *Writing* requires fine motor and visual-motor coordination and tactile-kinesthetic memory (Bannatyne 1966; Johnson and Myklebust 1967). Perception emerges from this list of functions involved in the process of reading and writing as the basis of learning reading and writing. This function must be seen as a complex of numerous integrated higher-order functions.

A useful psychophysiological definition of perception is that which sees it as the extraction of information from the environment (Gibson 1969). This implies that perception in an "active" process of search, selection, and organization of stimuli from the nearby environment which are related to precise intentions and tasks. Furthermore, perception includes a variety of processes other than those mentioned above: expectation, anticipation, attention, motivation, formulation, and verification of hypotheses in relation to the requirements of the task. The processes involved in the search and selection necessarily include motor components and the sequential organization of perceptual-motor patterns in relation to the task required (Birch and Lefford 1963; Luria 1973).

There is no test or school exercise which checks a single function: perception, language, personality, etc. On the contrary, the behavior which we see is the product of a complex set of interacting systems, none of which acts alone. These sensory systems interact in their turn with other systems, e.g., motor, linguistic, motivational, mnemonic, and programming. These are all in relation to the specific requirements of the task set in a socially determined context.

Naturally, the development and efficiency of a system are not determined only by the interaction of the subject and the environment, but also by the influence of one system on the other hierarchically organized. If this point of view is accepted, there are no school tests which involve only one of the processes in such a way as to study them individually and separately.

When the child is unable to reproduce a figure accurately it is thought to be due to distorted visual perception: If the figure is drawn by the child with an angle different from the model, the child is said to be unaware of spatial position or to be incapable of perceiving it accurately. If the relationship between two figures is not respected, it is said that the child is unable to perceive spatial relations between objects.

These interpretations are not able to explain which function, perceptive or motor-perceptive, is really being compromised. For example, in order to copy the drawing it is necessary that the visual process analysis and synthesis, which interact together, are coordinated by the kinesthetic functions described above. Copying is a motor activity guided by sight; it requires that kinesthetic information from the movement and posture of the trunk, head, arm, hand, etc. is continually and dynamically related to movement. Further, the child must organize these determined motor patterns and control the appropriate muscle tension related to the necessary succession of movements.

The inability to copy accurately could be based on defective function of the visual system, the kinesthetic system, or the motor system, or on the lack of integration between the visual-kinesthetic complex and the motor system. A further possibility is that these systems may be efficient while the systems for programming or verification of the performance may not be adequate.

This has led to the hypothesis that deficiencies in perceptive function are at the basis of reading and writing learning difficulties (Belmont 1980). This single-factor approach has led to a wide collection of plausible explanations of this disorder. Included are theories proposing: a deficiency in visual perception (Lyle and Goyen 1968, 1975; de Hirsch et al. 1966; Jansky and de Hirsch 1972; Silver and Hagen 1971; Rourke 1976; Satz et al. 1974), a dysfunction in cross-modal integration (Birch and Belmont 1964, 1965), difficulty in memory recall (Senf 1969; Senf and Feshback 1970; Senf and Freundl 1971), a difficulty in temporal order recall (Bakker 1967, 1972; Groenendaal and Bakker 1971), a disorder in cerebral dominance characterized by abnormalities in the degree of lateralization (Hynd et al. 1979), a delay in the maturation of lateralization and differentiation of motor, somatosensory, and linguistic processes (Satz and Sparrow 1970; Satz et al. 1971), and bilateral representation of spatial processing, normally thought of as a function of the right hemisphere, which interferes with the linguistic functions of the left hemisphere (Witelson 1976). The conclusions of most of these studies seem to indicate that dyslexia in not due to a deficit in or specific retardation of development but is the result of various interacting factors and that some higher levels of integration common to both multimodal and single-modal information are deficient in many dyslexic and dysgraphic children.

As we have seen, all these results indicate that higher central processes are involved in dyslexia. In reality all the above-listed studies examined only some aspects of the complex perception function, losing sight of the overall picture and attributing the explanation of dyslexia to disorders of some of the higher processes. Moreover there have been physiological studies such as that by Duffy et al. (1980) which, by recording cerebral electrical activity of normal and dyslexic children, have shown differences in the EEG spectra both at rest and during tests to activate the right (presentation of music and geometrical figures) and left hemispheres (reading excercises) alone or together (visual-verbal association).

These differences appear in the following areas: the frontomedial (supplementary motor area), the left frontal anterolateral (Broca's area), the left medial temporal (auditory association area), and the posterolateral parietal quadrant (Wernicke's area; parietal associative and visual associative areas). These studies show that numerous cerebral areas participate in the process of reading and that some of the brain areas are different in dyslexic and normal children, but they do not indicate which specific process is altered during reading.

To answer some of these questions, studies were conducted with sensory and cognitive evoked brain potentials (ERPs) [see Rosenthal et al. (1982) for a review of the topic]. Differences were seen in the amplitude and latency of time intervals of certain peaks in the evoked potentials. In particular, an increase in the latency of certain waves was generally interpreted as indicating a greater slowness of children in analyzing sensory information while a decrease in amplitude was interpreted as reduced "neural capacity."

Apart from some attempts to correlate dyslexic subgroups with some specific alterations in the ERPs (Rosenthal et al. 1982), it can be said that the majority of these studies suffer from the same limitations as the psychological studies in that they analyze a single aspect of the perception function complex, ignoring the fundamental notion that the perception function is principally an active process of extracting information from the environment. Especially when we face such complex functions we must be equipped with a method which can give an overall view and is at the same time able to analyze the different parts which make up the process. The method which we chose to study children with specific reading learning difficulties was to have them perform an integrated complex "task" (a motor-perceptive exercise) and to simultaneously record the performance, the electromyographic activity, and the brain electrical activity. The brain electrical activity accompanying the performance of this task is defined as movement-related brain macropotentials (MRBMs). The motor task consisted in starting the sweep of an oscilloscope trace by pushing a button with the thumb of the left hand and stopping it within 40–60 ms by pushing a button with the thumb of the right hand (Papakostopoulos 1978). The short time interval involved forced the subject to preprogram the task before it was carried out. The completion of such a task requires good bimanual coordination and the execution of ballistic movements. As visual feedback was provided in real time, the subject could adjust his strategy accordingly.

If we assume that reading is a complex and skillful process and consists of a set of modular subroutines serially and hierarchically organized, of which writing is the harmonic and integrated expression of a series of ballistic movements preprogrammed and correctable only after they have been executed and evaluated (kinesthetic feedback), a method which incorporates the study of motor performance, electromyographic activity, and MRBMs during the execution of the motor-perception ability task can supply useful information on those systems and subsystems which regulate and organize the motor-perceptive functions.

Recently Papakostopoulos (1978), in a critical review of the data from adults, proposed a taxonomy of such electrical phenomena. From observation of the

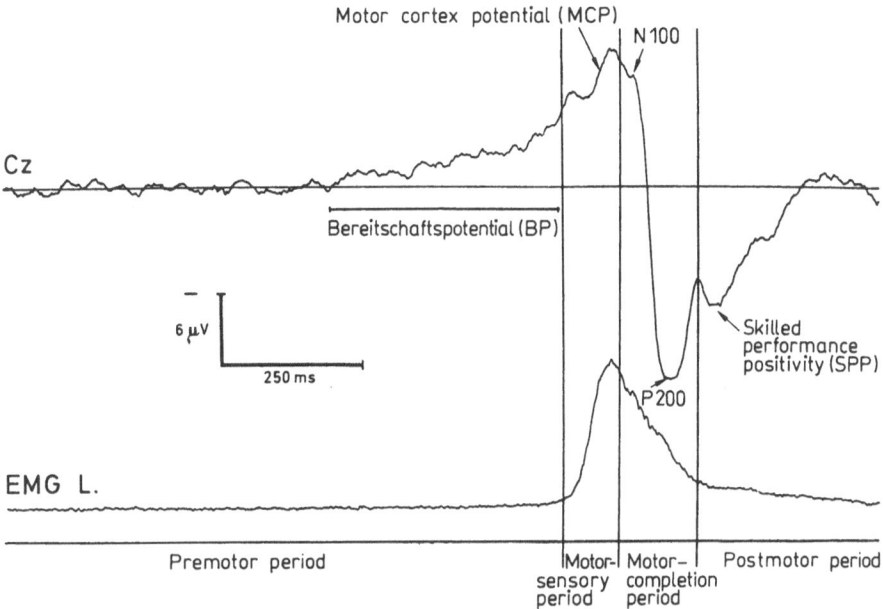

Fig. 1. Movement related brain macropotentials

myographic and brain electrical activity four periods can be distinguished: a premotor period, a motor-sensory period, a motor completion period, and a postmotor period (Fig. 1).

The premotor period is characterized by the presence of basic tonic muscular activity and the presence on the scalp of a phasic negative potential lasting 800–1200 ms. This potential is called the Bereitschaftspotential (BP) (Kornhuber and Deecke 1965; Vaughan et al. 1968). It is absent during passive movements. It has a low (5–7 μV) amplitude during simple tasks and a higher amplitude during more complex tasks (Papakostopoulos 1978). It is recorded prevalently in the frontal and central regions. It is absent in children younger than 6 years and the amplitude increases progressively with age, reaching adult values at adolescence (Chiarenza et al. 1983). The BP is believed to reflect the process of organization and selection of the strategy needed to carry out the task.

The sensory-motor period lasts about 200 ms and begins at the onset of phasic electromyographic activity. It coincides with the appearance on the cortex of the motor cortex potential (MCP), a negative potential which follows the negative slope of the BP. The MCP is absent during passive movements, present in simple voluntary motor actions, and increases in amplitude during complex motor actions (Papakostopoulos 1978). It is recorded prevalently from the precentral and central regions and is absent at the parietal regions. The MCP has been proposed as an index of sensory information from the muscle, skin, and tendon receptors (Papakostopoulos et al. 1975).

The motor completion period is characterized by the ending of the electromyographic phasic activity and by the presence of a negative cortical potential N100 and a positive potential defined as P200 (Vaughan et al. 1968). N100 is considered to be the response normally evoked by the oscilloscope trace and is partially suppressed in the central and precentral areas during movement. It has a latency of 100 ms and is an index of visual perception processes. P200 is a positive potential following N100 with a latency of about 200 ms from the beginning of the light stimulus. This potential is present during passive and active movements, both simple and complex. This potential is thought to be one of the components of the late somatosensory potentials (Chiarenza et al. 1983).

The postmotor period is marked by electromyographic tonic activity similar to that in the premotor period, by the appearance in the cortex of a positive potential with a latency of about 450 ms, denominated "skilled performance positivity" (SPP) (Papakostopoulos 1978, 1980), and by a slow negative potential labeled "post-action negativity" (PAN) with a latency of about 600 ms (Chiarenza et al. 1983, 1984). The SPP is recorded mainly in the parietal regions and appears towards the 9th year in the frontocentral region. The SPP is present only when the subject can evaluate the result of his performance. This potential is independent of the motor act and of the presence of any exteroceptive stimulation (Papakostopoulos 1980). SPP coincides with the subject's awareness of success or failure in the performance. The PAN is recorded mainly in the frontocentral regions. This potential decreases in amplitude with age and disappears by about the 10th year. It appears to be related to analysis and evaluation strategies different from those generating the SPP (Chiarenza et al. 1983, 1984).

Method and Material

The subject sat in an armchair in front of a Tektronix 5111 oscilloscope at a distance of 70 cm in a lighted and electrically shielded room.

The subject held a joystick-type push button in each hand. The excursion of the button was 5 mm. The task consisted in starting a sweep of the oscilloscope trace with the left thumb and stopping it in a predetermined part of the oscilloscope by pushing the other button with the right thumb. The speed of the trace was 10 ms/cm. The predetermined area corresponded to a time interval between 40 and 60 ms.

The time interval is measured and defined as "performance time." The distance from the target was also measured and defined as "performance shift." The number of performances reaching the target was measured and defined as "target performance." The number of performances shorter than 40 ms and longer than 60 ms was also measured.

After a verbal explanation of the task and before the placement of the electrodes, the subjects were allowed a short familiarization period and were asked to avoid eye movement or blinking during the execution of the task and to keep an interval of 7–20 s between any two attempts. The subjects were also asked to remain relaxed and to avoid muscular preparatory movements before pressing.

Silver chloride electrodes were fixed to the scalp with collodion in the prefrontal (Fpz), frontal (Fz), central (Cz), right precentral (RPC), left precentral (LPC), and parietal (Pz) regions. Each electrode was referred bilaterally to the mastoids. The surface electromyogram was recorded from the flexor muscles on the right and left forearms. The impedance of the electrodes was less than 3 kΩ. The time constant and high frequency were 4.5 s and 700 Hz for the EEG and 0.03 s and 700 Hz for the EMG respectively.

The EEGs and EMGs were recorded on magnetic tape for off-line analysis. The analysis begins with the acquisition of a $\pm 20 \mu V$ signal on each channel. During analysis, the arrival of the trigger signal, an electronic pulse generated by the left-hand button, starts the acquisition for each channel of 1600 points at a frequency of 500 Hz for 3.2 s. Of these points, 1100 precede the trigger and 500 follow it.

The first 500 points were averaged to give a baseline from which the amplitude of the potentials was measured. For every subject four blocks of 25 trials selected from those free of muscular artifacts, blinking, or eye movements were averaged and analyzed. The mean amplitude before movement, the peak amplitude during movement, and the rise time of the rectified electromyographic activity for the right and left forearms were measured. Further, the beginning of BP rise, called "BP onset," the total area of BP, and the mean amplitude of BP over the 200 ms prior to the beginning of movement were measured in the premotor period. In the sensory-motor period, the mean amplitude of the MCP, referred to the mean amplitude of the BP for 200 ms after the movement, and the latency of the MCP peak with respect to the beginning of the electromyographic phasic activity of the MCP peak were measured. In the motor completion period, the amplitudes of N100 and P200 from the baseline and with respect to the absolute amplitude of the MCP and their latency from the trigger were measured.

During the postmotor period, the mean amplitudes of SPP and PAN were taken as average values from the baseline over 200 ms centered around the main positive peak (SPP) and negative peak (PAN) in the latency band between 350 and 850 ms. SPP and PAN latencies were also measured from the trigger pulse.

There were 13 subjects aged 10 years, of whom nine were normal and four were dyslexic. All came from the same area of Milan and from the same school. All the children showed adequate visual acuity and normal hearing threshold.

To be considered as children with specific learning disabilities (LDs), they had to have an IQ of above 85 on both the Cattell and the WISC nonverbal performance test and to exceed the 5% tolerance limits on the reading and writing test in the Italian adaption of the Metropolitan Achievement Test (Faglioni et al. 1970). Further, in order to obtain a better clinical and psychological evaluation, apart from the teacher's report and the medical history supplied by the parents, the following psychological tests were done: the Lincoln Oseretsky Motor

Development Scale (Zucchi et al. 1959), the Bender Visual Motor Gestalt Test (Koppitz 1964), the Stamback test (Stamback 1965), the Goodenough test (Gesell and Amatruda 1974), and the Laterality test (Harris 1968).

Results

The neurological examination of the normal and LD children showed no classic signs of major or minor neurological damage. However, there were some consistent neurological signs in the group of LD children, even though they could not be grouped together as a well-defined clinical picture. The most frequent signs were disdiadochokinesia, motor clumsiness in fine manipulative activity, and synkinetic movements in the contralateral hand during the diadochokinesia test and finger opposition test. The quality of gross and fine movements was not optimal in terms of speed, adequacy, and fluidity. From the school reports of the LD children, the reading and writing difficulty was the problem most frequently noted, while signs of hyperactivity, impulsiveness, or attention disorders were rated as light. Comparison of the results of the psychological tests showed significant differences only on the WISC verbal tests (LDs = 100.25; normals = 134.44; $P < 0.05$) and the Oseretsky test, in which the LD subjects showed a lower developmental age than the normal subjects (LDs = 134.7; normals = 165.6; $P < 0.05$).

All the children completed their assigned motor-perceptive tasks, although with significant differences in terms of motor performance and MRBMs. The mean performance time was 99.44 ms for the children who had learning difficulties and 62.93 ms for the control group. The percentage of performances that could be defined as target performances was 14.25% and 26% respectively. Further, in LD subjects 66% of performances took longer than 60 ms compared with 47% in the normal children. The former were also less accurate, with a performance shift of 50 ms compared with 19.2 ms for the normal subjects. All these results were statistically significant ($P < 0.01$). Observing the performance during each of the four blocks of tests, it was apparent that the excercise produced a marked improvement in the LD children. In fact, their performance time fell from 112.03 ms in the first block to 62.61 ms in the fourth, while it fell from 67.5 ms to 60.47 ms among the normal children.

The accuracy of the LD children also showed the same improvement. The performance shift fell from 58.0 ms in the first block to 19.2 ms in the fourth, while it fell from 22.3 ms to 17.0 ms among the normal children. A comparison of the two groups of children shows significant differences ($P < 0.01$) in the first three blocks but not in the fourth for both performance time and performance shift.

The electromyographic activity of the two groups differed neither in amplitude of the electromyogram before and during the movement nor in the rise time of either left or right forearm muscles, while the onset of phasic activity was delayed in the dyslexic children.

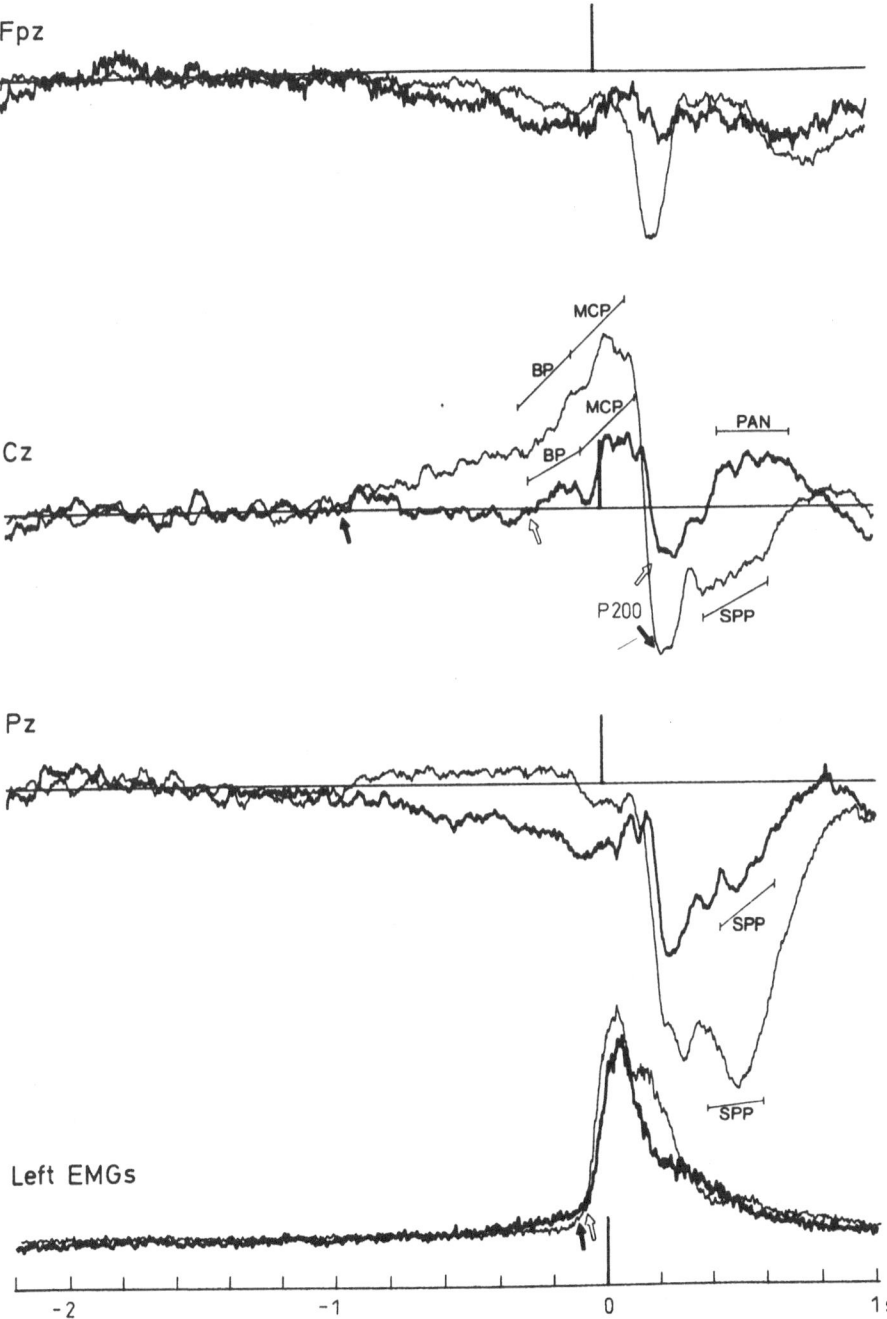

Fig. 2. Grand averages of MRBMs in normal (*light lines*) and dyslexic children (*heavy lines*). The divisions indicate the intervals of 200 ms over which the potentials shown above were measured. The *arrows left from zero* (*black* for the normal children and *white* for the dyslexic children) show the onset of BP and the onset of muscle phasic activity in the left electromyogram. (Chiarenza et al. 1986)

Consistent differences were found in MRBMs during all four motor periods (Fig. 2). The BP area in the dyslexic children in the premotor period was significantly reduced in the frontal central, precentral, and parietal regions (Fz, Cz, Pz, LPC: $P < 0.01$; RPC: $P < 0.05$). The BP amplitude 200 ms before the movement was reduced in Cz ($P < 0.05$) and Pz ($P < 0.01$). The BP onset was significantly delayed in the right and left precentral regions (RPC: $P < 0.05$; LPC: $P < 0.01$).

In the sensory-motor period the latency of MCP with respect to the EMG onset was greater, though not significantly so, in the frontal, central, and precentral regions of the dyslexic children. The amplitude of the MCP with respect to the BP amplitude was not different in the two groups.

In the motor completion period the latency of N100 was greater in the dyslexic children in the frontal, precentral, and central regions (Fz, Cz, RPC, LPC: $P < 0.05$) and increased further in the prefrontal and parietal regions (FPz, Pz: $P < 0.01$). There were no significant differences between the two groups in the latencies of P200 except in Fpz ($P < 0.05$). The amplitude of P200 with respect to the absolute amplitude of MCP was reduced in the LD children in all the brain areas (Fpz, Fz, Cz, Pz, RPC, LPC: $P < 0.05$). In the postmotor period, SPP was present in all brain areas of the normal subjects while in LD subjects it was only present in Pz and with a significantly reduced amplitude ($P < 0.05$). The latency of SPP in Pz was greater in children with learning difficulties than in normal children (LDs = 531.6 ms; normals = 495.2 ms). On the other hand, PAN was present in the frontal, central and precentral areas.

Discussion

The above results show that there are psychological and neurophysiological differences between normal children and children with learning difficulties. The most consistent differences are at the level of quality of movement—speed, fluidity, and adequacy—as shown by the neurological tests and the Oseretsky psychomotor examination. Children with learning difficulties showed lower than actual age in both motor development and in some tests of the neurological examination. These signs have been found previously, but their significance is still a topic of discussion since some are a matter of development and are present to a small extent in the normal population. Their presence could be interpreted as a sign of insufficient coordination and temporal control in motor sequencing.

Part of the preparation for movement is the activation of a central clock which controls the time sequencing of the motor action through afferent and efferent systems (Hirsch and Sterrich 1964; Rosenbaum and Patashnik 1980). The improvement of the performance of this clock depends on a greater synaptic efficiency of the central nervous system which in turn depends on age (Craik 1947) and on the presence of internal and external feedback on the accuracy of the performance.

Our subjects with learning difficulties showed themselves to be slower and less accurate and to have totalled a lower number of target performances in carrying out the motor-perceptual task. Further, a different trend in performance improvement was seen between the two groups of children during the experiment. The normal children achieved a better motor performance than the LD children and reached their peak in the second block, which they thereafter maintained. The children with learning difficulties, however, although they began with a much worse motor performance, improved steadily over the whole period of the experiment, reaching the level of the normal children only in the last 25 trials.

To these observations must be added the behavior observed during the performance of the task on closed circuit television and during the selection of the tests without artefacts. Compared with the normal children, those with learning difficulties had greater difficulty in controlling irrelevant and inappropriate movement, such as blinking immediately before or after the performance of the task or gross movement of the body or lower limbs. If all the results of the neurological tests, the Oseretsky test, and the performance of the motor-perception task are combined, it could be said that the setting of the central clock proceeds with difficulty. Further, these data seem to suggest that the control of the temporal sequencing of movements, based on the processing of internal and external feedback, does not occur in an appropriate way because, as we shall see later, MCP, N100, and SPP are altered in different ways in the children with learning difficulties.

It is known that children with learning difficulties have difficulty in executing simple motor tasks (Lewis et al. 1970; Pyfer and Carlson 1972; Bruininks and Bruininks 1977) and tests of bimanual coordination (Klicpera et al. 1981). Particularly insufficient motor performance has been observed in subjects with commissurotomy of the corpus callosum and it has been proposed that integrity of the callosum commissurae is essential to the performance of bimanual tasks (Kreuter et al. 1972; Preilowski 1972; Zaidel and Sperry 1977). It is not yet possible to ascertain to what extent these data agree with our observations. It is useful to point out that the myelination of the corpus callosum is completed approximately at the age of 10 years (Yakovlev and Lecours 1967).

In parallel with the motor performances, the MRBMs showed significant differences in normal and LD children.

The BP is a feature of the premotor period, when the ideomotor elements of the movement are being organized. In the LD children BP was reduced in amplitude in the parietal, central, and precentral areas and furthermore began only about 100 ms before the movement. It has recently been proposed that the BP could result from two components: the first begins 1.2 s before the movement and lasts for about 450/600 ms, and is followed by the second, characterized by a steep negative fall lasting 300/500 ms (Shibasaki et al. 1980). From the study of the development of the BP it was hypothesized that the first component might be linked to processes related to the representation of the action, while the second is the most operative part of the process and therefore the most automatic

(Chiarenza, in preparation). This first component was absent in the LD children while the second was considerably reduced in amplitude.

The sensory-motor period is characterized by the MCP. The MCP reflects processing of the kinesthetic reafferent information in the precentral and frontal areas related to the movement carried out. The MCP was of normal amplitude in LD children but the peak of this potential had a greater latency than in normal children. Experiments with animals (Dubrovsky and Garcia-Rill 1971) and observations of patients with posterior column lesions have shown that total or partial deafferentiation impedes the temporal control of a motor sequence. The increase in latency of the MCP may be interpreted as a difficulty of these children in processing the kinesthetic feedback.

The processing of the visual information, as represented by N100, the brain response evoked by the appearance of the light trace on the oscilloscope, was delayed in children with learning difficulties. It was delayed in the frontal central and precentral regions and even more so in the prefrontal and parietal regions. The P200 is thought to be one of the late components of the somatosensory potentials (Chiarenza et al. 1983); the reduction in amplitude of P200 seen in the various brain areas of the LD children but not in the normals could indicate a defect in the integration of the reafferent kinesthetic information.

The SPP was present only in the parietal regions in LD children and with reduced amplitude. There was increased latency compared with normal children. This may suggest difficulty for these children in the awareness and evaluation of their own performance. The PAN was present in the frontal, central, and precentral regions. This potential is recorded more often in children below the age of 10 years (Chiarenza et al. 1983). If, instead of the sequential averages, we look only at the averages of the MRBMs related to the target performance (Fig. 3), we see that there is SPP even in the LD children, in the frontocentral and precentral areas, even though the amplitude is considerably reduced compared with normal children. So the presence of PAN in the LD children could be due to a different strategy activated during the processing when the target was missed. These children seemed to give significance to the target performance only, without recognizing the failed performance.

Conclusions

Perception is an active process which requires the participation of various subsystems. The present data seem to indicate that during a motor-perceptual act, the different processes involved in the various brain areas are defective in children with learning difficulties. In particular, it can be stated that systems involved in the planning and programming of effective strategies are inadequate, and that, furthermore, those involved in verification and correction of errors are less efficient. These systems may be altered in themselves or may reflect deficiencies in those subsystems concerned with kinesthetic and visual process-

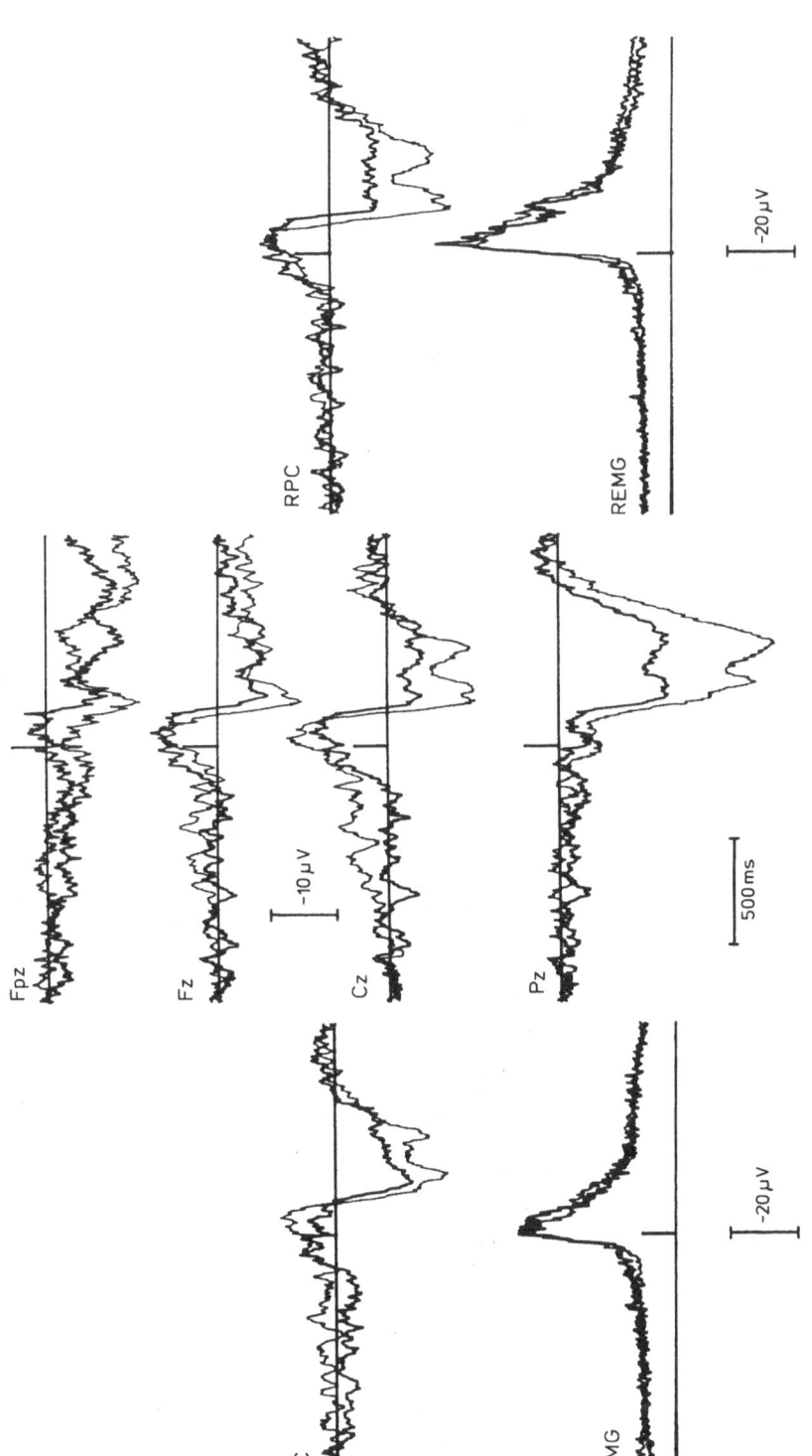

Fig. 3. MRBMs related to target performance of normal (*light lines*) and dyslexic children (*heavy lines*). Note the appearance of the SPP in the frontal, central, and precentral areas of the dyslexic children, even though the amplitude is lower than that of the normal children when they hit the target

ing, or they could be potentially adequate but not fully developed. Therefore, it can be said that dyslexia results from the defective integration and dysfunction of numerous processes which occur on different levels and at different times (Chiarenza et al. 1986).

Summary

Perception can be regarded as an active process of extracting information from the environment. It includes a variety of processes and is essential in learning to read and to write. Until now psychological and neurophysiological research has mainly examined isolated processes of perceptual function. The author suggests a method for evaluating the perceptual motor function by recording the MRBMs during the execution of a motor perceptual task.

The results obtained in a group of dyslexic children point to the conclusion that the subsystems for programming and verifying strategies are less efficient in these children; furthermore, those subsystems related to evaluation of kinesthetic and visual information are deficient.

It is hypothesized that dyslexia and dysgraphia are the results of a defective integration of the subsystems mentioned above.

References

Aaron PG, Bakker C (1982) The neuropsychology of dyslexia in college students. In: Malatesha RN, Hartlage LC (eds) Neuropsychology and cognition, vol 1. Sijthoff and Noordhoff, Alphen aan den Rijn, pp 128–146

Bakker DJ (1967) Temporal order, meaningfulness, and reading ability. Percept Mot Skills 24:1027–1030

Bakker DJ (1970) Temporal order perception and reading retardation. In: Bakker DJ, Satz P (eds) Specific reading disability: advances in theory and method. Rotterdam University Press, Rotterdam, pp 81–92

Bakker DJ (1972) Temporal order in disturbed reading. Rotterdam University Press, Rotterdam

Bannatyne AD (1966) The etiology of dyslexia and the color phonics system. In: Money J (ed) The disabled reader: education of the dyslexic child. Johns Hopkins University Press, Baltimore, pp 97–107

Bateman BC (1968) Interpretation of the 1961 Illinois test of psycholinguistic abilities. Special Child Publications, Seattle

Belmont, I (1980) Perceptual organization and minimal brain dysfunctions. In: Rie HE, Rie ED (eds) Handbook of minimal brain dysfunctions. A critical view. Wiley, New York, pp 253–271

Benton AL (1962) Dyslexia in relation to form perception and directional sense. In: Money J (ed) Reading disability: progress and research needs in dyslexia. Johns Hopkins University Press, Baltimore, pp 25–32

Birch HG (1962) Dyslexia and the maturation of visual function. In: Money J (ed) Reading disability: progress and research needs in dyslexia. Johns Hopkins University Press, Baltimore, pp 1–47

Birch HG, Belmont L (1964) Auditory-visual integration in normal and retarded readers. Am J Orthopsychiatry 34:852–861

Birch HG, Belmont L (1965) Auditory-visual integration, intelligence, and reading ability in school children. Percept Mot Skills 20:295–305

Birch HG, Lefford A (1963) Intersensory development in children. Monogr Soc Res Child Dev 28, Serial no 89

Boder E (1973) Developmental dyslexia: a diagnostic approach based on three atypical reading patterns. Dev Med Child Neurol 15:663–687

Bruininks VI, Bruininks RH (1977) Motor proficiency of learning disabled and non-disabled students. Percept Mot Skills 44:1131–1138

Chiarenza GA, Papakostopoulos D, Giordana F, Guareschi Cazzullo A (1983) Movement related brain macropotentials during skilled performances. A developmental study. Electroencephalogr Clin Neurophysiol 56:373–383

Chiarenza GA, Tengattini MB, Grioni A, Ganguzza D, Vasile G, Massenti A, Albizzati A, Papakostopoulos D, Guareschi Cazzullo A (1984) Long latency negative potentials durante un compito percettivo motorio. Caratteristiche evolutive in bambini normali. Riv Ital EEG Neurofisiol Clin 7:538–541

Chiarenza GA, Papakostopoulos D, Grioni A, Tengattini MB, Mascellani P, Guareschi Cazzullo A (1986) Movement related brain macropotentials during a motor perceptual task in dyslexic and dysgrafic children. In: McCallum WC, Zappoli R, Denoth F (eds) Cerebral psychophysiology studies in event related potentials (EEG Suppl 38). Elsevier, Amsterdam, pp 489–491

Craik KJW (1947) Theory of the human operator in control systems, Br J Psychol 38:56–61

de Hirsch K, Jansky J, Langford W (1966) Predicting reading failure. Harper and Row, New York

Duffy FH, Denckla MB, Bartles PH, Sandini G (1980) Dyslexia: regional differences in brain electrical activity by topographic mapping. Ann Neurol 7:412–420

Dubrovsky B, Garcia-Rill E (1971) Role of dorsal columns in sequential motor acts requiring precise forelimb projection. Exp Brain Res 18:165–177

Faglioni P, Gatti B, Paganoni AM, Robutti A (1970) Test di valutazione del linguaggio scritto. Manuale di istruzioni. Edizioni Organizzazioni Speciali, Firenze

Gesell A, Amatruda C (1974) Developmental diagnosis, 2nd edn. Hoeber, New York

Gibson EJ (1969) Principles of perceptual learning and development. Appleton-Century-Crofts, New York

Groenendaal HA, Bakker DJ (1971) The part played by mediation processes in the retention of temporal sequences by two reading groups. Hum Dev 14:62–70

Hynd GW, Obrzut JE, Weed W, Hynd CR (1979) Development of cerebral dominance: dichotic listening asymmetry in normal and learning disabled children. J Exp Child Psychol 28:445–454

Harris, AJ (1968) Harris test of lateral dominance, 3rd edn. Psychological Corporation, New York

Hirsch IJ, Sterrick CE (1964) Perceived order in different sense modalities. J Exp Psychol 67:103–112

Ingram TTS (1963) Delayed development of speech reference to dyslexia. Proc R Soc Med 56:199–205

Ingram TTS, Mason AW, Blackburn I (1970) A restrospective study of 82 children with reading disability. Dev Med Child Neurol 12:271–281

Jansky J, de Hirsch K (1972) Preventing reading failure—prediction, diagnosis and intervention. Harper and Row, New York

Johnson DJ, Myklebust HR (1967) Learning disabilities: educational principles and practices. Grune and Stratton, New York

Kinsbourne M, Warrington E (1966) Developmental factors in reading and writing backwardness. In: Money J (ed) The disabled reader: education of the dyslexic child. Johns Hopkins University Press, Baltimore, pp 77–83

Klicpera C, Wolff PH, Drake C (1981) Bimanual co-ordination in adolescent boys with reading retardation. Dev Med Child Neurol 23:617–625

Koppitz EM (1964) The Bender Gestalt test for young children. Grune and Stratton, New York

Kornhuber HH, Deecke L (1965) Hirnpotentialänderungen bei Willkürbewegungen und passiven Bewegungen des Menschen: Bereitschaftspotential und reafferente Potentiale. Pflügers Arch 284:1–17

Kreuter C, Kinsbourne M, Trevarthen C (1972) Are disconnected cerebral hemispheres independent channels? A preliminary study of the effects of unilateral loading on bilateral finger tapping. Neuropsychologia 10:453–461

Lewis FD, Bell DB, Anderson RP (1970) Relationship of motor proficiency and reading retardation. Percept Mot Skills 31:395–401

Lyle JG, Goyen J (1968) Visual recognition, developmental lag and strephosymbolia in reading retardation. J Abnorm Psychol 73:25–29

Lyle JG, Goyen J (1975) Effects of speed of exposure and difficulty of discrimination on visual recognition of retarded readers. J Abnorm Psychol 8:613–616

Luria AR (1973) The working brain. Penguin, London

Myklebust HR (1965) Development and disorders of written language: picture story language test. Grune and Stratton, New York

Papakostopoulos D (1978) Electrical activity of the brain associated with skilled performance. In: Otto DA (ed) Multidisciplinary perspectives in event-related brain potential research. US Environmental Protection Agency, Washington DC, pp 134–137

Papakostopoulos D (1980) A no stimulus no response event-related potential of the human cortex. Electroencephalogr Clin Neurophysiol 48:622–638

Papakostopoulos D, Cooper R, Crow HJ (1975) Inhibition of cortical evoked potentials and sensation by self-initiated movement in man. Nature 258:321–324

Preilowski BFB (1972) Possible contribution of the anterior forebrain commissures to bilateral motor coordination. Neuropsychologia 10:267–277

Pyfer HL, Carlson BR (1972) Characteristic motor development of children with learning disabilities. Percept Mot Skills 35:291–296

Rabinovitch RD (1968) Reading problems in children: definitions and classification. In: Keeney AH, Keeney VT (eds) Dyslexia: diagnosis and treatment of reading disorders. Mosby, St Louis, pp 28–48

Rosenbaum DA, Patashnik O (1980) A mental clock setting process revealed by reaction times. In: Stelmach GE, Requin J (eds) Tutorials in motor behavior. North–Holland, Amsterdam, pp 1–14

Rosenthal JH, Boder E, Callaway E (1982) Typology of developmental dyslexia: Evidence for its construct validity. In: Malatesha RN, Aaron PG (eds) Reading disorders, varieties and treatment. Academic, New York, pp 93–120

Rourke BD (1976) Issues in the neuropsychological assessment of children with learning disabilities. Can Psychol Rev 17:89–102

Satz P, Sparrow S (1970) Specific developmental dyslexia: a theoretical reformation. In: Bakker DJ, Satz P (eds) Specific reading disability: advances in theory and method. University of Rotterdam Press, Rotterdam, pp 17–40

Satz P, Rardin D, Ross J (1971) An evaluation of a theory of specific developmental dyslexia. Child Dev 42:2009–2021

Satz P, Friel J, Rudegeair F (1974) Differential changes in the acquisition of developmental skills in children who later became dyslexic. In: Stein DG, Rosen JJ, Butters N (eds) Plasticity and recovery of function in the central nervous system. Academic, New York, pp 88–98

Senf GM (1969) Development of immediate memory for bisensory stimuli in normal children, and children with learning disabilities. Dev Psychol 6:28–32

Senf GM, Feshback S (1970) Development of bisensory memory in culturally deprived dyslexic and normal readers. J Ed Psychol 61:461–470

Senf GM, Freundl PC, Silver A, Hagen R (1971) Memory and attention factors in specific learning disabilities. J Learn Dis 4:94–106

Shibasaki H, Barrett G, Halliday E, Halliday AM (1980) Components of the movement-related cortical potential and their scalp topography. Electroencephalogr Clin Neurophysiol 49:213–226

Stamback M (1965) Epreuves de niveau et de style moteurs. Actualités pédagogiques et psychologiques. Fasc 2:5–65

Vaughan GH, Costa LD, Ritter W (1968) Topography of the human motor potential. Electroencephalogr Clin Neurophysiol 25:1–10

Wepman JM (1962) Dyslexia: its relationship to language acquisition and concept formation. In: Money J (ed) Reading disability. Progress and research needs in dyslexia. Johns Hopkins University Press, Baltimore, pp 48–62

Witelson SF (1976) Abnormal right hemisphere specialization in developmental dyslexia. In: Knights RM, Bakker DJ (eds) Neuropsychology of learning disorders: theoretical approaches. University Park Press, Baltimore, pp 233–256

Yakovlev PI, Lecours AR (1967) Myelogenetic cycles of regional maturation in the brain. In: Minkowski A (ed) Regional development of the brain in early life. Blackwell, Oxford, pp 3–70

Zaidel D, Sperry RW (1977) Some long-term motor coordination effects of cerebral commissurotomy in man. Neuropsychologia 15:193–204

Zucchi M, Giuganino BM, Stella L (1959) Adattamento italiano della scala di sviluppo motorio di Oseretsky. Boll Psicol Sociol Appl 31–36

Biological Approaches to Classification and Treatment of Dyslexia

C. K. Conners

Current Problems in Diagnosis

A specific reading failure is considered to occur when a child is reading significantly below the level expected on the basis of IQ and age, and when alternative explanations are ruled out. This approach uses a positive criterion of discrepancy from expected reading age, and a varying number of negative or exclusionary criteria, to define the disorder. This concept of "specific reading retardation" was proposed by Rutter and Yule (1975) to differentiate backward readers with low intelligence from backward readers having more specific and selective deficits in reading which could not be attributed to general backwardness. Educational opportunity, cultural and emotional factors, sensory limitations, gross neurological abnormalities, and other medical conditions are generally regarded as contributory to reading failure and must be excluded in order to define the condition of dyslexia or specific reading retardation.

Within this group of specific disabled readers, further distinctions have been made in recent years between various subtypes, based upon patterns of reading and spelling errors (Boder 1971) or neuropsychological profiles (Mattis et al. 1975; Pirozzolo 1979; Petrauskas and Rourke 1979). A number of significant problems have arisen with these attempts at definition and subtyping.

First, the evidence that there are qualitative differences in the reading performance of backward and specific reading retardation has been seriously challenged (Share et al. 1987). Among the problems with the original concept proposed by Rutter and Yule (1975) is the failure to replicate the bimodal distribution of reading scores (a "hump" in the lower tail of the reading distribution), which was assumed to demarcate the specific and generally backward types of reading failure (Van der Wissel and Zegers 1985). It has been argued recently that those results reflect ceiling and floor effects in the reading tests (Share et al. 1987).

Second, even when the discrepancy model is employed, quite different results in terms of prevalence are obtained depending upon the statistical model utilized. For example, a "simple difference model" which computes the difference between IQ and reading standard scores produces a much higher prevalence figure than a regression model based upon the correlation between reading and IQ (Rispens 1987). This latter model has the further complication that the correlation of reading and IQ is much lower in younger than in older children. Multiple regression models (for example, using only those Wechsler subtest items which show correlations with reading) provide still different prevalence estimates.

Many studies of reading retarded children have failed to consider the confounding role of other diagnostic conditions such as attention deficit disorder, and the characteristics of reading samples have therefore varied according to the care with which such conditions are excluded (Denckla 1978; Denckla et al. 1985). The fact that reading retardation so frequently accompanies attentional and conduct disorders has not been adequately explained and there have been no successful attempts to show qualitative differences between the reading failure in attention deficit disorder and specific reading retardation (Conners 1987).

Finally, efforts at subtyping have suffered from a lack of replicability across samples, as well as differences in results depending upon the criteria and statistical methods employed for identifying the subtypes. Studies based upon clinically referred samples are invariably biased in unknown ways by the particular referral and demographic characteristics of the sample, leading to different proportions of subtypes across studies.

Problems in Treatment

The refractoriness of severe reading failure to treatment interventions is well known. For almost a century there have been concerted efforts to remediate reading failure by a variety of ingenious educational techniques. While early intervention and intensive special education can achieve substantial benefits in ultimate reading outcome, the gains are often extremely small and attained at the cost of intense effort by both child and teacher (Guthrie 1978). Hypothesized interactions between diagnostic subtypes and treatment responsiveness have not been confirmed, leaving most educational treatment planning at a strictly empirical level (Zigmond 1978).

Biological Models of the Reading Process

The discovery of fundamental genetic, neurodevelopmental, anatomical, or neurophysiological correlates of dyslexia might lead to a more satisfactory basis of classification and treatment. Such correlates could conceivably provide better measures of both normal growth processes and pathological disruptions related to reading failure.

Developmental Brain Changes

The Geschwind-Galaburda hypothesis that lateralized cerebral ectopies near the planum temporale are in some way crucial to dyslexia (Galaburda et al. 1985) suggests that at least some dyslexics might have experienced the disruption of normal processes of neuronal migration during early brain ontogenesis. At present, such neuroanatomical hypotheses of dyslexia are limited to postmortem studies, though new imaging techniques may eventually contribute data for such studies.

Recent large-scale studies of EEG spectra, phase, and coherence across the developmental span show distinct patterns of age-related development for left and right cerebral hemispheres, and distinct regional brain growth spurts which appear to coincide with piagetian stages of cognitive development (Thatcher et al. 1987). These growth spurts are in rough agreement with brain growth spurts identified from brain weight and head circumference (Epstein 1974). Such studies may provide the basis for important new leads in the identification of subtypes of reading disorder based upon *developmental* anomalies. For example, the growth spurt occurring in the left cerebral hemisphere between ages 4 and 6 years, which is evident in EEG power spectra (Thatcher et al. 1987), could provide the basis for identifying reading failures associated with a maturational delay of the left hemisphere relative to the right. Similarly, normative developmental EEG coherence and phase analyses may provide a means for identifying subtypes based upon neuronal conduction velocity or short vs long cortical axon functioning related to maturational changes in the frontoposterior direction (Thatcher et al. 1986).

The Neurobiology of Vision and Dyslexia

It is only quite recently that separate neural pathways for the processing of shape, color, and movement have been identified (Livingstone and Hubel 1984). A critical step was the discovery of large (magnocellular) fast-conducting cells which respond to low spatial frequencies and to peripheral movement, and smaller slow-conducting (parvocellular) neurons specifically tuned to high spatial frequencies and color and predominant in central vision. These cells maintain retinoptic organization from retinal ganglion cells, through the lateral geniculate bodies, and into areas V1 and V2 of the primary visual cortex. Event-related potentials readily distinguish stimulus selections based upon shape, color, or other attributes (Harter and Aine 1984).

Because of the differential sensitivity of the slow- ("X-type") and fast-conducting ("Y-type") fibers to spatial frequency (Blakemore 1975; Breitmeyer and Ganz 1976), it is possible to use certain perceptual phenomena associated with spatial frequency response of the visual pathways to test hypotheses relating dyslexia and functioning of these pathways. For example, the duration of visible persistence of gratings with different spatial frequencies is markedly different in dyslexics and normals (Badcock and Lovegrove 1981; Lovegrove and Brown 1978; Lovegrove et al. 1982; Slaghuis and Lovegrove 1985).

The data indicate that dyslexics have much longer persistence of low spatial frequency (two cycles per degree) but are no different from normal with higher frequencies (12 cycles/degree). These findings thus implicate the magnocellular neurons responsible for gross pattern vision and detection of peripheral movement. Classification of reading-disabled children according to parameters of spatial frequency response indicates that as many as 75% have this type of

abnormality. Not surprisingly, subtypes identified with these biological parameters do not agree with those based upon neuropsychological and behavioral
typing (Badcock and Lovegrove 1981).

The distribution of X- and Y-type cells in peripheral and central vision has
provided important hypotheses regarding other perceptual phenomena such as
lateral masking (Breitmeyer and Ganz 1976). A recent experiment by Geiger and
Lettvin (1987) reported that dyslexics recognize letters in peripheral vision better
than in foveal vision, and that the lateral masking effects of peripheral letters on
the recognition of central letters is diminished in dyslexics.

Eye Movements and Visual Attention in Dyslexia

Differences in the efficiency of central and peripheral vision may be the basis of
another well-known attribute of dyslexics, their abnormal eye movements.
Although many investigators have dismissed abnormal eye movements as
secondary phenomena associated with difficulty in decoding the meaning of the
visual text (Rayner 1978), substantial empirical arguments have been raised
against this point of view (Pavlidis 1981a, b, 1985). While there is no doubt that
the normal development of eye movement control is related to increased
efficiency of the reading process, there appear to be distinct abnormal qualitative
differences in the eye movements of dyslexics which do not disappear when they
read text below their reading level or when they are required to pursue nonlexical
targets.

The meaning of abnormal ocular pursuit and increased frequency of
retrogressive saccades in dyslexics may be clarified by recent experiments
involving single-unit recordings in the monkey (Goldberg 1982). In these
experiments it has been shown that areas in the posterior parietal cortex, as well
as in the frontal eye fields, are exquisitely responsive to targets in the field of vision
which will subsequently become the target of an eye movement. The parietal
areas are particularly interesting in that activation of those cells occurs when the
animal pays attention to a peripheral target, whether or not it subsequently
makes an eye movement towards the target. This cortical area thus appears to
reflect activity of an important *visual attention mechanism*, perhaps functionally
responsible for directing saccades in order to foveate visual stimuli which have
attentional value.

Since Y-cells are more heavily represented in the periphery of vision, and since
they act reciprocally to inhibit activity of X-cells, this parietal attention
mechanism would be important in the priming of visual pathways prior to the
next saccade towards peripheral material, and could thus explain the attentional
priming effects reported for letters by Posner (1980). It has been shown that
visual-spatial, not lexical or semantic features are responsible for the priming
effects of parafoveal stimuli (Rayner et al. 1980; Inhoff and Rayner 1980).

Similarly, it has been proposed that the inhibitory activity of the Y-cells on the
X-cells responsible for detailed pattern vision in the foveal region is important in
preventing confusability of stimuli which would occur from forward masking

effects if the long-lasting images created in foveal vision were not terminated prior to a saccade (Breitmeyer and Ganz 1976). The lateness of the myelination of the parietal cortices (Yakovlev and Lecours 1967), their documented role in certain acquired dyslexias involving confusability of letters (Pirozzolo 1979; Pirozzolo and Rayner 1978; Shallice and Warrington 1977), and the evidence from electrophysiological studies which frequently implicate parietal abnormalities in dyslexics (Conners 1970; Duffy et al. 1980; Dykman et al. 1979; Johnstone et al. 1984; Preston et al. 1977), raise the possibility that some dyslexias may be based on a visual attentional deficit, either in the tectopulvinar visual pathway or at its cortical terminus in the parietal area. The finding of abnormal eye movements in children with attention deficits who also have learning disorders (Bala et al. 1981) is consistent with this attentional hypothesis.

Quantitative EEG Studies of Dyslexics

A number of differences between dyslexics and normals have been reported for quantitative EEG measures, including increased low frequencies (delta and theta) with a concomitant decrease in high frequencies, increased 3–7 Hz theta, decreased alpha, and decreased beta (19–24 Hz). With the exception of studies by Fein et al. (1983, 1986), most of the findings have not been replicated. The use of small samples, the tendency to obscure developmental changes by lumping together children across a wide age range, the failure to eliminate behavioral disorders from the samples, and the use of relative power measures which artifactually produce interdependencies among frequency bands mean that much of this work must be interpreted with caution. At least three carefully conducted studies, however, have confirmed a decrease in beta activity in dyslexics during passive, eyes-closed conditions (Dykman et al. 1982; Fein et al. 1986). While decrease in beta activity is not specific to dyslexics, also appearing as a finding in hyperactive children, it is consistent with the notion of a failure to sustain in aroused and attentive state. It has been compared with the sensory-motor rhythms in animals which become apparent during states of highly focused attention. As suggested by Fein et al. (1986), decreased beta activity appears to be a meaningful basis for subtyping of dyslexia.

Event-Related Potentials

Abnormalities of parietal visual event-related potentials (ERPs) in dyslexics and correlation of ERPs with reading progress were identified quite early (Conners 1970) and replicated with words as stimuli (Preston et al. 1977). A number of studies have found either latency or amplitude components, particularly positivities in the 300–900 ms latency range, which discriminate poor readers from controls (Dainer et al. 1981; Dykman et al. 1979; Holcomb et al. 1985, 1986; Johnstone et al. 1984; Lovrich and Stamm 1983). However, reviews of ERPs and reading indicate considerable lack of agreement with regard to ERP scalp topography and type of effects (Harter et al. in press; Otto and Squires 1986).

Much of this disagreement may be due to several factors: (a) failure to separate concomitant disorders, such as attention deficit disorder, (b) use of a wide age range, (c) differences between passive and task-relevant effects, and (d) use of linguistic vs nonlinguistic stimuli.

Most of these problems were controlled in a recent careful study by Harter and colleagues (in press). They compared four groups of children composed of the 2×2 fourfold classification of children with and without attention deficits, and children with and without dyslexia. Several effects differentiated normal and poor readers, and poor readers from children with attention deficits. A smaller and later positivity over the left central and parietal regions for a wave at about 200–400 ms was found in the dyslexics. The amplitudes of the children with attention deficits, in contrast, were larger. This wave was associated with the ability to differentiate between relevant and nonrelevant stimuli, or "selective neural processing." The effects were more central and frontal over the right hemisphere for the children with attention deficits, and more central-occipital over the left hemisphere for the dyslexics.

Importantly, the magnitude of the central-occipital positivity is highly correlated with reading ability. In the study by Harter et al. the correlation was 0.57. This compares with the values of 0.65 reported by Ollo and Squires (1986) and 0.61 reported by Conners (1970). The study by Harter et al. indicates that there are deficits in dyslexia associated with selective neural processing (possibly reflecting a specific kind of selective attention deficit) as well as deficits unrelated to the content of the stimuli (such as letter relevance). Thus, there may be more than one neurophysiological subtype definable on the basis of ERP responses. We have proposed elsewhere (Conners 1987) that one type may reflect selection of information related to phonological coding, and primarily involving the geniculostriate pathway, and another type, activity in the tectopulvinar pathway and involving visual selective attention and priming effects.

Regional brain activation of left frontotemporal and supplementary motor area activity during the reading process has been identified from cerebral blood flow measures taken during reading (Lou et al. 1984), and agrees in this respect with electrophysiological topographical mapping studies which also identify areas of frontal involvement as well as left parieto-occipital differences between normal and dyslexic readers (Duffy et al. 1980). Intersubject variations in these patterns, however (Duffy et al. 1980), means that it is likely there will prove to be several subtypes rather than a single archetypal pattern of dysfunction relative to normals or other pathological groups.

Biological Treatments of Dyslexia

Successful biologically based treatments of dyslexia, such as pharmacotherapy, could point to particular biological mechanisms associated with dyslexia, particularly if the mechanism of action of the treatment is known. Although an empirically effective treatment could occur even if the basic disorder was unrelated to the mechanism of action, such treatments could be a basis for powerful hypotheses regarding underlying mechanisms.

Several recent studies have suggested that the nootropic drug piracetam has a significant impact upon reading progress in severe dyslexics (Conners and Reader 1987; Conners et al. 1987; Wilsher et al. 1987). In a randomized, double-blind study carried out over an entire academic year (Wilsher et al. 1987), there were significant drug effects on both reading comprehension and speed. A special study which was part of this multicenter collaborative trial showed that piracetam had significant effects on a very early visual ERP component which appeared to reflect a priming effect when the subject was signalled that a letter was about to occur (Conners et al. 1987). Later components were also affected and this was interpreted as evidence of effects on selective neural processing or processing negativity.

Summary and Conclusions

Persistent problems in classification and treatment of dyslexia are closely interrelated. Without a rational classification system it is not possible to study the subtype by treatment interactions, which most observers feel is crucial to successful pedagogic planning and effectiveness. Classification has been hampered by the lack of knowledge of the underlying biological parameters which determine crucial steps in the acquisition of reading skills, and lack of appropriate biological markers for segregating children into appropriate subgroups for analysis. A simple differentiation of "specific reading retardation" from poor reading at the lower end of the IQ scale seems to be in trouble.

One proposal for isolating a *developmental subtype* is to utilize recent electrophysiological data which strongly suggest differential rates of maturation for left and right cerebral hemispheres. Of particular significance is the confirmation that rapid growth spurts in brain function occur between 4 and 6 years of age. Quantitative EEG methods may provide a useful way of tracking these changes longitudinally and relating them to progress in learning to read.

Another subtype suggested here reflects problems of selective visual attention, probably due to delayed development of parietal mechanisms which are closely related to control of eye movements. Since visual ERP waveforms are capable of differentiating activity in foveal and peripheral vision (Cohn and Hurley 1985), a direct test of this hypothesis seems possible. A study which segregates subjects on the basis of abnormal eye movement patterns and examines ERPs to foveal and peripheral stimuli might provide confirmation of this suggestion of subtyping and its relation to neural pathways.

Recent ERP studies also indicate that there are selective attention deficits for meaningful or relevant vs nonmeaningful or irrelevant stimuli which are specific to dyslexics, and clearly different from the generalized attentional deficits associated with attention deficit/hyperactivity disorders. These ERP measures may provide the means for defining those types of poor readers with higher-order processing deficits related to phonological awareness or lexical-semantic aspects of processing.

Sufficient sophistication has now developed with these methods that the time seems ripe for employing them as means of subdividing the globally defined pool of poor readers. The concept of "specific reading retardation" needs buttressing from an understanding of the basic neurobiology underlying the reading process. Further development of successful nootropic drugs may eventually provide complementary information regarding the chemical pathways involved in crucial reading processes.

References

Badcock D, Lovegrove W (1981) The effects of contrast, stimulus duration, and spatial frequency on visible persistence in normal and specifically disabled readers. J Exp Psychol 7:495–505

Bala SP, Morris AG, Atkin A, Gittelman R, Kates W (1981) Saccades of hyperactive and normal boys during ocular pursuit. Dev Med Child Neurol 23:323–336

Blakemore C (1975) Central visual processing. In: Gazzaniga MS, Blakemore C (eds) Handbook of psychobiology. Academic, New York

Boder E (1971) Developmental dyslexia: prevailing diagnostic concepts and a new diagnostic approach. In: Myklebust H (ed) Progress in learning disabilities. Grune and Stratton, New York

Breitmeyer BG, Ganz L (1976) Implications of sustained and transient channels for theories of visual pattern masking, saccadic suppression, and information processing. Psychol Rev 83:1–36

Cohn R, Hurley CW (1985) Differential visual evoked cortical responses to direct and peripheral stimulation in man. Electroencephalogr Clin Neurophysiol 61:157–160

Conners CK (1970) Cortical visual evoked responses in children with learning disorders. Psychophysiology 7:418–428

Conners CK (1987) Dyslexia and the neurophysiology of attention. In: Pavlidis GT (ed) Perspectives on dyslexia vol. 1, Wiley, New York

Conners CK, Reader MJ (1987) The effects of piracetam on reading achievement and visual event-related potentials in dyslexic children. Child Health Dev 5:75–90

Conners CK, Reader MJ, Reiss A, Caldwell J, Caldwell L et al. (1987) The effects of piracetam upon visual event-related potentials in dyslexic children. Psychophysiology 24:513–521

Dainer KB, Klorman R, Salzman LF, Hess DW, Davidson PW, Michael RL (1981) Learning disordered children's evoked potentials during sustained attention. J Abnorm Child Psychology 9:79–84

Denckla MB (1978) Critical review of "electroencephalographic and neurophysiological studies in dyslexia". In: Benton AL, Pearl D (eds) Dyslexia: an appraisal of current knowledge. Oxford University Press, New York

Denckla MB, Rudel R, Chapman C, Krieger J (1985) Motor proficiency in dyslexic children with and without attentional disorders. Arch Neurol 42:228–231

Duffy FH, Denckla MB, Bartels PH, Sandini G (1980) Dyslexia: regional differences in brain electrical activity by topographic mapping. Ann Neurol 7:412–420

Dykman RA, Ackerman PT, Oglesby DM (1979) Selective and sustained attention in hyperactive, learning-disabled and normal boys. J Nerv Ment Dis 167:288–297

Dykman RA, Holcomb PJ, Oglesby DM, Ackerman PT (1982) Electrocortical frequencies in hyperactive, learning-disabled, mixed, and normal children. Biol Psychiatry 17:675–685

Epstein HT (1974) Phrenoblysis: special brain and mind growth periods. I: Human brain and skull development. Dev Psychobiol 7:207–216

Fein G, Galin D, Johnstone J, Yingling CD, Marcus M, Kiersch ME (1983) EEG power spectra in normal and dyslexic children, 1. Reliability during passive conditions. Electroencephalogr Clin Neurophysiol 55:399–405

Fein G, Galin D, Yingling CD, Johnstone J, Davenport L, Herron J (1986) EEG spectra in dyslexic and control boys during resting conditions. Electroencephalogr Clin Neurophysiol 63:87–97

Galaburda AM, Sherman GF, Rosen GD, Aboitiz F, Geschwind N (1985) Developmental dyslexia: four consecutive patients with cortical anomalies. Ann Neurol 18:222–233

Geiger G, Lettvin JY (1987) Peripheral vision in persons with dyslexia. N Engl J Med 316:1239–1243

Goldberg ME (1982) Moving and attending in visual space: single-cell mechanisms in the monkey. In: Potegal M (ed) Spatial abilities: development and physiological foundations. Academic, New York

Guthrie JT (1978) Principles of instruction: a critique of Johnson's "Remedial approaches to dyslexia". In: Benton AL, Pearl D (eds) Dyslexia: an appraisal of current knowledge. Oxford University Press, New York

Harter MR, Aine CJ (1984) Brain mechanisms of visual selective attention. In: Parasuraman R (ed) Varieties of attention. Academic, New York

Harter MR, Anllo-Velnto L, Wood FB, Schroeder MM (1988) Separate brain potential characteristics in children with reading disability and attention deficit disorder: color and letter relevance effects. Brain Cognition 7:115–140

Holcomb PJ, Ackerman PT, Dykman RA (1985) Cognitive event-related brain potentials in children with attention and reading deficits. Psychophysiology 22:656–667

Holcomb PJ, Ackerman PT, Dykman RA (1986) Auditory event-related potentials in attention and reading disabled boys. Int J Psychophysiology 3:263–273

Inhoff AW, Rayner K (1980) Parafoveal word perception: a case against semantic preprocessing. Percept Psychophysiology 27:457–464

Johnstone J, Galin D, Fein G, Yingling C, Herron J, Marcus M (1984) Regional brain activity in dyslexic and control children during reading tasks: visual probe event-related potentials. Brain Lang 21:233–254

Livingstone MS, Hubel DH (1984) Anatomy and physiology of a color system in the primate visual cortex. J Neurosci 4:309–356

Lou HC, Henriksen L, Bruhn P (1984) Focal cerebral hypoperfusion in children with dysphasia and/or attention deficit disorder. Arch Neurol 41:825–829

Lovegrove W, Brown C (1978) Development of information processing in normal and disabled readers. Percept Mot Skills 46:1047–1054

Lovegrove WJ, Martin F, Bowling A, Blackwook M, Badcock D, Paxton S (1982) Contrast sensitivity functions and specific reading disability. Neuropsychologia 20:309–315

Lovrich D, Stamm JS (1983) Event-related potential and behavioral correlates of attention in reading retardation. J Clin Neuropsychol 5:13–37

Mattis W, French JH, Rapin I (1975) Dyslexia in children and young adults: three independent neuropsychological syndromes. Dev Med Child Neurol 17:150–163

Otto C, Squires N (1986) Event-related potentials in learning disabilities. In: Cracco R, Bodis-Wollner I (eds) Frontiers in clinical neurosciences: evoked potentials. Liss, New York

Pavlidis GT (1981a) Sequencing, eye movement and the early objective diagnosis of dyslexia. In: Pavlidis GT (ed) Dyslexia research and its applications to education. Wiley, London

Pavlidis GT (1981b) Do eye movements hold the key to dyslexia? Neuropsychologia 19:57–64

Pavlidis GT (1985) Eye movements in dyslexia: their diagnostic significance. J Learn Disabil 18:42–50

Petrauskas RJ, Rourke BP (1979) Identification of subtypes of retarded readers: a neuropsychological, multivariate approach. J Clin Neuropsychol 11:17–37

Pirozzolo FJ, Rayner K (1978) The neural control of EMs in acquired and developmental reading disorder. In: Avakian-Whitaker H, Whitaker HA (eds) Advances in neurolinguistics and psycholinguistics. Academic, New York

Pirozzolo JF (1979) The neuropsychology of developmental reading disorders. Praeger, New York

Posner MI (1980) Orienting of attention. Q J Exp Psychol 32:3–25

Preston MS, Guthrie JT, Kirsch I, Gertman D, Childs B (1977) VERs in normal and disabled adult readers. Psychophysiology 14:8–14

Rayner K (1978) Eye movements in reading and information processing. Psychol Rev 85:618–660

Rayner K, McConkie GW, Zola D (1980) Integrating information across eye movements. Cogn Psychol 12:206–226

Rispens J (1987) In search of diagnostic criteria. Paper presented at the 3rd world congress of dyslexia. Crete, Greece

Rutter M, Yule W (1975) The concept of specific reading retardation. J Child Psychol Psychiatry 16:181–197

Shallice T, Warrington EK (1977) The possible role of selective attention in acquired dyslexia. Neuropsychologia 15:31–41

Share DL, McGee R, McKenzie D, Williams S, Silva PA (1987) Further evidence relating to the distinction between specific reading retardation and general reading backwardness. Br J Dev Psychol 5:35–44

Slaghuis WL, Lovegrove WJ (1985) Spatial-frequency-dependent visible persistence and specific reading disability. Brain Cogn 4:219–240

Thatcher RW, Krause PJ, Hrybyk M (1986) Cortico-cortical associations and EEG coherence: a two-compartmental model. Electroencephalogr Clin Neurophysiol 64:123–143

Thatcher RW, Walker RA, Giudice S (1987) Human cerebral hemispheres develop at different rates and ages. Science 236:1110–1113

Van der Wissel A, Zegers FC (1985) Reading retardation revisited. Br J Dev Psychol 3:3–9

Wilsher CR, Bennett D, Chase CH, Conners CK, DiIanni M, Feagans L, Hanvik LJ et al. (1987) Piracetam and dyslexia: effects on reading tests. J Clin Neuropsychol 7:230–237

Yakovlev P, Lecours P (1967) Myelogenetic cycles of regional maturation in the brain. In: Minkowski A (ed) Regional development of the brain in early life. Blackwell, Oxford, pp 3–70

Zigmond N (1978) Remediation of dyslexia: a discussion. In: Benton AL, Pearl D (eds) Dyslexia: an appraisal of current knowledge. Oxford University Press, New York

The Special Case of Down Syndrome*

G. A. Chiarenza

Introduction

Down syndrome is one of the main causes of mental retardation. Given its high incidence (1.6 cases per 1000 live births) and the social importance of reeducation and of enabling those affected to enter a working environment, the syndrome has become an object of research with regard to retardation and learning failure.

Macroscopic neuropathological findings appear indicative of the peculiarity of Down syndrome compared with other forms of mental retardation: there is a slight weight loss and brain size reduction; the fronto-occipital diameter is shorter, and the frontal lobes, brain stem, and cerebellum are smaller. The cortical structure appears to be simplified: the main sulci are less deep, while the secondary ones are fewer; the gyri are wider and the cortex is thinner (Colon 1972).

These macroscopic characteristics are due to a structural organization at the microscopic level consisting in (a) poor myelinization or demyelinization of nervous fibers, particularly of arcuate fibers, which connect primary sensory cortex with association areas (Benda 1969), (b) a severe cellular neuronal loss, including cholinergic neurons of Meynert's nucleus, and (c) abnormalities in the dendritic formation process (Ball and Nuttal 1980; Colon 1972; Balazs and Brooksbank 1985; Ohara 1972). The formation of dendritic spines seems to be normal throughout gestation but then drastically decreases during the postnatal period. This process, which can be attributed to an early growth standstill (Takashima et al. 1981), explains the reduced numbers of dendritic spines in adults (Balazs and Brooksbank 1985).

However, most histopathological microscope investigations deal with the relation between early onset of mental deficiency in Down syndrome and dementia in Alzheimer's disease. These two conditions present similar histo-pathological patterns that are characterized by cerebral atrophy and by the occurrence of senile plaques, neurofibrillar bodies, and granulovacuolar degeneration (Crapper et al. 1975; Ball and Nuttal 1980; Balazs and Brooksbank 1985). In subjects with a normal karyotype the occurrence of senile plaques and

*This project was supported in part by CIRAH, Centro Italiano Autosufficienza Handicappati.

neurofibrillar bodies has been found to increase with age and to correlate positively with the degree of intellectual decay. In contrast, it is not at all certain that in Down syndrome the early occurrence of and the rapid increase in such changes are associated with early mental deterioration (Ropper and Williams 1980), although it is acknowledged that it is difficult in Down syndrome to detect clinical signs of a demential process given that mental retardation is in any case a clinical characteristic. In fact, it has been shown that the most significant age-related clinical signs in Down syndrome are much subtler and more selective and concern the short-term memorization of visual stimuli and the occurrence of frontal inhibitory release reflexes (Crapper et al. 1975; Ropper and Williams 1980). So it seems that dementia and neuropathological alterations in Down syndrome dissociate: the high incidence of such abnormalities in itself is not sufficient to explain the sporadic appearance of a demential pattern in elderly Down syndrome subjects.

Investigations aiming to show in Down syndrome the early occurrence of biochemical, cerebral, and noncerebral reactions typical of the normal aging process have been carried out. In the process, abnormalities in the metabolism of nucleic acids and particularly of free oxygen radicals have been detected. A basic enzyme in the metabolism of free oxygen radicals is superoxide dismutase, which is codified by chromosome 21 and the activity of which is increased by 50% in the red blood cells of Down syndrome subjects (Balazs and Brooksbank 1985). An increased production of peroxides with consequent lipoperoxidation of biological membrane lipids would alter the fluidity of the double lipidic layer, causing a biochemical and functional disorganization in biological membranes. It seems likely that in Down syndrome the formation of both biochemically and functionally anomalous cellular membranes takes place at an early ontogenic stage. However, neither the increased production of intracellular free radicals nor the loss of membrane fluidity are peculiar to Down syndrome. In fact both events occur in the normal aging process (Hansford 1983), but in subjects affected by Down syndrome they occur earlier and more clearly, being present even during the initial postnatal period (Balazs and Brooksbank 1985).

Biochemical investigations also provide information about the involvement of various receptor systems, at both central and peripheral level. In particular, in Down syndrome as well as in Alzheimer's disease the cholinergic system is certainly involved, as shown both by the decrease in cortical choline acetyl-transferase (CAT) activity and by the neuronal reduction of Meynert's nucleus, which is considered to be the origin of most cholinergic projections towards the neurocortex. However, since the CAT activity decrease exceeds by far the neuronal loss of the basal Meynert's nucleus, the subcortical damage is likely to be secondary to primary cortical damage (Perry et al. 1985). Histochemical investigations carried out on subjects affected by Alzheimer's disease show it to be probable that just the senile plaques might be the site of damage of cholinergic nerve endings (Kitt et al. 1984; Price et al. 1982).

The remarkable sensitivity of subjects affected by Down syndrome to the peripheral effects of atropine, a selective blocker of the muscarine receptor, points to a generalized anomaly of the cholinergic transmitter system (Harris and Goodman 1968).

The dopaminergic neurons of the substantia nigra projecting towards the basal ganglia and of the ventral tegmentum area projecting towards the frontal and limbic cortex also appear to be seriously affected, probably by secondary reverse degeneration from primitive cortical damage (Mann et al. 1987). The low blood concentration of serotonin in patients with Down syndrome and the reduced plasmatic activity of dopamine-β-hydroxylase may be peripheral signs of a change in the catecholaminergic metabolism (Wetterberg et al. 1972).

In addition to neuroanatomical and biochemical studies, investigations have been carried out by recording the cerebral electrical activity. It has been shown that the diffuse neuronal loss, the senile plaques, and the neurofibrillar degeneration may bring about a change in synaptic transmission: in fact, the involved neurons would lose the ability to produce postsynaptic potentials (Crapper et al. 1975).

Despite the neuropathological findings, electroencephalography in Down syndrome has proved to be only moderately helpful in describing cerebral development. Ellingson and Peters (1980) found significant retardation in the maturation of brain electrical activity in trisomy 21 infants which was correlated with delayed early behavioral development but not with the presence of conventional signs of EEG abnormality. Electroencephalography is also of little use for the description of cerebral aging in Down syndrome subjects (Callner et al. 1978), since from the fourth to the sixth decades of life 75% of these subjects have a normal EEG, while the rest show a diffuse but nonspecific slowing of the cortical electrical activity (Ellingson et al. 1973). Moreover, there are few EEG characteristics that specifically correlate with any form of mental retardation (Bigum et al. 1970).

More recently, investigations have been carried out by recording the sensory evoked potentials (EPs) and cognitive potentials (event-related potentials = ERPs), allowing cognitive components of mental deficiency to be distinguished from the merely perceptive ones. Despite the absence of specific EEG patterns characterizing the brain in Down syndrome, studies of the early and late components of EPs and ERPs have yielded consistently abnormal findings which confirms at the neurofunctional level that such patients are unique in that they differ substantially in this respect from both normal subjects and subjects affected by other forms of mental retardation.

Brainstem auditory evoked potentials (BAEPS) reflect the brainstem function at the pontomesencephalic level. Comparing Down syndrome and normal subjects, waves II and III and the IV–V complex show shorter latencies in the former in response to monaural (Ferri et al. 1986; Gigli et al. 1984; Gliddon et al. 1975) and binaural stimulation (Squires et al. 1980). Only the latency of wave I is prolonged in correspondence with the auditory deficit that is particularly frequent in Down syndrome, and is often serious and bilateral (Balkany et al. 1979). The central conduction time measured from the interpeak latency (IPL) I–V is shorter than in normal subjects. It should be pointed out that the IPL I–V is prolonged in subjects affected by "idiopathic" mental retardation and in subjects with cerebral malformations (Chiarenza and Radelli, in preparation).

The reduction in the brain stem conduction time has been related by some authors to the shorter brain stem length, measured as the inion–C7 distance, and to its perpendicular insertion in the brain (Squires et al. 1980), identified in anatomopathological studies (Benda 1969; Burger and Vogel 1973); others, in contrast, have related the reduction to the degree of mental retardation (Ferri et al. 1986) or to a generalized abnormality of the nerve conduction speed also present at a peripheral level (Scott et al. 1982).

But the most constant and significant finding concerning EPs in Down syndrome is the greater amplitude and the longer latency of late components as compared with findings in both normal subjects and subjects affected by other forms of mental retardation. This particular aspect emerges with all types of evoked potential (Gliddon et al. 1975; Galbraith et al. 1970; Straumanis et al. 1973; Dustman and Callner 1979), irrespective of age and with reduced intra- and interindividual variability (Bigum et al. 1970).

On this basis it has been assumed that the increase in the amplitude of EPs in Down syndrome is due to a failure of cerebral inhibitory processes. This inhibitory deficit, which lowers the neuronal threshold of discharge, might cause an increase in amplitude of cortical evoked responses through a neurophysiological disinhibition mechanism. The inhibitory failure is probably due to the reduced activity of the reticulothalamic sensory gating system, which would result in absent or abnormal inhibition of the sensory stimuli afferent to the cortex. In fact, experimental block of the nonspecific thalamocortical system produces an increase in the amplitude of visual and auditory EPs (Skinner and Lindsley 1971). In support of the assumption that reduced activity of the reticulothalamic sensory gating system occurs, there are some investigations showing that subjects affected by Down syndrome do not present the phenomenon of "habituation" in cerebral EPs (Schafer and Peeke 1982). Habituation, defined as the decreasing response to repeated stimuli, acts as a fundamental adaptive mechanism of central origin allowing the organism not to react to insignificant external stimuli in an environmental situation characterized by continuous sensory stimulation. The inability of Down's syndrome subjects to display this elementary form of "learning" could be an important neurobiological substrate of mental deficiency. According to Luria (1963, 1973), subjects affected by mental retardation have a reduced "plasticity" because of a failure of central inhibitory processes that does not allow them to adopt mental states adequate to the continuously evolving requirements of the external environment.

The amplitude decreases in subjects not affected by mental retardation could indicate that they perceive the relative lack of significance in a series of identical stimuli more promptly than do Down syndrome subjects. Therefore the latter are thought to have no ability to inhibit selectively the meaningless-stimulus-related information (Dustman and Callner 1979).

Furthermore, in Down syndrome there is not the typical decrease in amplitude of cerebral potentials evoked by self-induced stimuli, whose time span is therefore known (Schafer and Peeke 1982): these subjects can be considered unable to perceive the sequential order in a succession of external events, and for that reason to reduce the response not only to insignificant stimuli but also to foreseeable ones. Such habituation requires an unimpaired short-term memory function and the correct utilization of the fundamental time parameters, similarity and timing of sequential events, that control voluntary actions.

The lack of age-related changes in evoked responses in Down syndrome suggests furthermore that the development of central inhibitory processes stops very early. This harmonizes with the hypothesis that the rate of cortical development in Down syndrome quickly decreases after the first months of life (Barnet and Lodge 1967).

The ERPs in Down syndrome have shown that the main differences with regard to the control groups do not concern the components N1–P1, but rather the components N2–P3, which have a longer latency and a lower amplitude in Down syndrome (Karrer and Ivins 1976b; Squires et al. 1979; Lincoln et al. 1986). Therefore differences in ERPs between Down syndrome subjects and control subjects cannot be attributed to a different sensory perception, because latencies and amplitudes of the component N1 do not differ in the two groups. The appearance of P3 is usually related to the recognition of an external event. The increase in P3 latency would represent a greater slowness, in Down syndrome subjects, of the processes of stimulus recognition for decisional purposes. In contrast the P3 amplitude is influenced by dimensions such as subjective probability, stimulus meaning, and the proportion of information lost during transmission owing to equivocation or inattention (Johnson 1986). The lower amplitude of P3 in Down syndrome could reflect a failure in one of these dimensions and in the processes of memory and formation of expectancy patterns (Squires et al. 1979). This assumption is confirmed by studies carried out on contingent negative variation (CNV) in subjects affected by mental retardation and in a control group; there were no significant differences between the groups as regards amplitudes, but such differences were seen in respect of latency and CNV rise time recorded at Cz, both of which were prolonged in the mentally retarded subjects (Karrer and Ivins 1976a,b). Down syndrome subjects take more time to develop the CNV, which then tends to increase during the test. As the CNV is considered an "expectancy" potential that can reflect both an orientation process following a warning stimulus and the expectancy and/or preparation for the motor response after an imperative stimulus, this behavior could indicate the inability to evaluate and utilize correctly the time succession of external events in programming a motor-perceptual act. From a behavioral point of view this results in an increase in reaction times, which are steadily and markedly longer than those of control subjects.

Until now there have been few controlled investigations of the electrical cerebral activity of subjects with severe mental deficiency in relation to complex tasks consisting in programming a temporal sequence of self-paced and goal-directed actions. It occurred to me and my co-workers that in Down syndrome subjects we might employ the same method used in respect of learning-disabled children, which consists in the study of motor performances, electromyographic activity, and movement-related macropotentials in order to evaluate motor-perceptual skills and cognitive processes related to the selection and evaluation of operative strategies.

Material and Method

Down syndrome subjects aged from 18 to 25 (average 23.10), with a mean IQ of 63 on the Wechsler Intelligence Scale (WISC), a mean mental age of 10.6 years on the Termann Merril Scale, and a mean age of 10.2 years on the Psychosocial Development Scale were examined. In addition two control groups were tested. The first control group, matched to the retardates' mental age, consisted of nine normal children (group C), with a mean age of 10 years and a mean IQ of 123.3 on the WISC. The second control group, matched to the retardates' chronological age, consisted of nine young adults (group A) of normal intelligence, with a mean age of 25.9 years. None of the examined subjects had visual or auditory deficiencies or severe neurological abnormalities.

The employed method is reported on pp. 131–146.

Results

The Down syndrome subjects required a longer training period than the control groups to learn the correct bimanual sequence of movements.

Performance

The mean performance time was 178.50 ms in the Down syndrome subjects, 62.93 ms in group C, and 57.79 ms in group A. The target performance rate was 13.82% in the Down syndrome subjects, 26.00% in group C, and 32% in group A. The Down syndrome subjects were also less accurate, their performance shift being 134.15 ms, whereas that of the children was 19.27 ms and that of the normal adults, 14.00 ms. All these differences were significant by Student's t-test ($P < 0.01$). Group A showed a higher target performance rate than group C ($P < 0.05$) and was also more accurate ($P < 0.01$) (Table 1). Moreover the Down syndrome subjects showed a higher rate of performance above 60 ms (58.1%) than either group C (47.5%) ($P < 0.01$) or group A (39.8%) ($P < 0.01$). The difference was also significant between the two control groups ($P < 0.05$). In the Down syndrome group 32.30% of performances were above 200 ms, against 0.25% in the adults and 0.22% in the children (Fig. 1).

Table 1. Means and SD of the performance of the Down syndrome subjects and of the two control groups. In this and in Table 2, a superscript a or b indicates a significant correlation between the Down syndrome subjects and the control group in question. The asterisks indicate a significant difference between the two control groups

Block		Performance time (ms)		Performance shift (ms)		Target performance (%)	
		$\bar{x}$	SD	$\bar{x}$	SD	$\bar{x}$	SD
I	A	70.30[b]	45.90	22.01[b]	38.86	31.11[b]	46.39
	C	67.54[b]	37.87	22.31[b]	26.59	22.22[a]	41.66
	D	260.76	278.98	214.14	268.30	12.00	32.58
II	A	56.21[b]	27.67	12.93[b]*	17.13	32.44[b]	46.92
	C	64.32[b]	53.35	21.28[b]	45.67	23.11[a]	42.24
	D	161.41	224.54	117.48	215.79	13.71	34.49
III	A	51.20[b]	22.64	9.84[b]*	11.81	34.22[a]	47.55
	C	59.40[b]	45.35	16.40[b]	38.24	31.55[a]	46.57
	D	119.47	110.63	76.26	98.04	17.24	37.93
IV	A	53.45[b]	25.19	11.20[b]*	14.67	30.22[a]	46.02
	C	60.47[b]	33.20	17.07[b]	22.20	27.11[a]	44.55
	D	128.42	122.59	86.15	109.12	13.18	34.02
Total	A	57.79[b]*	32.51	14.00[b]**	23.67	32.00[b]*	46.67
	C	62.93[b]	43.18	19.27[b]	34.49	26.00[b]	43.88
	D	178.50	220.13	134.15	210.22	13.82	34.54

A, adults; C, children; D, Down syndrome subjects
a/*, $p < 0.05$; b/**, $p < 0.01$

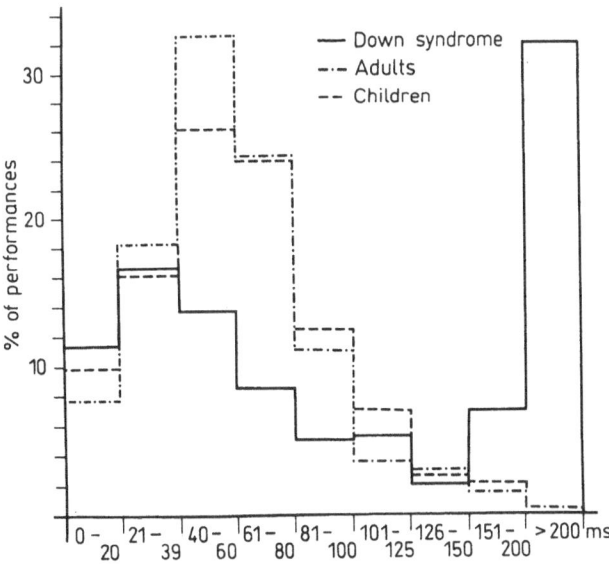

Fig. 1. Performance times of Down syndrome subjects, adults, and children in the nine time intervals

In the Down syndrome subjects practice produced only a partial improvement in the performance time, which remained steadily higher in all four blocks in comparison with the two control groups ($P < 0.01$). In fact, their performance time decreased from 260.7 ms in the first block to 128.4 ms in the fourth, whereas it dropped from 70.3 ms to 53.4 ms in group A and from 67.5 ms to 60.4 ms in group C.

The performance accuracy in the Down syndrome subjects showed the same course: the performance shift decreased from 214.1 ms in the first block to 86.1 ms in the fourth. In group A it fell from 22.0 ms to 11.2 ms and in group C from 22.3 ms to 17.0 ms. Comparison of the performance shift in the Down syndrome subjects in the four blocks with that of the two control groups was steadily significant ($P < 0.01$). Moreover, in comparing the two control groups, the adults were found to improve their accuracy with practice more than did the children: their performance shift did not differ from that of children in the first block, whereas it considerably diminished in the second one ($P < 0.05$) and retained this advantage in the following blocks. The target performance rate in Down syndrome subjects was steadily lower than that of the other two groups: 12% in the first block and 13% in the fourth, against 31% ($P < 0.01$) and 30% ($P < 0.05$), respectively, in group A and 22% ($P < 0.05$) and 27% ($P < 0.05$), respectively, in group C. No significant differences were found in the target performance rate between the two control groups in any of the four blocks, although the adults had a significantly higher total target performance rate than the children ($P < 0.05$) (Table 1).

Electromyography

The EMG of the left forearm muscles group did not show any significan' difference in the three groups as regards the amplitude before and during th(movement and the rise time. In the Down syndrome subjects the EMG of the right arm related to the arrest of the sweep was not different in amplitude before and during the movement in comparison with the other two groups. In contrast, the rise time was slower in comparison with both control groups, with a significant difference ($P < 0.05$) between the Down syndrome subjects and group C.

Movement-Related Brain Macropotentials

Remarkable differences were found in movement-related brain macropotentials (MRBMs) during all four motor periods (Fig. 2).

In the premotor period the Bereitschaftspotential (BP) was present as a negative deflection in the frontal, central, and precentral regions in both control groups; the amplitude and the area of BP did not differ in a statistically significant way, except for Pz ($P < 0.01$), where the amplitude value was higher in adults. In the Down syndrome subjects it was absent or showed a reduced amplitude

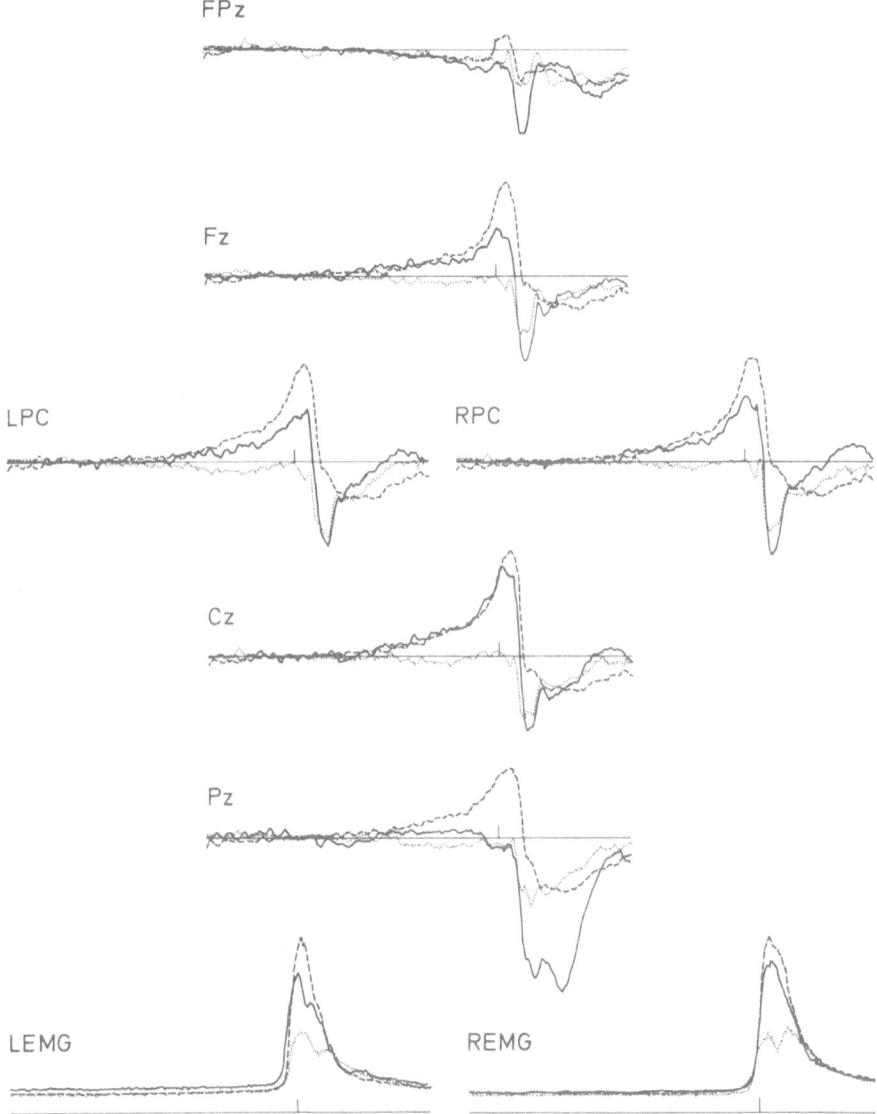

Fig. 2. Grand averages of rectified EMGs and MRBMs of Down syndrome children (....... *traces*), adults (----- *traces*), and normal children (—— *traces*). In this and in Fig. 3 the *vertical bar* in each trace indicates the instance of the computer trigger and a calibration signal of 5 μV. Negativity is upwards. The time scale is 3200 ms

at all recording sites. So there were statistically significant differences in BP between the Down syndrome subjects and the control groups at all recorded cerebral areas, except for FPz in the comparison between Down syndrome subjects and group A, and for FPz and Pz in the comparison between Down syndrome subjects and group C. The results of the BP area comparison in Down syndrome subjects and in adults were the same as those for the amplitude (Pz: $P < 0.01$; Fz, Cz, RPC, and LPC: $P < 0.05$). In Down syndrome subjects and in children, too, results of the BP area comparison proved statistically similar to those regarding the BP amplitude, except for Fz and LPC (Table 2).

The BP onset did not differ between the two control groups, whereas it could not be recorded and measured in the Down syndrome subjects because of the poor or insignificant amplitude of the potential.

In the sensory-motor period the motor cortex potential (MCP) was markedly present in control subjects in the central and precentral regions. In the Down syndrome subjects the MCP was absent or showed a significantly reduced amplitude in all cerebral areas compared with group A (Fz, Cz, Pz, RPC, and LPC: $P < 0.01$; FPz: $P < 0.05$) and in the frontal, central, and precentral areas compared with group C (Fz: $P < 0.05$; Cz, RPC, LPC: $P < 0.01$). The two groups of normal subjects were also different from each other: the MCP amplitude was larger in group A in all cerebral regions (Fz, Pz, RPC, and LPC: $P < 0.01$; FPz, Cz: $P < 0.05$). The latency of the MCP with respect to the EMG onset did not differ in adults and control children, whereas in the Down syndrome subjects it was impossible, because of the reduced MCP amplitude, to find a peak whose latency could be measured with respect to the EMG onset. The absence or reduction in MCP, as well as in other potentials dealt with later, should not be attributed to a jitter effect related to the performance variability, since, as shown in Fig. 3, the grand average of rectified EMGs and MRBMs related to the target performance of all the three groups confirms the absence of MCP in the Down syndrome subjects.

The latency of N100 was always shorter in the Down syndrome subjects at all recording sites, in comparison with both group C and group A. The difference was significant only for FPz and Pz ($P < 0.05$) between the Down syndrome subjects and group C; the two control groups did not differ from each other. Since the measurement of the N100 amplitude depends on the amplitude of BP, it was impossible to carry out a statistical comparison.

In the motor completion period the latency of P200 was not significantly different in the Down syndrome subjects and the adults. The children presented a P200 latency significantly higher than that of the Down syndrome subjects in the frontal (Fz: $P < 0.01$) and in the central and left precentral regions ($P < 0.05$). In the children, P200 had a larger amplitude in all cerebral regions; the differences, with respect to both the adults and the Down syndrome subjects, were statistically significant for all recording sites ($P < 0.01$). The lowest amplitude values were found in the Down syndrome subjects.

In the postmotor period, skilled performance positivity (SPP) was present as a positive deflection in all cerebral regions in all three groups, but with a

Table 2. Means and SD of the MRBMs of the Down syndrome subjects and the two control groups.

			FP_z x̄	SD	F_z x̄	SD	C_z x̄	SD	P_z x̄	SD	RPC x̄	SD	LPC x̄	SD
BP	Amp.	A	1.65	4.34	-6.52[b]	5.01	-8.33[b]	6.22	-7.10[b]**	4.74	-8.96[b]	4.73	-8.36[b]	4.88
		C	3.15	4.89	-4.41[a]	6.99	-9.25[b]	7.59	-0.68	7.66	-6.51[b]	6.56	-5.02[a]	6.50
		D	1.67	6.22	0.58	5.05	-0.16	4.89	1.13	4.18	0.06	4.39	0.89	5.18
	Area	A	370.95	1332.94	-1090.41*	1073.42	-1657.03[a]	1421.66	-1641.59[b]	1241.66	-1468.81[a]	1086.28	-1519.90[a]	1227.94
		C	941.39	1261.16	-1054.68	1661.59	-1771.81[a]	1641.04	-568.91	1872.87	-1342.44[a]	1333.52	-954.48	1345.34
		D	1061.77	2152.90	508.00	1567.65	318.55	1865.01	1063.66	1261.20	436.25	1541.84	1158.62	2303.43
MCP	Amp.	A	0.86*	5.22	-13.95[b]**	6.50	-15.07[b]	7.78	-10.94[b]**	4.70	-15.49[b]**	5.75	-14.57[b]**	6.06
		C	3.64	5.94	-6.09[a]	8.50	-11.45[b]	10.98	3.19	8.32	-7.33[b]	9.45	-6.27[b]	10.40
		D	2.09	9.59	1.27	8.40	0.53	6.81	1.81	4.87	0.88	6.34	1.98	7.47
N_{100}	Lat.	A	113.70	20.12	120.27	19.53	118.34	21.61	123.02	25.25	116.83	22.33	119.72	20.45
		C	121.70[a]	20.21	119.48	19.66	118.47	17.89	127.61[a]	25.18	117.57	18.81	116.91	16.95
		D	104.16	8.50	109.30	14.16	109.05	7.92	112.71	5.06	108.33	10.80	111.22	11.29
	Amp.	A	7.77**	3.22	15.86**	5.89	17.82**	10.38	14.53**	5.30	17.17**	7.10	15.91**	7.10
		C	12.62	4.64	22.19[b]	6.98	26.04[b]	7.60	22.03[b]	10.57	25.34[b]	7.29	23.32[b]	6.46
		D	7.59	6.30	11.12	7.16	13.60	7.53	11.64	5.42	13.36	6.99	12.55	8.01
P_{200}	Lat.	A	192.81*	15.48	210.61	21.88	211.16*	15.91	219.00	18.91	211.50	19.25	213.83	18.95
		C	210.44	23.50	223.05[b]	18.80	266.55*	26.76	232.51	27.38	219.27	27.05	227.00*	28.37
		D	201.06	23.50	195.73	25.87	201.73	31.19	210.34	33.41	210.56	28.21	201.82	27.20
	Amp.	A	7.70	5.54	8.08	6.21	8.18	7.96	10.24**	5.25	7.50	5.68	7.80	5.53
		C	6.97	7.42	5.84	9.56	7.65	12.90	25.24[b]	10.86	5.08	9.71	6.09	9.92
		D	6.97	13.25	4.05	11.24	5.60	11.38	9.28	8.31	4.91	9.58	5.15	11.02
SPP	Lat.	A	525.80	126.32	507.33	102.18	501.88	98.90	497.82	79.89	516.05	99.34	509.72	97.08
		C	537.09	99.28	539.41	99.84	501.74	65.34	495.27	45.70	516.75	81.92	510.70	79.83
		D	532.55	88.71	521.10	72.88	532.38	77.31	508.85	66.15	515.70	48.99	507.82	62.05

A, adults; C, children; D, Down syndrome subjects
a/*, $P < 0.05$; b/**, $P < 0.01$

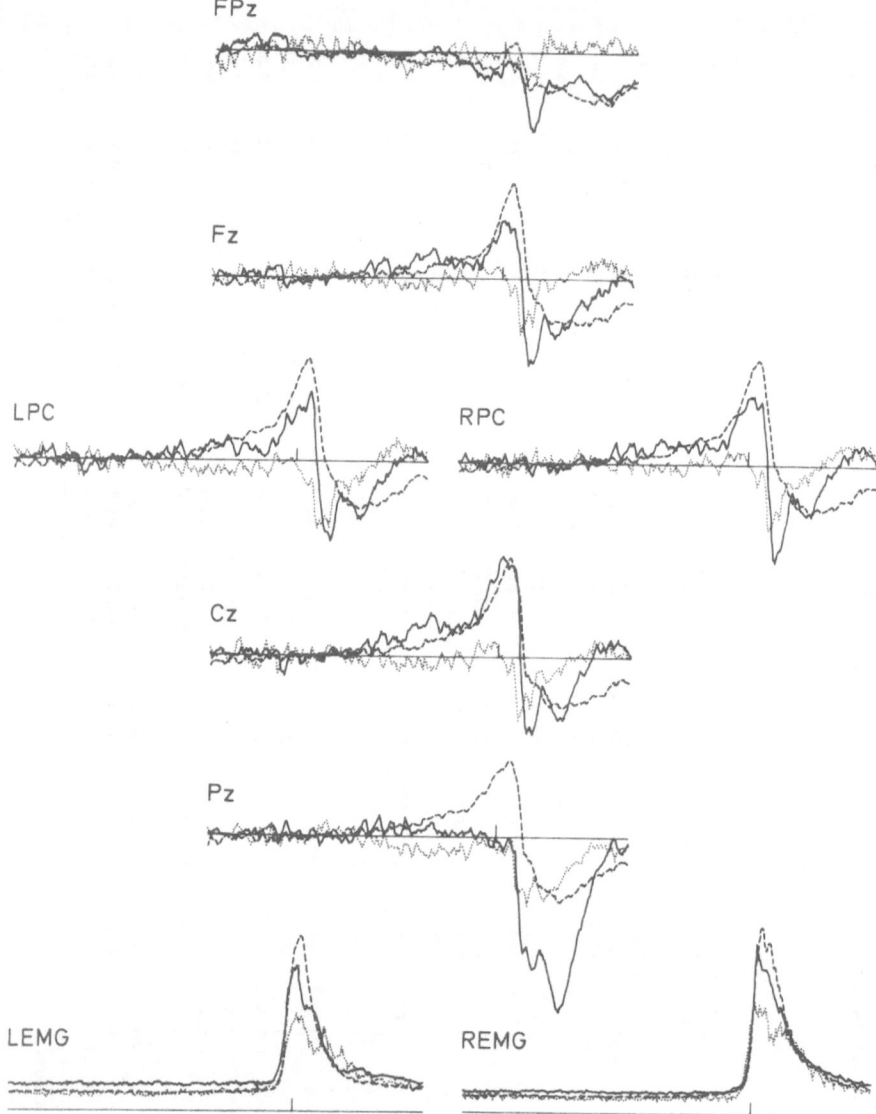

Fig. 3. Grand average of rectified EMGs and MRBMs related to target performances of Down syndrome subjects (....... *trace*), adults (---- *trace*), and children (—— *trace*)

lower amplitude in the Down syndrome subjects. The latencies of this potential were not significantly different in the three groups. The amplitude of SPP measured from the baseline was not significantly different in the three groups, except for Pz in the comparison between Down syndrome subjects and children (Table 2). When comparison was made with SPP amplitude values measured as the difference from the P200 peak, the SPP amplitude of the adults was significantly greater at all recording sites ($P < 0.01$) as compared with the Down syndrome subjects and the children, except for Pz in the comparison with the children. The SPP amplitude of the children was not significantly different from that of the Down syndrome subjects.

Discussion

The motor-perceptual task lies in carrying out ballistic bimanual and self-initiated movements. Its successful performance basically depends on a correct and accurate temporal sequence of movements. The limited temporal range of actions forces a temporal and motor programming of the whole task. Moreover, as subjects can evaluate the result of each test in real time owing to the visual feedback, they are also able to compare each time the obtained result with the preprogrammed motor strategy and to change it in the most suitable way to reach the target.

The preparation of a movement sequence like that required by such a bimanual task involves the development of a central clock which controls the temporal course through afferent and efferent systems (Hirsch and Sterrick 1964; Rosenbaum and Patashnik 1980). The performance improvement of this clock depends in part on the presence of proprioceptive and exteroceptive feedback as regards the accuracy of the performance. However, it does not eliminate the need for motor programming (Rosenbaum 1983), which plays a fundamental role in the organization of the temporal sequence of movements, which depends, in turn, on a higher age-related synaptic efficiency of the central nervous system (Craik 1947). In fact, the interval between two consecutive movements has been found to be often shorter than the time required for the proprioceptive and exteroceptive feedback of the first one to act as a trigger for the second one (Lashley 1951).

The performances of the Down syndrome subjects in our investigation show that the development of this central clock proceeds with difficulty. Down syndrome subjects, in fact, meet with great difficulties in carrying out the bimanual movement in the correct temporal sequence. They were found to be steadily slower in executing the task in comparison with both control groups. They showed the highest rate of trials with performance times above both 60 and 200 ms. Moreover, their performance time decreased greatly from the first block to the fourth, but the target performance rate remained practically unchanged and below that of the controls.

The children showed the highest increase in the target performance rate as the task proceeded, even though they did not reach the values of the adults. The adults did not improve with practice the target performance rate, which was already high in the first block, but they showed a greater increase in their performance accuracy as compared with children.

The performance accuracy of the Down syndrome subjects showed the same behavior as the target performances; the performance shift remained at steadily higher values throughout the task.

Therefore the adults were able to carry out their task more speedily and accurately throughout the test. The children proved able to utilize experience, progressively improving their performance, even though they did not reach the level of the adults. The Down syndrome subjects were permanently below the other two groups as regards accuracy and speed. Not only were the Down syndrome subjects slower and less accurate, but they were also unable to benefit from practice, unlike the other two control groups which were similar as regards mental or chronological age.

These findings agree with several previous studies showing that motor-perceptual functions of Down syndrome subjects appear to be greatly impaired in comparison with subjects with the same chronological or mental age (Cratty 1969). The difficulty in maintaining equilibrium or in executing tasks requiring it (Pesch and Nagy 1978), the markedly prolonged reaction time (Berkson 1960), and the inability to carry out rapid movement sequences (Frith and Frith 1974) are some of the motor-perceptual functions in regard of which Down syndrome subjects display worse performances than subjects affected by other forms of mental retardation.

Furthermore they meet with particular difficulties in carrying out tasks involving a temporal component and when the sequence of movements must be programmed so as to make the resulting action coincide with an external event (Henderson et al. 1981a). To make this possible, in fact, it is necessary for the movement to be programmed according to precise spatial and temporal parameters. The specific problem of motor programming in Down syndrome subjects seems to lie solely in the temporal component and not in the spatial one (Henderson et al. 1981b).

In parallel to motor performances, MRBMs also showed significant differences between Down syndrome sybjects and control subjects. The BP is characteristic of the premotor period (Kornhuber and Deecke 1965), when the organization of ideokinetic elements for the execution of the movement takes place (Chiarenza et al. 1982, 1983). Its clinical and neurophysiological character-istics make it an important index of cortical maturation. The BP appears, in fact, at about the age of 7 years in the frontocentral regions, and it progressively increases in amplitude until in adolescence it reaches that seen in adults (Chiarenza 1986a). It is absent or has a low amplitude in various clinical situations: chronic schizophrenia (Chiarenza et al. 1985), Parkinson's disease (Deecke et al. 1977), dyslexia–dysgraphia (Chiarenza et al. 1986), and learning disabilities (Chiarenza et al. 1982). In the current study the BP was present in the two control groups, with a greater amplitude in adults, whereas it was absent or greatly reduced in all cerebral regions of the Down syndrome subjects.

Warren and Karrer (1984) showed that during the execution of unskilled movements the BP is missing or appears as a positive deflection in young adults affected by mental retardation. The absence of BP in Down syndrome subjects could therefore indicate the presence of a programming failure of motor-perceptual performance, both simple and complex.

It has been assumed that the BP is a cholinergic potential: its absence in Down syndrome subjects would therefore agree with the microscopic and histochemical findings showing a remarkable deficit of the central cholinergic system in Down syndrome subjects (Perry et al. 1985; Kitt et al. 1984; Price et al. 1982).

During the sensory-motor period the MCP was absent or had a reduced amplitude in the Down syndrome subjects. This potential is considered to be an index of reafferent sensory activity: it represents the elaboration in precentral and frontal regions of the kinesthetic information related to the executed movement (Papakostopoulos et al. 1975; Papakostopoulos and Crow 1984). Since suitable proprioceptive information is of fundamental importance for the preparation and correct execution of movements, the lack of elaboration of this sensory feedback, expressed by the MCP, could be responsible for the poor capacity for temporal organization which Down syndrome subjects show in carrying out complex motor acts. Animal experiments (Dubrovsky and Garcia-Rill 1973) and observations on patients with damaged posterior spinal columns have in fact shown that total or partial deafferentation prevents the temporal control of a motor sequence. Furthermore, it is important to observe that in the same experimental situation subjects over 60 also show slower performances and a steadily reduced MCP amplitude (Papakostopoulos and Banerji 1980). Since macro- and microscopic investigations are in agreement in proving, in Down syndrome subjects, an early onset of anatomopathological signs of cerebral aging, the reduced MCP amplitude could indicate a poor cortical reactivity to the reafferent sensory information. The lower MCP amplitude in children compared with adults could reflect, in contrast, a condition of relative immaturity of the sensory reafferent activity, with which lower speed and accuracy of execution as regards performances would correspond.

The N100 wave is considered the cerebral response evoked by the appearance of the sweep on the oscilloscope. Its latency is shorter in Down syndrome subjects, but in this study it was significantly reduced only for FPz and Pz in comparison with the children's group. This result is in agreement with the studies of BAEPs by Squires et al. (1981) and in contrast with those of cortical ERPs by Bigum et al. (1970), Marcus (1970), and Gliddon et al. (1975). These discrepancies can be attributed mainly to the different experimental paradigms used by these authors, their experiments being externally paced.

The shorter latency of N100 in the Down syndrome subjects in the task performed in this study could be attributed to a deficit of the central mechanisms responsible for the perceptive elaboration of sensory input. An analogous interpretation has been suggested for the flat recovery function of wave V of the BAEPs of the Down syndrome subjects (Squires et al. 1981; Otto et al. 1984). P200 is considered to be one of the late components of somatosensory potentials (Chiarenza et al. 1983). Its latency was greater in children than in the other two groups; there was no significant difference between Down syndrome subjects and adults. As the latency of P200 has been found to decrease with age (Chiarenza et al. 1983), these findings would indicate that in Down syndrome subjects the neuronal systems subtended by this potential have reached a maturation comparable with that in normal adults, whereas they are still relatively immature in 10-year-old children. The lower amplitude of P200 in Down syndrome subjects could be an index of reduced elaboration of reafferent sensory input. This result is in agreement with other investigations of late sensory and cognitive components of cerebral evoked potentials in Down syndrome (Squires et al. 1979).

The presence of SPP on all recorded brain areas in Down syndrome subjects and the fact that its amplitude was similar, except for Pz, to that in subjects with the same mental age, but lower than that in subjects with the same chronological age, could indicate that Down syndrome subjects are able to recognize and evaluate the results of their motor-perceptual performances but that they do not manage to use such experience to improve their performances. In fact, the SPP is present only when, besides elaborating movement strategies, the subject can also evaluate from time to time the result of his performances and utilize the acquired knowledge to change or influence future actions (Papakostopoulos 1978; Chiarenza 1986a). If the possibility of evaluation is lacking, the SPP does not appear (Papakostopoulos 1980; Papakostopoulos et al. 1986). This potential has peculiar developmental characteristics: it is always present in the parietal regions and appears at the age of 9–10 years in the frontocentral regions. With age its amplitude in these areas increases until it reaches the adult amplitude in adolescence, whereas the latency decreases (Chiarenza et al. 1983). Children under 9–10 are likely to be unable to elaborate complex strategies based on formal and probabilistic thinking (Chiarenza 1986b). These results agree with, and extend to self-paced tasks, the previous observations of Karrer and Ivins (1976b), Squires et al. (1979), and Lincoln et al. (1986) with P300 experiments.

There is evidence to support the assumption that all surface-positive slow potentials, such as SPP and P300, originate from the hyperpolarizing inhibition of pyramidal neurons and the electrotonic diffusion of postsynaptic inhibitory potentials to apical dendrites (Creutzfeldt et al. 1966). The physiological tone of the cholinergic component of the ascending activating reticular system projecting towards thalamus and cortex is thought to play a predominant role in the genesis of such positive potentials (Marczynski 1978).

Since the SPP is considered a potential produced by cholinergic systems (Marczynski 1978), its low amplitude might reflect the cortical and subcortical cholinergic deficit shown by histochemical and microscopic studies in Down syndrome subjects (Perry et al. 1985; Price et al. 1982). Furthermore, these findings would suggest the SPP and BP are generated by different neuronal systems, both because they are differently distributed on the scalp and because they are differently involved in Down syndrome.

In conclusion, the pathological structural organization of the CNS, from brain stem to cortex, may be responsible both for the bad performance of the

Down syndrome subjects on this motor-perceptual task and for the abnormalities of MRBMs that are its electrophysiological equivalent.

Conclusions

Analysis of the performances during the execution of a motor-perceptual, self-initiated task shows that Down syndrome subjects have great difficulty in organizing correct temporal sequence of ballistic movements. Moreover, they are much slower in performing the task because of a defective timing of motor sequences. From a neurophysiological point of view, these behavioral aspects express themselves in a reduced preparation of the movement (absent or very low BP), a lack of elaboration of the reafferent somatosensory information (absence of MCP), and a reduced capacity for evaluating the outcome of the performance (presence of low SPP).

References

Balazs R, Brooksbank WLB (1985) Neurochemical approaches to the pathogenesis of Down's syndrome. J Ment Defic Res 29:1–14

Balkany TJ, Downs MP, Jafek BW, Krajicek MJ (1979) Hearing loss in Down's syndrome. Clin Pediatr (Phila) 18:116–118

Ball MJ, Nuttal K (1980) Neurofibrillary tangles, granulovacuolar degeneration and neuron loss in Down syndrome: quantitative comparison with Alzheimer dementia. Ann Neurol 7:462–465

Barnet, AB, Lodge A (1967) Click evoked EEG responses in normal and developmentally retarded infants. Nature 214:252–255

Benda CE (1969) Down's syndrome. Grune and Stratton, New York

Berkson G (1960) An analysis of reaction time in normal and mentally deficient young men. J Ment Defic Res 4:51–77

Bigum HB, Dustman RE, Beck C (1970) Visual and somato-sensory evoked responses from mongoloid and normal children. Electroencephalogr Clin Neurophsiol 28:576–585

Burger PC, Vogel FS (1973) The development of the pathologic changes of Alzheimer's disease and senile dementia in patients with Down's syndrome. Am J Pathol 73:457–468

Callner DA, Dustman RE, Madsen JA, Schenkenberg T, Beck EC (1978) Life span changes in the averaged evoked responses of Down's syndrome and nonretarded persons. Am J Ment Defic 82:398–405

Chiarenza GA (1986a) Development of sensory motor and cognitive processes. Movement related brain macropotentials in children. In: Gallai V (ed) Maturation of the CNS and evoked potentials. Elsevier, Amsterdam, pp 236–246

Chiarenza GA (1986b) Electrophysiology of skilled performances in children. Ital J Neurol Sci (Suppl) 5:155–162

Chiarenza GA, Radaelli L (1988) New parameters in the evaluation of the brain stem acoustic evoked potentials in children with language disorders. (in preparation)

Chiarenza GA, Papakostopoulos D, Guareschi Cazzullo A, Giordana F, Giammari Aldè G (1982) Movement related brain macropotentials during skilled performance task in children with learning disabilities. In: Chiarenza GA, Papakostopoulos D (eds) Clinical application of cerebral evoked potentials in pediatric medicine. Excerpta Medica, Amsterdam, pp 259–292

Chiarenza GA, Papakostopoulos D, Giordana F, Guareschi Cazzullo A (1983) Movement related brain macropotentials during skilled performances. A developmental study. Electroencephalogr Clin Neurophysiol 56:373–383

Chiarenza GA, Papakostopoulos D, Dini M, Cazzullo CL (1985) Neurophysiological correlates of psychomotor activity in chronic schizophrenics. Electroencephalogr Clin Neurophysiol 61:218–228

Chiarenza GA, Papakostopoulos D, Grioni A, Tengattini MB, Mascellani P, Guareschi Cazzullo A (1986) Movement related brain macropotentials during a motor perceptual task in dyslexic and dysgraphic children. In: Mc Callum WC, Zappoli R, Denoth F (eds) Cerebral psychophysiology: studies in event related potentials (EEG Suppl 38). Elsevier, Amsterdam, pp 489–491

Colon EJ (1972) The structure of the cerebral cortex in Down's syndrome. A quantitative analysis. Neuropädiatrie 3:362–376

Craik KJW (1947) Theory of the human operator in control systems. Br J Psychol 38:56–61

Crapper D, Dalton AJ, Skopitz M, Scott JW, Hachinski VC (1975) Alzheimer degeneration in Down syndrome. Electrophysiologic alterations and histopathologic findings. Arch Neurol 32:618–623

Cratty BJ (1969) Motor activity and the education of retardates. Lea and Febiger, Philadelphia

Creutzfeldt OD, Lux HD, Watanabe S (1966) Electrophysiology of cortical nerve cells. In: Purpura DP, Yahr MD (eds) The thalamus. Columbia University Press, New York, pp 35–55

Deecke L, Englitz HG, Kornhuber HH, Schmitt G (1977) Cerebral potentials preceding voluntary movement in patients with bilateral or unilateral Parkinson akinesia. In: Desmedt JE (ed) Attention, voluntary contraction and event related cerebral potentials. Prog Clin Neurophysiol 1:151–163

Dubrovsky B, Garcia-Rill E (1973) Role of dorsal columns in sequential motor acts requiring precise forelimb projection. Exp Brain Res 18:165–177

Dustman RE, Callner DA (1979) Cortical evoked responses and response decrement in nonretarded and Down's syndrome individuals Am J Ment Defic 83:391–397

Ellingson RJ, Peters JF (1980) Development of EEG and daytime sleep patterns in trisomy-21 infants during the first year of life: longitudinal observations. Electroencephalogr Clin Neurophysiol 50:457–466

Ellingson RJ, Eisen JD, Ottersberg G (1973) Clinical electroencephalographic observations on institutionalized mongoloids confirmed by karyotype. Electroencephalogr Clin Neurophysiol 34:193–196

Ferri R, Bergonzi P, Colognola SA, Musumeci S, Sanfilippo P, Tomassetti A, Viglianesi A, Gigli GL (1986) Brainstem evoked potentials in subjects with mental retardation and different karyotypes. In: Gallai V (ed) Maturation of the CNS and evoked potentials. Elsevier, Amsterdam, pp 369–374

Frith U, Frith CD (1974) Specific motor disabilities in Down syndrome. J Child Psychol Psychiatry 15:293–301

Galbraith GC, Gliddon JB, Busk J (1970) Visual evoked responses in mentally retarded and nonretarded subjects. Am J Ment Defic 83:341–348

Gigli GL, Ferri R, Musumeci SA, Tomassetti P, Bergonzi P (1984) Brainstem auditory evoked responses in children with Down's syndrome. In: Berg JM (ed) Perspectives and progress in mental retardation, vol 2. Biomedical aspects. University Park Press, Baltimore, pp 277–286

Gliddon JB, Galbraith GC, Busk J (1975) Effect of preconditioning visual stimulus duration on visual-evoked responses to a subsequent test flash in Down's syndrome and nonretarded individuals. Am J Ment Defic 80:186–190

Hansford RG (1983) Bioenergetics in aging. Biochim Biophys Acta 726:41

Harris WS, Goodman RM (1968) Hyper-reactivity to atropine in Down's syndrome. N Engl J Med 279:407

Henderson SE, Morris J, Ray S (1981a) Performance of Down syndrome and other retarded children on the Cratty Gross-Motor test. Am J Ment Defic 85:416–424

Henderson SE, Morris J, Frith U (1981b) The motor deficit in Down's syndrome children: a problem of timing? J Child Psychol Psychiatry 22:233–245

Hirsch IJ, Sterrick CE (1964) Perceived order in different sense modalities. J Exp Psychol 67:103–112

Johnson R (1986) A triarchic model of P300 amplitude. Psychophysiology 23:367–384

Karrer R, Ivins J (1976a) Steady potentials accompanying perception and response in mentally retarded and normal children. In: Karrer R (ed) Developmental psychophysiology of mental retardation. Thomas, Springfield, pp 361–417

Karrer R, Ivins J (1976b) Event-related slow potentials in mental retardates. In: McCallum WC, Knott JR (eds) The responsive brain. Wright, Bristol, pp 154–157

Kitt CA, Price DL, Struble RG (1984) Evidence for cholinergic neurites in senile plaques, Science 226:1443–1445

Kornhuber HH, Deecke L (1965) Hirnpotentialänderungen bei Willkürbewegungen und passiven Bewegungen des Menschen: Bereitschaftspotential und reafferente Potentiale. Pflügers Arch 284:1–17

Lashley KS (1951) The problem of serial order in behavior. In: Jeffres LA (ed) Cerebral mechanisms in behavior. Wiley, New York

Lincoln AJ, Courchesne E, Kilman BA, Galambos R (1986) Auditory ERPs and information processing in Down's syndrome children. In: McCallum WC, Zappoli R, Denoth F (eds) Cerebral psychophysiology: studies in event-related potentials (EEG Suppl 38). Elsevier, Amsterdam, pp 492–495

Luria AR (1963) The mentally retarded child. Pergamon, New York

Luria AR (1973) The working brain. Penguin, London

Mann DMA, Yates PO, Marcyniuk B (1987) Dopaminergic neurotransmitter systems in Alzheimer's disease and in Down's syndrome at middle age. J Neurol Neurosurg Psychiatry 50:341–344

Marcus MM (1970) The evoked cortical response: a technique for assessing development. Calif Ment Health Res Dig. 8:59–72

Marczynski TJ (1978) Neurochemical mechanisms in the genesis of slow potentials: a review and some clinical implications. In: Otto DA (ed) Multidisciplinary perspectives in event-related brain potential research. US Environmental Protection Agency, Washington DC, pp 25–35

Ohara PT (1972) Electron microscopical study of the brain in Down's syndrome. Brain 95:681–684

Otto B, Karrer R, Halliday R, Horst RL, Klorman R, Squires N, Thatcher RW, Fenelon B, Lelord G (1984) Developmental aspects of event-related potentials: aberrant development. In: Karrer R, Cohen J, Tueting P (eds) Brain and information: event-related potentials. Ann NY Acad Sci 425:319–337

Papakostopoulos D (1978) Electrical activity of the brain associated with skilled performance. In: Otto DA (ed) Multidisciplinary perspectives in event-related brain potential research. US Environmental Protection Agency, Washington DC, pp 134–137

Papakostopoulos D (1980) A no stimulus, no response event-related potential of the human cortex. Electroencephalogr Clin Neurophysiol 48:622–638

Papakostopoulos D, Banerji N (1980) Movement related brain macropotentials during skilled performance in Parkinson's disease. Electroencephalogr Clin Neurophysiol 49:93

Papakostopoulos D, Crow HJ (1984) The precentral somatosensory evoked potential. In: Karrer R, Cohen J and Tueting P (eds) Brain and information: event-related potentials. Ann NY Acad Sci 425:256–262

Papakostopoulos D, Cooper R, Crow HJ (1975) Inhibition of cortical evoked potentials and sensation by self-initiated movement in man. Nature 258:321–324

Papakostopoulos D, Stamler R, Newton P (1986) Movement related brain macropotentials during self-paced skilled performance with and without knowledge of results. In: McCallum WC, Zappoli R, Denoth F (eds) Cerebral psychophysiology: studies in event-related potentials (EEG Suppl 38). Elsevier, Amsterdam, pp 261–262

Perry KE, Curtis M, Dick DJ, Candy JM, Atack JR, Bloxham CA, Blessed G, Fairbairn A, Tomlinson BE, Perry RH (1985) Cholinergic correlates of cognitive impairment in Parkinson's disease: comparisons with Alzheimer's disease. J Neurol Neurosurg Psychiatry 48:413–421

Pesch RS, Nagy DK (1978) A survey of the visual and developmental-perceptual abilities of the Down's syndrome child. J Am Optom Assoc 9:1031–1037

Price DL, Whitehouse PJ, Struble RG, Coyle JT, Clark AW, Delong MR, Cork LC, Hedreen JC (1982) Alzheimer's disease and Down's syndrome. Ann NY Acad Sci 396:145–151

Ropper AH, Williams RS (1980) Relationship between plaques, tangles and dementia in Down syndrome. Neurology 30:639–644

Rosenbaum DA (1983) Central control of movement timing. Bell Syst Tech J 62:1647–1657

Rosenbaum DA, Patashnik O (1980) A mental clock setting process revealed by reaction times. In: Stelmach JE, Requin J (eds) Tutorials in motor behavior. North-Holland, Amsterdam, pp 1–14

Schafer EWP, Peeke HVS (1982) Down syndrome individuals fail to habituate cortical evoked potentials. Am J Ment Defic 87:332–337

Scott BS, Petit TL, Becker LE, Edwards BAV (1982) Abnormal electric membrane properties of Down's syndrome DRG neurons in cell culture. Dev Brain Res 2:257

Skinner JE, Lindsley DB (1971) Enhancement of visual and auditory evoked potentials during blockade of the nonspecific thalamocortical system. Electroencephalogr Clin Neurophysiol 33:1–6

Squires N, Aine C, Buchwald J, Norman R, Galbraith G (1980) Auditory brain stem response abnormalities in severely and profoundly retarded adults. Electroencephalogr Clin Neurophysiol. 50:172–185

Squires NK, Galbraith GC, Aine CJ (1979) Event related potential assessment of sensory and cognitive deficits in the mentally retarded. In: Lehmann D, Callaway E (eds) Human evoked potentials applications and problems. Plenum, New York, pp 397–413

Squires NK, Buchwald J, Liley F, Strecker J (1981) Brain stem evoked potential abnormalities in retarded adults. In: Courjon J, Mauguière F, Revol M, (eds) Clinical applications of evoked potentials in neurology. Raven, New York, pp 129–139 (Advances in neurology, vol. 32)

Straumanis JJ, Shagass C, Overton DA (1973) Somatosensory evoked responses in Down syndrome. Arch Gen Psychiatry 29:544–549

Takashima S, Becker LE, Armstrong DL, Chan FW (1981) Abnormal neuronal development in the visual cortex of the human fetus and infant with Down syndrome. A quantitative and qualitative Golgi study. Brain Res. 225:1

Warren C, Karrer R (1984) Movement-related potentials in children. In: Karrer R, Cohen J, Tueting P (eds) Brain and information: event-related potentials. Ann NY Acad Sci 425:489–495

Wetterberg L, Gustavson KH, Backstrom M, Ross SB, Froden O (1972) Low dopamine-beta-hydroxylase activity in Down's syndrome. Clin Genet 3:152–153

Speech and Language

Evaluation of Speech Performance in Children and Its Implications for Developmental Psychopathology

H. Amorosa

Introduction

Numerous studies have shown that children with developmental speech and language problems are at a much higher risk of suffering psychiatric disorders than the population as a whole (Cantwell and Baker 1985; Beitchman et al. 1986; Amorosa et al. 1986; Esser and Schmidt 1987; Mayr, in press). Why there is a high correlation, with up to 50% of children with speech and language problems showing behavioral disorders, is unclear. The increase does not seem to be due to an increase in any one particular disorder as compared to the distribution seen in children of the same age but without speech/language problems. In a small proportion of children it seems likely that the psychiatric disorder is a reaction to the speech and language problem, whereas it is extremely rare for the opposite to be true, i.e., for a psychiatric disorder to be the cause of a speech and language problem. In general, however, the most likely explanation for the correlation is that a common factor is responsible for both the abnormal behavior and the abnormal speech and language development. One such factor has been seen in early interference with normal brain development (Lou et al. 1984).

The prospective longitudinal studies from the Rostock group (Meyer-Probst and Teichmann 1984) have shown that besides environmental influences, pre-, peri-, and postnatal risk factors likely to interfere with normal brain development are an important determinant of later behavior and speech/language development. In our own study of children with severe impairments of speech and language we found not only a large number of children with signs of motor coordination problems (Noterdaeme et al. 1988) and other symptoms commonly used to diagnose minimal brain dysfunction, but also a much higher incidence of minimal brain dysfunction in those children with both a disorder of speech and language development and psychiatric symptoms (Berger et al., in press). In a longitudinal study of preterm and term-born babies, Largo (1987) found a significant correlation between the perinatal optimality score and the ages at which various stages of language development were reached.

In contrast to these studies, in an epidemiological study of 8-year-old boys, Esser et al. (1983) and Esser and Schmidt (1987) could not find evidence for a connection between minimal brain dysfunction and behavioral disorders or speech and language problems, although they did find a relationship between speech/language disorders and behavioral disorders.

The differences in findings can be related in part to differences in the populations studied. Since Esser et al. (1983) and Esser and Schmidt (1987) excluded all children from special classes, one has to assume that children with only minor problems of speech and language development were most prevalent in their group. The studies by Noterdaeme et al. (1988) and Berger et al. (in press), however, included only patients with severe disorders.

A further problem, and one that is often overlooked, in all studies on the relationship of behavioral disorders, developmental speech/language impairment, and underlying disorders of brain development is that of global ratings of neurological abnormality, risk factors for brain development, psychiatric diagnosis, and speech/language deficit. Global ratings necessarily deemphasize individual differences and neglect heterogeneity within groups. From animal studies on disorders of brain development (Prechtl 1981; Isaacson and Spear 1984; Almli 1984) we know that depending on the time of insult, often even within a short time span, very different brain systems can be affected, the susceptibility of brain tissue to damaging agents being dependent on the specific developmental state of that tissue. If we assume that the situation is similar in humans, we have to analyze specific functions in detail and relate our findings to studies of the development of specific brain systems to infer stages of development where damage is most likely to occur. Global ratings of abnormalities averaged over groups therefore are not sufficient to analyze the relationship between early interference with brain development and later behavioral or cognitive outcome in individual children.

A detailed assessment is thus necessary not only for therapeutic planning for a specific child but also for an understanding of the more general question of the relationships between that child's behavioral disorder, speech/language disorder, and underlying problems of brain development.

The present report contributes to one aspect of this assessment, the detailed analysis of fine motor coordination problems in children with specific developmental speech and language disorders. We chose disorders of fine motor coordination because they are common in this group of children, but also because motor functions allow a much closer look at brain functioning than does most other behavior. We hoped the study of disorders of fine motor coordination would yield some insights into disorders of brain functioning and also into the time at which interference with normal development could have occurred.

Problems of fine motor coordination are common among children with developmental speech and language disorders and have been mentioned repeatedly in clinical descriptions (Luchsinger and Arnold 1970; Seeman 1974; Wirth 1983). Fine motor coordination of both speech movements and finger movements are involved. Usually the children start to talk late. Their utterances are unintelligible even into their school years. Later their articulation improves, but they are often described as having sloppy pronunciation. They are considered to be clumsy, and their handwriting is often very irregular.

In our studies with a group of $4\frac{1}{2}$- to 8-year-old children with unintelligible speech we found that the children had more problems adjusting the speech breathing system to the length of a planned utterance than children matched for age but with normal speech development: we observed more breaks for breathing at syntactically inappropriate places, more instances of talking on an inspiratory airstream, and more variation in loudness. All these symptoms point to insufficient control of the subglottal air pressure, which in turn is controlled by the speech breathing system (Amorosa, in press).

The second system involved in speech is the phonatory system, which controls the functions of the vocal folds. The tension and degree of opening of the vocal folds is important for the intonation of an utterance and the voice characteristics. In the same children with unintelligible speech just mentioned, we found many signs of insufficient fine motor control of the vocal folds. Clinical signs of insufficient control are a tense and rough voice, preutterance vocalizations, intraphonemic disruption, voice tremor, and pitch breaks. They were much more common in these children than in the age-matched children with normal speech. The voice disorders persisted in spite of improvements in intelligibility over a 16-month period (Amorosa et al. 1990). Martin (1981) reports similar problems.

Articulatory problems are the most common symptoms in children with speech and language abnormalities. Often they are considered to be of a linguistic–cognitive nature, but studies on simple syllable repetitions (Amorosa 1989) and oral movements (Gabriel et al. 1976a, b) support the idea that fine motor coordination problems play an important part in unintelligible speech.

Studies of finger movements are also indicative of a deficit in fine motor coordination. Bäumler (1989) showed that children with specific developmental disorders of speech and language performed less well than normal children of the same age on a "test battery" of fine motor tasks [(Hamster 1980) apparatus available from Dr. G. Schufried, 2340 Mödling, Austria]. Holding the hands steady and performing fast repetitive movements were especially difficult. Noterdaeme et al. (1988) showed that the isolation of movements creates problems. For example, when the children were instructed to move only their wrist or fingers the children in the experimental groups moved their whole arm, whereas those in the control group moved just their wrist or fingers.

All studies mentioned so far very clearly point to disorders of fine motor coordination in children with specific disorders of speech and language development. Most of these studies included younger children of preschool or beginning school age. Since it has been said often that these fine motor abnormalities are symptoms of delayed development, we were interested in investigating fine motor performance on speech and hand movement tasks in children between 10 and 12 years of age who still had symptoms of specific developmental disorders of speech and language. We conducted two studies. In the first study we investigated speech motor performance on repetitions of short utterances, using the variability of timing as a measure of performance. In the second study we compared the tapping performance of older language-impaired children with that of age-matched and younger children with normal speech and language development.

Method

Subjects

The subjects were ten children (four girls and six boys) who were diagnosed by
two speech therapists as having a specific developmental disorder of speech and
language (according to ICD-9). The children were far below age and IQ level
on language tests; this was especially true for expressive language, less so for
language comprehension. They still had articulatory abnormalities, but only
minor ones. The age range was 119–151 months (mean 132.4). All children had
a nonverbal IQ above 85 and normal hearing. In addition, we studied two
groups of ten children each with normal speech and language development.
These children, too, had nonverbal IQs above 85 and normal hearing, but they
differed in age. One group had an age range of 121–139 months (mean: 128.1),
the second group an age range of 61–82 months (mean: 71.9).

Procedure

The children were tested individually in a quiet room in their school.

Experiment 1

Among other tasks, the children were asked to repeat the utterances /ipi/, /ipi
saw ipi/, and /ipi saw ipi with ipi/ five times each after the examiner had said

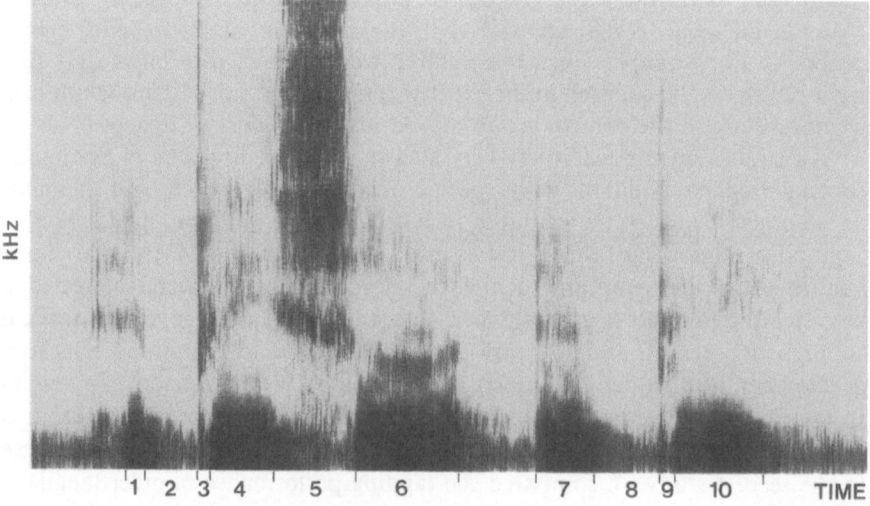

Fig. 1. Sonagram of the utterance |ipi saw ipi| (initial (1,7) and final (4,10)|i|; closure (2,8) and VOT
(3,9) for the |p| in |ipi;|s| (5) and |a| (6) in |saw|)

them (Di Simoni 1974). It seemed important to give the model each time to exclude any memory problems. The utterances were tape recorded with an Uher 4200 tape recorder and a Sennheiser directional microphone. Broad band spectrograms were taken with a Kay-Elemetrics sonagraph. Measurements were made for the utterances /ipi/ and /ipi saw ipi/ only because too many of the children in the experimental group were unable to repeat the longer utterance fluently. The following measurements were made by two raters independently: the length of the initial and final /i/, the closure and voice onset time (VOT) for the /p/ in /ipi/, the duration of the /s/ and /a/ in /saw/, the overall duration of the utterance, and the length of pauses between words. There were 65 measurements for each child. Figure 1 shows the measurements taken for the utterance /ipi saw ipi/.

Interrater agreement for measurements (within 2 mm) was 88%. The correlation was 0.95 (Pearson r). A discriminant analysis (SPSS) was used to determine differences between groups. Since we were interested in the variability of performance, the standard deviation of the duration was entered as the dependent variable.

Experiment 2

We used the finger tapping procedure as described by Wolff and Hurwitz (1976). The children had to tap with their index fingers on two metal plates (18 × 12 cm) mounted about 20 cm apart from each other on a board. The contact between the finger and the plate activated a 100 ms square wave signal that was recorded on a two-track tape recorder for each hand separately. The speed of tapping was given by a metronome, set at 120 beats per minute. The metronome was turned on 5 s before the child began tapping and was left on for another 15 s after the child began tapping. Then it was turned off, but the child continued to tap for another 30 s. The variability of tapping was assessed only for the last 30 s without the metronome, the measure of variability being the standard deviation of the interval between taps. Each child performed four trials, always in the same order: tapping with one hand only, tapping with both hands simultaneously, and tapping with both hands in alternation. An analysis of variance with repeated measures was used to analyze group differences.

Results

Figure 2 shows the scatter plot of the discriminant analysis with two discriminant functions of the speech task. Table 1 gives the details of the discriminant analysis. The stepwise procedure was used with the selection rule to minimize Wilk's lambda. All of the children in the older control group were classified correctly, as were nine of the ten children in the younger control group and eight of the ten children in the experimental group. The scatter plot shows the close vicinity

Table 1. Canonical discriminant functions and classification results for the speech task with two canonical discriminant functions and stepwise inclusion of variables

Canonical discriminant functions

Function	Eigenvalue	Percent of variance	Canonical correlation
1	3.30204	66.55	0.8761005
2	1.66001	33.45	0.7899754

After function	Wilks's lambda	Chi-squared	df	Sign.
0	0.0873862	54.842	20	0.0000
1	0.3759389	22.012	9	0.0088

Classification results

		Predicted group membership		
Actual group	No. of cases	1	2	3
1	10	9	0	1
2	10	0	10	0
3	10	0	2	8

Percent of "grouped" cases correctly classified: 90%

1, younger control group; 2, older control group; 3, group with speech/language disorders

Table 2. Results of the analysis of variance with three repeated measures and group membership as independent variable of the tapping task

Effect	SS	df	MS	F	P
A	523557.6573	2	261778.829	4.473	0.0205
Error 1	1580099.1837	27	58522.1920		
R	627372.7483	2	313686.374	7.985	0.0012
A × R	511290.0267	2	127822.507	3.254	0.0182
Error 2	2121365.8554	54	39284.553		

A, group membership; R, tasks (single hand, bimanual, alternating tapping)

Table 3. The means and standard deviations for the tapping task

Codes	Level	Mean	SD	No.
1	V1	201.4	241.0791	
	V2	151.7	133.8881	
	V3	275.6	206.4834	10
2	V1	47.2	18.9315	
	V2	43.9	21.2679	
	V3	79.9	52.4478	10
3	V1	85.0	39.6260	
	V2	103.6	73.5288	
	V3	491.4	531.0886	10

1, younger control group; 2, older control group; 3, group with speech/language disorders; V1, single hand tapping; V2, bimanual tapping; V3, alternating tapping

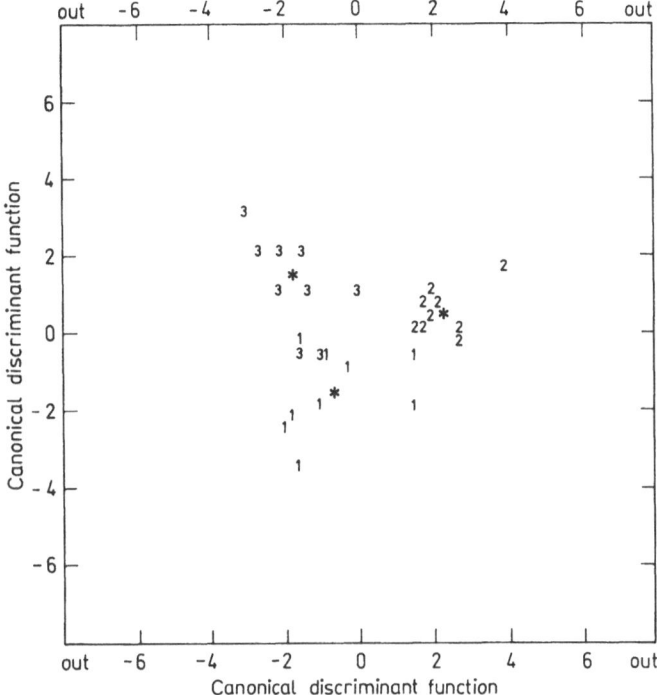

Fig. 2. The scatter plot of the discriminant analysis with the two canonical discriminant functions as the x and y axes. *1*, children from the younger control group; *2*, children from the older control group; *3*, children from the group with speech/language disorders. *indicates a group centroid

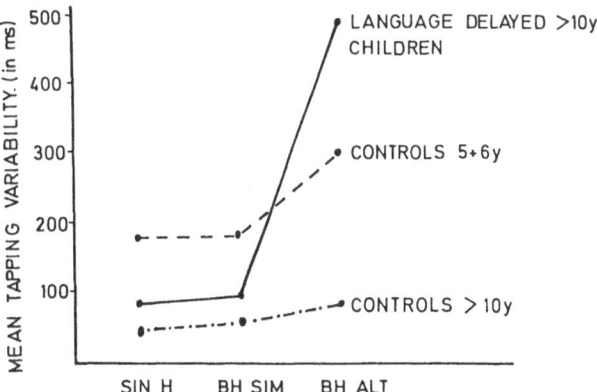

Fig. 3. The mean tapping variability for three conditions and three groups. *SIN H*, single hand trials, left and right hand collapsed; *BH SIM*, both hands simultaneously; *BH ALT*, both hands alternating

within the group of older children with normal speech and language development. The variables entered first in the stepwise selection were the standard deviation of /a/ in /saw/ and the closure duration for /p/ in /ipi/.

Figure 3 shows the results for the tapping tasks, Table 2, the results of the analysis of variance with repeated measures, and Table 3, the group means.

Since there was no significant difference in tapping variability between the preferred and nonpreferred hand, we collapsed these data. The older normal children showed very little difference between the various conditions, whereas the younger children showed much more variability on the alternating tapping task. The children with the speech/language disorders showed even higher variability when tapping alternately with the two index fingers, although they were of the same age as the older control group.

Discussion

The results of our study for both the speaking and the tapping tasks show that children with developmental speech and language disorders differ from children of the same age and from younger children with normal speech and language development in the variability of their timing. The children in the experimental group showed much greater variability on the repetition of short utterances and on tapping alternately with their index fingers.

The findings for the speech task are similar to those of Dames (1986). Dames assessed sentence repetitions with different degrees of linguistic complexity and found increased variability of duration in the children with speech and language disorders for all sentences. Moreover, Amorosa (1989) found higher intraindividual variability in younger children with unintelligible speech when they repeated simple syllables such as /pa/, /ta/, and /ka/.

Our tapping results are comparable to those of Badian and Wolff (1977). These authors found that older children with reading disabilities were much more variable in their tapping on alternating tapping tasks only and did not differ from an age-matched control group on simultaneous tapping of both hands. The results do not support the hypothesis that learning-disabled children are like younger children in their motor performance. Even though the performance of the experimental group was almost as steady as that of the older control group in single hand and simultaneous bimanual trials, their alternating bimanual tapping was much more variable than that of the younger control group.

Our results are comparable to those reported in the literature. Using low variability as an indication of skilled movements they allow the conclusion that even at 10–12 years, children with specific developmental speech and language disorders have problems with fine motor coordination of speech and finger movements. These deficits can be demonstrated in the variability of repetitions of phrases or bimanual alternating movements.

We know from observations of children with speech and language disorders and from the high variability within the group that they are not homogeneous in terms of fine motor coordination. As can be seen in the results of the present study, some of the children were slow and variable even on simple movements, i.e., single hand tapping, but the majority showed deficits only when more complex movements were tested or when automatization was required. This result is interesting in light of our data on the repetition of simple syllables (Amorosa 1989). There too, among those children with speech/language deficits we found some who were as fast as the children in the control group in the repetition of the same syllable. But they were unable to repeat the sequence /pataka/ even at a slow pace. The conclusion can thus be drawn that fine motor control of speech and finger movements is not only less refined in children with specific speech and language disorders but is qualitatively different and heterogeneous in its origin.

Studies by Orgogozo and Larsen (1979) and Roland et al. (1980) have shown that sequential finger movements increase the blood flow in the supplementary motor area, whereas repetitive simple finger movements do not. The premotor cortex also seems to be important for the execution of variable motor sequences such as speech (Wise 1984). Roland et al. (1980) assume that the supplementary motor area is involved in planning motor sequences, while the cerebellum seems important for the execution of learned automatized movements. But we also know that the lateral cerebellum is involved in updating the evolving movements depending on the sensory input from the moving body parts (Evarts 1984). In addition it is involved in motor equivalence, a very important factor in speech and finger movements, which implies that, for example, lip closure for a /p/ can be accomplished with many combinations of jaw, upper lip, and lower lip movements.

These are only a few of the brain areas involved in the execution of movements of the hands and speech. If we assume that one of the areas in the brain just mentioned does not function as well in the children with speech and language problems as in normal children, we would expect their performance to be more variable. The reason for variability would be different for different disorders. For example, a child with an intermediate cerebellum that functions less well could have problems updating movements in response to information from muscles, while if there is a dysfunction in the supplementary motor area one would expect movement planning to be less efficient. At the moment we are unable to decide whether just one or several areas are usually insufficient for skilled complex movements in children with speech/language deficits. But only if we are able to separate the group into more homogeneous subgroups can we analyze whether damage to specific areas could have occurred. For further studies on the functional organization of the brain with the new imaging techniques it will be important to have a better understanding of the different brain areas involved in tasks that seem very similar at first sight, such as bimanual simultaneous and alternating tapping. This may then lead to conclusions about whether behavioral disorders are more likely to occur with certain fine motor symptoms, and allow a more detailed analysis of interconnections between early interference with brain development, psychiatric symptoms, and specific learning disorders.

References

Almli CR (1984) Early brain damage and time course of behavioral dysfunction: parallels with neural maturation. In: Finger S, Almli CR (eds) Early brain damage, vol 2. Neurobiology and behavior. Academic, Orlando, pp 99–116

Amorosa H (1989) Die Untersuchung kindlicher Sprechbewegungsstörungen mit Hilfe der akustischen Analyse. Postdoctoral thesis, University of Munich

Amorosa H (in press) Disorders of vocal signalling in children. In: Papoušek H, Jürgens U, Papoušek M (eds) Origins and development of nonverbal vocal communication: evolutionary, comparative and methodological perspectives. Cambridge University Press, Cambridge

Amorosa H, von Benda U, Wagner E (1986) Die Häufigkeit psychiatrischer Auffälligkeiten bei 4- bis 8-jährigen mit unverständlicher Spontansprache. Z Kinder Jugendpsychiatr 14:289–295

Amorosa H, von Benda U, Wagner E (1990) Voice problems in children with unintelligible speech as indicators of deficits in fine motor coordination. Folia Phoniatr (Basel) pp 64–70

Badian NA, Wolff PH (1977) Manual asymmetries of motor sequencing in boys with reading disability. Cortex 13:343–349

Bäumler C (1989) Untersuchung über Störungen der Feinmotorik bei sprachentwicklungsgestörten Kindern. Doctoral dissertation, University of Munich

Beitchman JH, Nair R, Clegg M, Ferguson B, Patel PG (1986) Prevalence of psychiatric disorders in children with speech and language disorders. J Am Acad Child Psychiatry 25:528–535

Berger F, Amorosa H, Scheimann G (in press) Psychiatrische Auffälligkeiten bei sprachunanfälligen Kindern mit und ohne Minimale Cerebrale Dysfunktion. Z Kinder Jugendpsychiatr

Cantwell DP, Baker L (1985) Psychiatric and learning disorders in children with speech and language disorders: a descriptive analysis. In: Gadow KD (ed) Advances in learning and behavioral disabilities, vol 4. JAI, Greenwich CT, pp 29–47

Dames K (1986) Einfluß der Syntax auf die Zeitstruktur der Nachsprechleistungen sprachentwicklungsgestörter und sprachunauffälliger Kinder. In: Kegel G, Arnhold T, Dahlmeier K, Schmid G, Tischer B (eds) Sprechwissenschaft und Psycholinguistik: Beiträge aus Forschung und Praxis. Westdeutscher, Opladen, pp 145–216

Di Simoni FG (1964) Influence of utterance length upon bilabial closure duration for /p/ in three-, six- and nine-year-old children. J Acoust Soc Am 55:1353–1354

Esser G, Schmidt M (1987) Minimale Cerebrale Dysfunktion—Leerformel order Syndrom? Enke, Stuttgart

Esser G, Lehmkuhl G, Schmidt M (1983) Die Beziehung von Sprechstörungen und sprachlichem Entwicklungsstand zur zerebralen Dysfunktion und psychiatrischen Auffälligkeiten bei 8-jährigen Grundschülern. Sprache Stimme Gehör 2:59–62

Evarts EV (1984) Hierachies and emergent features in motor control. In: Edelman GM, Gall WE, Cowan WM (eds) Dynamic aspects of neocortical function. Wiley, New York, pp 557–579

Gabriel P, Chilla R, Kozielski P (1976a) Zur sprachlichen Entwicklung des Vorschulkindes 1: Artikulationsstörungen und Zungenmotilität. Folia Phoniatr (Basel) 28:17–25

Gabriel P, Chilla R, Kozielski P (1976b) Zur sprachlichen Entwicklung des Vorschulkindes 2: Geschlechtsdifferenzen bei Artikulation und Zungenmotilität. Folia Phoniatr (Basel) 28:26–33

Hamster W (1980) Die motorische Leistungsserie. Manual. Available from the author at the Department of Neuropsychology, Neurological Hospital of the University of Tübingen

Isaacson RL, Spear LP (1984) A new perspective for the interpretation of early brain damage. In: Finger S, Almli CR (eds) Early brain damage, vol 2. Neurobiology and behavior. Academic, Orlando, pp 73–98

Largo RH (1987) Influence of pre-, peri-, and postnatal events on language development during the first five years of life In: Rauh H, Steinhausen H-C (eds) Psychobiology and early development. Elsevier, Amsterdam, pp 171–184 (Advances in Psychology, vol 46)

Lou HC, Henriksen L, Bruhn P (1984) Focal cerebral hypoperfusion in children with dysphasia and/or attention deficit disorder. Arch Neurol 41:825–829

Luchsinger R, Arnold GE (1970) Handbuch der Stimm- und Sprachheilkunde, 3rd edn, vol 2. Die Sprache und ihre Störungen. Springer, Berlin Heidelberg Vienna New York

Martin JAM (1981) Voice, speech, and language in the child: development and disorder. Springer, Berlin Heidelberg Vienna New York (Disorders of human communication, vol 4)

Mayr T (in press) Verhaltensauffälligkeiten bei Vorschulkindern mit unterschiedlich schweren Sprech- und Sprachstörungen-eine epidemiologische Studie. Heilpädagog Forsch

Meyer-Probst B, Teichmann H (1984) Risiken für die Persönlichkeitsentwicklung im Kindesalter. Thieme, Leipzig

Noterdaeme M, Amorosa H, Ploog M, Scheimann G (1988) Quantitative and qualitative aspects of associated movements in children with specific developmental speech and language disorders and in normal pre-school childen. J Hum Move Stud 15:151–169

Orgogozo JM, Larsen B (1979) Activation of the supplementary motor area during voluntary movement in man suggest it works as a supramotor area. Science 206:847–850

Prechtl HFR (1981) The study of neural development as a perspective of clinical problems. In: Connolly KJ, Prechtl HFR (eds) Maturation and development: biological and psychological perspectives. Heinemann, London, pp 198–215 (Clinics in developmental medicine, no 77/78)

Roland PE, Larsen B, Lassen NA, Skinhoj E (1980) Supplementary motor area and other cortical areas in organization of voluntary movements in man. J Neurophysiol 43:118–136

Seeman M (1974) Sprachstörungen bei Kindern. VEB Verlag Volk und Gesundheit, Berlin

Wirth G (1983) Sprachstörungen-sprechstörungen-kindliche Hörstörungen: Lehrbuch für Ärzte, Logopäden und Sprachheilpädagogen, 2nd edn, revised. Deutscher Ärzte-Verlag, Cologne

Wise SP (1984) The nonprimary motor cortex and its role in the cerebral control of movement. In: Edelman GM, Gall WE, Cowan WM (eds) Dynamic aspects of neocortical function. Wiley, New York, pp 525–555

Wolff PH, Hurwitz I (1976) Sex differences in finger tapping: a developmental study. Neuropsychologia 14:35–41

Nonfluent Speech Disturbances— Psychobiological Interactions

H. S. Johannsen and H. Schulze

Introduction

In their chronic form, disorders of speech flow, in particular stuttering [category 307.00 of the international DSM III classification of psychological and psychiatric disturbances (1984)], constitute a disorder of communication which seriously impairs the social and functional integration of the affected person and is frequently associated with a series of long-lasting psychological problems for the person concerned but also for his/her environment.

Nonfluency of speech occurs at some time in almost all children as a developmental variant in preschool age. In the vast majority of these children, it is lost very quickly without faulty social or psychological developments or appreciable impairment of communication. In about 5% of all children, however, this nonfluency of speech is manifested to such an extent that one must refer to stuttering or to transition to chronic stuttering. Even in children diagnosed as stuttering, about 80% experience a spontaneous remission when their development is followed into adolescence. Chronic stuttering persists in about 1% of the adult population.

The theoretical controversy as to whether stuttering in the usual case is to be considered as a manifestation of an intra- or interpersonal psychological disorder or whether different organic/physiological features constitute the core of the problem (Cooper 1977) has shifted in favor of an integrative view against the background of new research data on stuttering in early childhood. Accordingly, stuttering in early childhood must be considered as a superficial phenomenon of various mutually interacting psychobiological factors. The current theoretical state of discussion, which derives from the research results, and the resulting conclusions for the prevention and early treatment of stuttering are described synoptically and discussed below.

State of Research

The fact that stuttering is the disorder of communication on which probably the most publications are available can be considered as an indication that it has so far turned out to be exceedingly difficult to do justice to this multifaceted phenomenon, especially when the levels of clinical presentation of the individual case and theoretical consideration are departed from and research data are exploited to describe and explain the phenomenon. The intensified scientific concern with the genesis and the early developmental phases of stuttering has yielded some important findings which improve our understanding of the disorder, give rise to new research questions, provide fruitful impulses for clinical practice of prevention and early treatment, and justify an interim review of the state of research.

Multimodal Idiographic View of Early Childhood Stuttering

Research into and theoretical discussion of stuttering in former years and decades were dominated by the idea that this disorder is etiologically homogeneous, that it has a common but not yet recognized cause. From this search for the cause of stuttering, there arose a multiplicity of monocausal, mostly mutually competitive, etiological theories: stuttering as a social neurosis, as a manifestation of an intrapsychic conflict, as a class of learned behavior, as a servomechanical defect, etc. However, none of them proved to be generally applicable to the multifaceted phenomenon of stuttering.

Experimental and clinical research on stuttering in the last 40 years involved a large number of studies which attempted to identify the variables on which stuttering subjects differ from nonstuttering subjects. Despite all the contradictions in these research results, which were in some cases due to methodological deficiencies or noncomparability of the methods of investigation, an important trend could be discerned. No characteristics separating the two groups of subjects (apart from quantitative and qualitative aspects of stuttering symptoms) could be found, but there were significant differences in individual subjects and subpopulations from among the stutterers as compared with normal subjects (cf. Adams 1985; Andrews et al. 1983, see below). A logical conclusion from these research results was the abandonment of the conception of stuttering as a homogeneous disorder and instead of this to postulate the existence of subgroups within the overall population of stutterers (e.g. Bloodstein 1987; Conture 1982; Preus 1981; Schwartz and Conture 1988; Van Riper 1982). These and other models constitute a rough orientational context, but are not documented empirically. The longitudinal studies necessary for this purpose are not available. It hence remains very doubtful (as confirmed by clinical practice) whether prognostically valid predictions can be made with regard to the further course of the disorder on the basis of the subgroup allocation, and whether meaningful conclusions can be drawn from this for the choice and establishment of therapy contents and methods. In addition, the subgroups postulated by various researchers show overlaps, so that an exact allocation to one of the subgroups may be impossible in the individual case.

This automatically led to the view (which is generally held today) that stuttering is in principle or at least momentarily only conceivable on the basis of the individual case. This idiographic view regards stuttering as a multifactorial, multimodal phenomenon in which every stutterer acquires his/her disorder individually on the basis of completely different factors (Fig. 1). In the further time course, the significance of individual variables may change, the variables may mutually influence each other, and the stuttering itself may in turn have repercussions for individual variables. Besides the possibility of integrating different and in some cases rival views and positions, a multifactorial and idiographic model for the exploration of stuttering explains the interindividual variability of the disorder better than earlier theories. Similar model concepts have also been developed for stuttering in children by other authors (Cooper and Cooper 1985; Myers and Wall 1982; Shine 1984; Starkweather 1987). The factor groups specified, with their subvariables, constitute hypotheses which

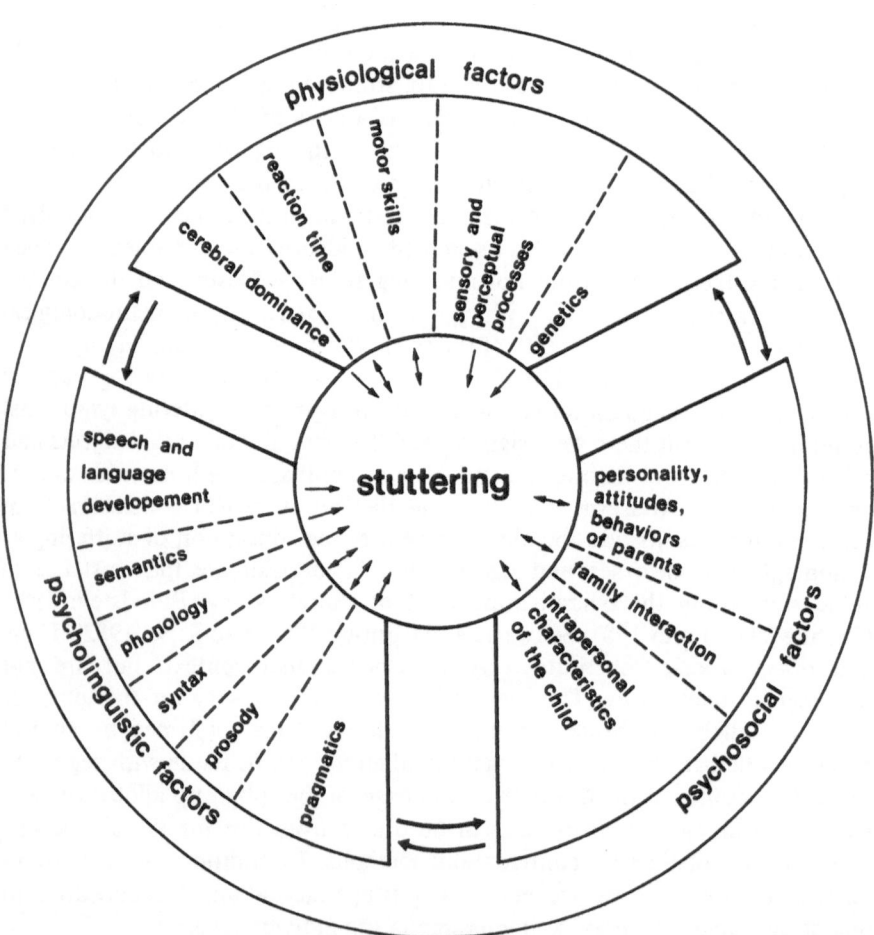

Fig. 1. Descriptive model of the interaction between physiological, linguistic, and psychosocial variables which are important for the onset, the development, and the maintenance of childhood stuttering

were derived from current experimental and clinical research work, and can be regarded as scientifically substantiated. Their significance for the genesis and maintenance of stuttering is also underscored by the investigations of Riley and Riley (1984) in a group of stuttering children, whose results, obtained by factorial analysis, are subsumed in their nine-component model for the development of stuttering in children.

The Moment of Stuttering

The bundle of factors with their subvariables can on their own explain the conditions under which stuttering arises; they offer an account of the onset of stuttering. A second important question with regard to the theory of stuttering consists in explaining the nature of discrete instances of stuttering behavior or the moment of stuttering. The concepts and models available on this are subdivided by Bloodstein (1987) into the breakdown, suppressed need, and anticipatory struggle hypotheses. With regard to the question to be investigated in this book, the model of Zimmermann (1980) is presented below as an example. According to this model, stuttering is to be viewed as a disorder of movement; as an organically orientated model, it serves as representative for other models (see also Moore and Haynes 1980; Kent 1984; MacKay and MacDonald 1984; Yeudall 1985).

Zimmermann assumes that the speech structures vary only within certain limits with regard to their movement parameters and with regard to the chronological and spatial relationships of the organs of articulation. If these limits of variability are exceeded, the brain stem reflexes which participate via feedback impulses also change. This may disturb the system of articulation in its natural and adequate course and thus interfere with fluent speech and lead to a collapse of the system, in the form of either oscillations or standstill. However, it must be critically noted that while the conception of Zimmermann can explain repetitions and blockages very well, it cannot explain so well the parallel occurrence of revisions and interjections that appear to reflect uncertainties in formulating and planning utterances rather than disintegration of movement (Hubbard and Yairi 1988).

Models both on the genesis and maintenance of stuttering and on the moment of stuttering only result in a valid concept, if it is assumed that the specified variability of the movement parameter is influenced by the factors shown in Fig. 1 and that these factors lead to the consequence that the critical variability limits are exceeded.

Stuttering in a Psycholinguistic Model

A rate of speech with an average of 15 phonemes per second and correspondingly fewer syllables or words can only be attained when parts of the speech and speaking process are preprogrammed and take place automatically. In the moment of speaking, the psychomotor performance of the speaker consists more in the content than in sentence structure, in the choice of words, and in articulation. These parts proceeding automatically and in parallel, which are designated as modules, comprise the formulation with grammatical and phonological encoding, articulation and understanding with recognition of sounds, and phonological and grammatical decoding (Levelt 1987). The author regards stuttering as a disorder in the module articulator; a word is completely planned phonologically, but stutterers cannot bring more than the initial consonants across their lips. The program of articulation of the word must be started anew several times before the word can be articulated fluently.

This assumption can explain the different nonfluency types in stuttering, part-word repetitions, and blockages, but not other phenomena such as interjections, word repetitions, phrase repetitions, or dysrhythmic phonation.

Research Results: Physiological Factors

The hope that the overall group of stutterers might be sharply discriminated from fluently speaking persons by some specific results has not so far been fulfilled, apart from the phenomenon of stuttering itself. Thus it was always only subgroups within the stutterers which differed from control subjects with regard to the characteristics investigated. In respect of psycholinguistic and psychosocial factors which play a role in the etiology of stuttering, we can only refer to appropriate synoptic representations (Bloodstein 1987; Schulze and Johannsen 1986; Wall and Myers 1984). However, it must be emphasized that these groups of factors are of equivalent importance in the present view of the genesis and maintenance of stuttering. A few factors from the physiological/organic field will be emphasized. Special features which can only be observed during the stuttering event always entail the difficulty in appraisal as to whether they are the cause or only the consequence of stuttering. A recognizable trend in the method of studying the phenomenon of stuttering is to investigate also the fluent speech of a stuttering person. Since the aforementioned difficulty arises in investigations of older stutterers, children who are as young as possible, and who moreover only stutter for a short time, must be considered in order to obtain results which can be interpreted unequivocally.

Physiological factors have always played a more or less major role in the discussion with regard to the etiology of stuttering. In recent years, they have come increasingly to the fore, because ever more subtle and precise possibilities of investigating such variables have been developed. The objects of such investigations in recent years have been in particular activities of the central nervous system such as cerebral dominance for speech and speaking, reaction time, and sensory and perceptual processes, as well as questions of respiration, phonation, and articulation and their coordination.

The concept of viewing stuttering as a consequence of inadequately pronounced cerebral dominance for speech and speaking was initially formulated as a hypothesis (Stier 1911; Sachs 1924; Orton 1927; Travis 1931), and then investigated with diverse approaches (Wada test, dichotic listening, tachisto-scopy, evoked potentials, regional cerebral blood flow, EEG activity). The overall results obtained with the different methods of investigation can be summarized with the observation that no discriminatory characteristics between stutterers and nonstutterers as overall groups have as a rule been found, although such characteristics have been found for individual stutterers compared with the fluently speaking control group. Individual investigations on stuttering children can also be interpreted in the same terms (Blood and Blood 1984; Cimorell Strong 1983).

In the measurements of reaction time, stutterers' ability to initiate or to terminate phonation and their manual reaction time in response to a visual or auditory signal were at the focus of interest. In the investigations on voice initiation and termination time, it was apparent that stutterers were slower than reference subjects. This was variously interpreted, on the one hand as an indication of a disturbance in the region of the larynx (Rieber and Wollock 1977), then as a manifestation of disturbed coordination of respiration, phonation, and articulation (Till et al. 1983), and finally as a sign of a poorly integrated central nervous system (Wall and Myers 1984). Cullinan and Springer (1980) carried out their investigations in children, and distinguished between those stuttering children who had difficulties in their early speech development and those without such special features. The first group was slower in voice initiation and termination times, while the second differed only slightly, and not signifi-cantly, from the control group. Other investigators (Murphy and Baumgartner 1981) found no differences between stuttering children aged 4–6 years without disturbances of articulation or speech development and fluently speaking control subjects. The slight and not significant prolongation of the voice reaction times of the stuttering children in the investigation of Cullinan and Springer (1980) was underscored even more in older than in younger children, so that the above explanations have to be questioned at the moment and the differences established might possibly also be a consequence of stuttering. For the time being, the investigation results on the manual reaction time do not show any unequivocal direction, since there are roughly the same number of results showing longer reaction times in stutterers as there are not showing any differences between stutterers and nonstutterers.

As was already the case with the method of dichotic listening, differences were found between a subgroup of the stutterers and control subjects which indicate an altered central auditory function in stutterers. Other results confirm this thesis. Stuttering only rarely occurs in the congenitally deaf, and it may disappear on the occurrence of adventitious deafness (Wingate 1970). Alterations (DAF) or abolition (masking white noise) of the acoustic feedback in some cases alter the speech behavior of stutterers dramatically (synopsis and literature in Bloodstein 1987). The theory which has been developed from this (Van Riper 1971, 1982; Fiedler and Standop 1978, 1986)—that the transition from an acoustic feedback to kinesthetic-tactile-proprioceptive channels which takes place in the normal development of speech and speaking does not occur, that competitive feedback with its effects on the coordination of the output signals for the articulation musculature thus initially gives rise to developmental stuttering, and that a chronic course results in additionally unfavorable environmental factors—has not been proved so far.

Variations in speech breathing are so large in fluently speaking children under 7 years that differences in stuttering children can only be established with difficulty. The behavior of the larynx in stuttering and also in fluently spoken speech segments of a stutterer can now be investigated more readily owing to the development of techniques of investigation with high accuracy and speed of registration. In particular, the electroglottographic investigations of Conture (Adams et al. 1985), which show markedly raised tension in the laryngeal musculature and variations in the position of the vocal cords in stuttering as compared with nonstuttering reference children, are to be emphasized, since they were obtained in children. They can confirm in children findings obtained in adults by fiber laryngoscopy (Conture et al. 1985) and by electromyography (Freeman 1977; Freeman and Ushijima 1978).

Differences between stutterers and nonstutterers have also been found for the field of articulation with various techniques. In particular, emphasis should be placed on the study by Stromsta (1965), who, by means of spectrographic investigations of formant transitions, established special features of coarticulation in stuttering children as a discriminating characteristic compared with children with developmental nonfluency. Unfortunately, such a longitudinal study, which is so important for the prediction of the further development of stuttering in nonfluent preschool children, has not been repeated up to now.

The application of models of genetic transmission to the distribution of stuttering in families with stutterers (Kidd et al. 1981; Cox et al. 1984) has led to the well-founded assumption that inheritance plays a role in a proportion of stutterers, but that only a predisposition to stuttering is inherited and that probably, if anything, several genes rather than one alone are responsible for this. Differences in the sex distribution of stuttering are explained in that girls require a greater predisposition in order to become stutterers (Andrews and Harris 1964). However, it is also unequivocal that the inherited predisposition alone does not give rise to stuttering, but that other factors must also be present. This is shown in particular by research on twins (Howie 1981) according to which in twinpairs with at least one stuttering twin, a second dizygotic twin has a 32% risk of stuttering and a second monozygotic twin a 77% risk, whereas siblings of a stutterer have only a 20% risk.

Consequences for Prevention and Early Treatment

Proceeding from a multivariantet idiographic understanding of stuttering in early childhood, it is consistent that greater emphasis be placed on diagnostic and differential diagnostic aspects in order to establish the appropriate dosage and content of therapy or the likely effect of counseling, and thus enable effective intervention as early as possible in the developmental course of the disorder. General guidelines on the indication for therapy, e.g., waiting until the age of school entry has been reached before starting therapy, are regarded as wrong, just as it is wrong to restrict as a matter of principle the content of stuttering treatment in preschool children to the consideration of psychosocial factors.

Differential Diagnostics: Developmental Nonfluency vs Stuttering

The differential diagnostic question as to whether a nonfluently speaking preschool child has only a (temporary) developmental stutter or whether there are indications of a possible chronic course was not raised at all for years or was unimportant because there were no potential consequences. So long as the diagnosogenic and semantogenic theory of the genesis of stuttering of W. Johnson (1942, 1959) was regarded as correct (according to this theory the two groups differ neither in qualitative nor quantitative characteristics of their nonfluency, but solely on the basis of the perception and evaluation of this nonfluency by their parents), the question did not arise. Because stuttering children were not directly treated, because a high rate of remission can be reckoned with even in the children classified as stutterers, and because the development of an awareness of the disorder by the child was feared as a result of therapy, the question was unimportant; the answering of the question with an appropriate allocation of the children to one of the two groups would not have affected the interventional strategies. However, since stuttering infants are increasingly included in therapeutic procedures, differential diagnosis is a prerequisite for all further decisions. Its significance must be seen against the background that the fundamental assumption of the theory of Johnson must be regarded as refuted: in fact, differences are indeed present between children who have only a nonfluent development and those who are already chronic stutterers (Bernstein 1981; Bloodstein 1987; Myers and Wall 1981; Starkweather 1982; Ingham 1984).

Various attempts have been made to answer the differential diagnostic question of the allocation of a nonfluent child to the group of developmentally nonfluent or stuttering children. On the one hand rather simple criterion catalogs and on the other hand more complex instruments are applied. The simpler approaches (a) single out quantitative features of the stuttering symptoms as discrimination criteria, e.g., incidence and frequency of repetitions, duration of vowel lengthening, and blockage (Costello 1983; Rustin 1982), (b) use qualitative characteristics of speech production, e.g., defective coarticulation (Stromsta 1986), or (c) decide on the basis of the presence or absence of certain experimental and attitudinal characteristics in the child, e.g., the awareness of a disorder (Hood 1978; Selmar 1981). As against these simple criterion catalogs, multidimensional approaches predominate which use quantitative and qualitative characteristics of stuttering symptoms for differential diagnosis, but besides this also evaluate concomitant symptoms, duration and course, attitudes of the child and/or the

parents, genetic factors, motor activity, speech development, and visual and
auditory perception in different constellations (Adams 1980, 1984; Cooper and
Cooper 1985; Conture 1982; Culp 1984; Curlee 1980; Gregory and Hill 1980;
Johnson 1980; Riley 1984; Stocker 1980). The value of these differential diagnostic
characteristics catalogs is restricted in that without exception they so far derive
from the clinical experience of the investigator and in that they have not yet been
substantiated by long-term clinical study involving observation of a large
number of children over several years. The objective of all efforts to improve the
differential diagnostic appraisal of the speech nonfluency of a child must be to
reduce wrong decisions with regard to the classification of children as developmentally nonfluent or as stutterers since a whole series of unfavorable consequences may result from erroneous classification (Schulze and Johannsen 1986).

Diagnostics: Preparatory to Therapy

If the probability of chronic stuttering is high in a nonfluently speaking child, or if
chronic stuttering is already present, the contents and therapy methods,
motivational, or organizational conditions for a program of preventional therapy
must be established in a separate diagnostic procedure.

The contents of therapy depend on the individual constellations present
which are important for the genesis, the course, and the current maintenance of
stuttering (Gregory 1986; Wall and Myers 1984). On the one hand, the social,
emotional–affective, cognitive, psycholinguistic, and motor conditions on the
part of the child and constituent abilities which must be present for fluent
speaking on a specific developmental level must be clarified in differential
diagnosis. On the other hand, the demands which the child makes on him/herself
or which are imposed on him/her from the outside by significant reference
persons must be identified (Starkweather 1987). In addition, a detailed analysis of
stuttering behavior and secondary symptoms provides additional indications for
the design of therapy (Cooper and Cooper 1985; Costello and Ingham 1985).
Since several factors are as a rule involved (in accordance with our present-day
understanding of stuttering in early childhood) and since, moreover, these may
alter over time, the diagnostics preparatory to therapy must be broad in scale and
are relatively time consuming (Adams 1984; Graichen 1985; Myers and Wall
1981; Riley and Riley 1984). The existing multivariate theoretical model concepts
(cf. Fig. 1) constitute the theoretical frame of reference for the dimensioning of
diagnostics into which the data obtained can subsequently be rationally
integrated for the purpose of planning therapy. The investigation inventory
essentially consists of psychological and neurological batteries for investigation
of speech, voice attentional behavior, motor activity, feelings, and attitudes as
well as parent–child interactions. Such investigation batteries have been
compiled by various study groups and provide assistance in establishing the
indication for therapy and in laying down the optimal sequence of measures and
the overall therapy design, so that as individual and as problem-oriented a
treatment of the child as possible can take place with inclusion of his/her

significant reference persons. A detailed description of these diagnostic concepts is to be found in Schulze and Johannsen (1986, Chap. C.II).

Intervention Strategy

The multifactorial background of the disorder which is usually present necessitates a multimodal concept of therapy even in very young stutterers. In principle, all therapy methods may be used which have proved to be effective in the treatment of adolescent and adult stutterers (Ham 1986; Ingham 1985; Peins 1984). However, the principles of therapy are in some cases completely different, since they are oriented to the state of development of the child and must consider his/her social, cognitive, motor, linguistic, and emotional–affective basic conditions individually (Culp 1984). The therapeutic approaches which are to be considered in principle can be subdivided into direct and indirect treatment approaches.

Direct treatment of stuttering may consist in changing the speech pattern of the child as a whole or in some aspects, e.g., by alteration of the speaking tempo, the speech melody, breathing, coarticulation, or voice onset (e.g. Cooper 1984, 1985; Culp 1984; Shine 1984, 1985; Stromsta 1986). A further possibility consists in systematically extending the speech components already fluently spoken by methods of operative behavioral modification (in terms of behavioral shaping). At the same time it is part of the principle of the procedure that the linguistic complexity of what is spoken is raised in finely graduated working steps, and that the advances in therapy be maintained by transfer and generalization steps and specific follow-up measures so that the speaking behavior can be stabilized in the long term (Costello 1980; Guitar 1982; Ryan and Van Kirk-Ryan 1983). The parents or other important reference persons frequently take over some therapeutic functions in the transfer and generalization phase of the treatment (Bopp and Schulze 1979; De Vries 1982). Therapeutic approaches which change the pattern of stuttering itself or the immediately preceding and subsequent behaviors, cognitions, and feelings (cf. Ham 1986; Luper and Mulder 1964; Van Riper 1973) tend to constitute the exception in very young children because of their restricted capacity for introspection with regard to currently occurring feelings and cognitions and their restricted metalinguistic abilities.

Indirect therapy is referred to when the interventions are not targeted directly on the behavioral class of stuttering or on the speaking pattern. Some forms of indirect therapy completely dispense with treating the child him/herself and concentrate exclusively on the work with significant reference persons (Schaar 1980; Zwitman 1978), while others also include the stuttering children in the therapy. The objective is to create psychological, physiological, and linguistic conditions on the part of the child so that fluent speech can be developed and stabilized in the long term. Points of approach for such therapies may consist in attentional behavior, auditory perception, motor activity (in particular the motor activity of the mouth), vocabulary, grammar, and pragmatics (synoptic presentation and discussion by Schulze and Johannsen 1986, Chap. II). Furthermore, a focal point of therapy may consist in altering unfavorable general or stuttering-specific attitudes and emotional-affective forms of coping in the child so long as they are relevant to the problem of stuttering. As already indicated, another focal point of the indirect approaches to therapy consists in work with the parents and significant reference persons of the child. This approach, which is the traditional form of intervention in the stuttering preschool child, entails programs of counseling and parent training which attempt to alter behaviors, attitudes, and feelings towards the child and his stuttering.

As shown by current synoptic papers on the prevention and early treatment of stuttering (Luper 1982; Prins 1983; Schulze and Johannsen 1986, 1987; Wall and Myers 1984), an abundance of treatment concepts for young stuttering children and their significant reference persons exist today which allow justice to be done to the individual problem constellations in a specific way. It also becomes evident that the controversy of the 1970s as to whether young children may be treated directly (cf. Cooper 1977) has largely been resolved, at least in English-speaking countries. Many therapists in these countries do not engage in this controversy on the general level; they decide in the individual case and practise mainly multimodal therapeutic approaches, as shown by a comparative review of 26 therapeutic approaches for stuttering preschool children (Schulze and Johannsen 1986, p. 156 ff). The analysis furthermore shows that the fundamental renunciation of direct therapy measures tends to constitute the exception today. The clinical experience available so far, which in some cases comprises a time span of 10 years (Culp 1984; Gregory and Hill 1980), documents that direct therapy of stuttering preschool children is possible and successful when certain rules regarding indication are observed. An unfavorable awareness of the disorder is by no means aroused (as formerly alleged) when the therapist orientates his treatment to the stage of development of the child and considers his/her cognitive, motor, linguistic, and emotional-affective basic preconditions individually (Cooper 1984; Costello 1983; Culp 1984; Guitar 1982).

Research Perspectives

The research data obtained in the last decade in stuttering children have doubtless led to a better understanding of the variables important for the etiology of stuttering and have also had an innovative effect on the overall practice of prevention and therapy via the development of integrative models. However, a critical examination of the research data must also lead to the conclusion that these are frequently only particular findings obtained from cross-sectional studies with a small sample size and often without control group comparison. So far, replication studies have not been able to confirm the findings in every case. Synoptic empirical testing of the single hypotheses derived from these studies and further specification of the problems to be investigated must be carried out in an interdisciplinary research approach and might be implemented in the context of a broad-scale cross-sectional study with control group comparison. Thus, for example, the research data available so far on the significance of psycholinguistic variables support the hypothesis that the overall development of speech, and in particular the development of grammar and semantics, is important for the development of fluent speaking. However, the specific role the psycholinguistic variables play, for example, in connection with cognitive and articulomotor abilities in the development of chronic stuttering or the significance of pragmatic abilities has not yet been adequately clarified.

As stated by many international experts, the most fruitful approach will consist in a methodologically demanding longitudinal study on a large random sample of preschool children with nonfluent speaking to explore the developmental dynamics of stuttering (Gregory 1986; Van Riper 1982). The fundamental scientific relevance of such an approach consists in the possibility of obtaining specific data on the developmental characteristics depending on different physiological, psychosocial, and psycholinguistic variables using the most modern methods of investigation in child stuttering (cf. Conture 1987) and weighting the significance of the variables for the development of stuttering. The clinical relevance of such an approach consists in obtaining reliable and valid prognosis parameters to determine the probability of a chronic course of stuttering or spontaneous remission and thus be able to establish the indication for therapy more reliably. In addition, the rules of differential indication for the content of therapy might be rendered more precise and a systematic evaluation of direct and indirect therapy methods could be carried out.

Exploration of the processes which accompany the development of normal fluent speaking is also regarded as an important approach in basic research today (cf. Starkweather 1980, 1987) from which further important components in the understanding of pathological nonfluent speaking are to be expected.

References

Adams MR (1980) The young stutterer: diagnosis, treatment and assessment of progress. Semin Speech Lang Hear 4:289–298
Adams MR (1984) The differential assessment and direct treatment of stuttering. In: Costello J (ed) Speech disorders in children. College Hill Press, San Diego, pp 261–290
Adams MR (1985) The speech physiology of stutterers: present status. Semin Speech Lang 6:177–190
Adams MR, Freeman FJ, Conture, EG (1985) Laryngeal dynamics of stutterers. In: Curlee RF, Perkins WH (eds) Nature and treatment of stuttering: new directions. Taylor and Francis, London, pp 89–129
Andrews G, Harris M (1964) The syndrome of stuttering. Spastics Society Medical Education and Information, London
Andrews G, Craig A, Feyer M-M, Hoddinott S, Howie P, Neilson M (1983) Stuttering: a review of research findings and theory circa 1982. J Speech Hearing Disord 48:226–246
Bernstein NE (1981) Are there constraints on childhood disfluency? J Fluency Disord 6:341–350
Blood GW, Blood IM (1984) Central auditory function in young stutterers. Percep Mot Skills 59:699–705

Bloodstein O (1987) A handbook on stuttering, 4th edn. National Easter Seal Society, Chicago
Bopp W, Schulze H (1979) Einbeziehung von Eltern als Kotherapeuten in die Verhaltensmodifikation des Stotterns bei Grundschülern. In: Wendlandt W (ed) Verhaltenstherapeutische Gruppenprogramme in der pädagogischen Praxis. Schwann, Düsseldorf, pp 97–126
Cimorell Strong JM (1983) Language facilitation. A complete cognitive therapy program. University Park Press, Baltimore
Conture EG (1982) Stuttering. Prentice Hall, Englewood Cliffs
Conture EG (1987) Studying young stutterers' speech productions: a procedural challenge. In: Peters HFM and Hulstijn W (eds) Speech motor dynamics in stuttering. Springer, Berlin Heidelberg Vienna New York, pp 117–140
Conture EG, Schwartz HD, Brewer DW (1985) Laryngeal behavior during stuttering: a further study. J Speech Hear Res 28:233–240
Cooper EB (1977) Controversies about stuttering therapy. J Fluency Disord 2:75–86
Cooper EB (1984) Personalized fluency control therapy: a status report. In: Peins M (ed) Contemporary approaches in stuttering therapy. Little Brown, Boston, pp 1–37
Cooper EB (1985) The Cooper personalized fluency control therapy revised (PFC-R). Workshop at the Oxford dysfluency conference, Oxford
Cooper EB, Cooper CS (1985) Cooper personalized fluency control therapy revised. DLM Teaching Resources, Allen
Costello, JM (1980) Operant conditioning and the treatment of stuttering. Semin Speech Lang Hear 1:311–325
Costello JM (1983) Current behavioral treatments for children. In: Prins D, Ingham RJ (eds) Treatment of stuttering in early childhood. Methods and issues. College Hill Press, San Diego, pp 69–112
Costello JM, Ingham RJ (1985) Assessment strategies for stuttering. In: Curlee RF, Perkins WH (eds) Nature and treatment of stuttering. Taylor and Francis, London, pp 303–334
Cox, NJ, Kramer PL, Kidd KK (1984) Segregation analyses of stuttering. Genet Epidemiol 1:245–253
Cullinan WL, Springer MT (1980) Voice initiation times in stuttering and nonstuttering children. J Speech Hear Res 23:344–360
Culp DM (1984) The preschool fluency development program: assessment and treatment. In: Peins M (ed) Contemporary approaches in stuttering therapy. Little Brown, Boston, pp 39–72
Curlee RF (1980) A case selection strategy for young disfluent children. Semin Speech Lang Hear 1:277–287
De Vries U (1982) Behandlung von stotternden Vorschulkindern mit Einbeziehung der Eltern. In: Prävention und Sprachbehinderung. Fachtagung der Arbeiterwohlfahrt, 5–7 Nov 1981, Oldenburg, AWO-Bezirksverband Weser-Ems, Oldenburg, pp 157–169
Fiedler P, Standop R (1978) Stottern: Wege zur integrativen Theorie und Behandlung. Urban and Schwarzenberg, Munich
Fiedler P, Standop R (1986) Stottern. Ätiologie, Diagnose, Behandlung. Urban and Schwarzenberg, Munich
Freeman FJ (1977) The stuttering larynx. An electromyographic study of laryngeal muscle activity accompanying stuttering. City University, New York
Freeman FJ, Ushijima T (1978) Laryngeal muscle activity during stuttering. J Speech Hear Res 21:538–562
Graichen J (1985) Organismische Fehlregulation als direkte Ursache von Redeflußstörungen (Stottern) in neurologischer Differentialdiagnostik. Sprache Stimme Gehör 9:34–40
Gregory HH (1986) The problem of stuttering: where are we in 1986? S Afr J Commun Disord 33:3–7
Gregory HH, Hill D (1980) Stuttering therapy for children. Semin Speech Ling Hear 1:351–363
Guitar, B (1982) Fluency shaping with young stutterers. J Child Commun Disord 6:50–59
Ham R (1986) Techniques of stuttering therapy. Prentice Hall, Englewood Cliffs
Hood, SB (1978) The assessment of fluency disorders. In: Singh S, Lynch J (eds) Diagnostic procedures in hearing, language and speech. University Park Press, Baltimore, pp 529–632
Howie PM (1981) Intrapair similarity in frequency of disfluency in monozygotic and dizygotic twin pairs containing stutterers. Behav Genet 11:227–237
Hubbard CP, Yairi E (1988) Clustering of disfluencies in the speech of stuttering and nonstuttering preschool children. J Speech Hear Res 31:228–233
Ingham RJ (1984) Stuttering and behavior therapy: current status and experimental foundations. College Hill Press, San Diego

Ingham RJ (1985) Stuttering treatment outcome evaluation: closing the credibility gap. Semin Speech Lang 6:105–123

Johnson LJ (1980) Facilitating parental involvement in therapy of the disfluent child. Semin Speech Lang Hear 1:301–309

Johnson W (1942) A study of the onset and development of stuttering. J Speech Disord 7:251–257

Johnson W (1959) The onset of stuttering. University of Minnesota Press, Minneapolis

Kent RD (1984) Stuttering as a temporal programming disorder. In: Curlee RF, Perkins WH (eds) Nature and treatment of stuttering: new directions. College Hill Press, San Diego, pp 283–301

Kidd KK, Heimbuch RC, Records MA (1981) Vertical transmission of susceptibility to stuttering with sex-modified expression. Proc Natl Acad Sci USA 78:606–610

Levelt WJM (1987) Hochleistung in Millisekunden—Sprechen und Sprache verstehen. In: Jahrbuch 1987 der Max-Planck-Gesellschaft. pp 61–77

Luper HL (1982) Intervention with young stutterers. J Child Commun Disord 6 (1):3–4

Luper HL, Mulder RL (1964) Stuttering: therapy for children. Prentice Hall, Englewood Cliffs

MacKay DG, MacDonald M (1984) Stuttering as a sequencing and timing disorder. In: Curlee RF, Perkins WH (eds) Nature and treatment of stuttering: new directions. College Hill Press, San Diego, pp 261–282

Moore WH, Haynes WO (1980) Alpha hemispheric asymmetry and stuttering. Some support for a segmentation dysfunction hypothesis. J Speech Hear Res 23:229–247

Murphy M, Baumgartner JM (1981) Voice initiation and termination time in stuttering and nonstuttering children. J Fluency Disord 6:257–264

Myers FL, Wall MJ (1981) Issues to consider in the differential diagnosis of normal childhood nonfluencies and stuttering. J Fluency Disord 6:189–195

Myers FL, Wall MJ (1982) Toward an integrated approach to early childhood stuttering. J Fluency Disord 7:47–54

Orton ST (1927) Studies on stuttering. Arch Neurol Psychiatry 18:671–672

Peins M (ed) (1984) Contemporary approaches in stuttering therapy. Little Brown, Boston

Preus A (1981) Identifying subgroups of stutterers. Universitetsforlaget, Oslo

Prins D (1983) Continuity, fragmentation and tension: hypotheses applied to evaluation and intervention with preschool disfluent children. In: Prins D, Ingham RJ (eds) Treatment of stuttering in early childhood. Methods and issues. College Hill Press, San Diego, pp 41–42

Rieber RW, Wollock J (1977) The historical roots of the theory and therapy of stuttering. J Commun Disord 10:3–24

Riley GD (1984) Stuttering prediction instrument for young children. CC Publications, Tigard

Riley GD, Riley JA (1984) Component model for treating stuttering in children. In: Peins M (ed) Contemporary approaches in stuttering therapy. Little Brown, Boston, pp 123–172

Rustin L (1982) Early intervention in the treatment of stuttering. North Irel Speech Lang Forum 8:7–14

Ryan BP, Van Kirk-Ryan B (1983) Programmed stuttering therapy for children: comparison of four establishment programs. J Fluency Disord 8:291–321

Sachs MW (1924) Zur Ätiologie des Stotterns. Klin Wochenschr 36:113–115

Schaar E (1980) Zur Frühbehandlung des Stotterns—Ein Trainingsprogramm unter besonderer Berücksichtigung neuropsychologischer Aspekte. University of Würzburg

Schulze H, Johannsen HS (1986) Stottern bei Kindern im Vorschulalter. Theorie, Diagnostik, Therapie. Phoniatrische Ambulanz der Universität Ulm. University of Ulm

Schulze H, Johannsen HS (1987) Therapie des Stotterns bei Kindern im Vorschulalter—Zum Stand der Diskussion in der anglo-amerikanischen Fachliteratur. Sprachheilarbeit 32:97–108

Schwartz HD, Conture EG (1988) Subgrouping young stutterers: preliminary behavioral observations. J Speech Hear Res 31:62–71

Selmar JW (1981) The early identification and treatment of children with problems of non-fluency. Bull Coll Speech Ther

Shine RE (1984) Assessment and fluency training with the young stutterer. In: Peins M (ed) Contemporary approaches in stuttering therapy. Little Brown, Boston, pp 173–216

Shine RE (1985) Systematic fluency training for young children, revised edn. CC Publications, Tigard

Starkweather CW (1980) A multiprocess behavioral approach to stuttering therapy. Semin Speech Lang Hear 1:327–338

Starkweather CW (1982) Stuttering and laryngeal behavior: a review. ASHA monographs no 21. American Speech and Hearing Association, Rockville

Starkweather CW (1987) Fluency and stuttering. Prentice Hall, Englewood Cliffs

Stier E (1911) Untersuchung der Linkshändigkeit und die funktionellen Differenzen der Hirnhälften. Fischer, Jena

Stocker B (1980) The Stocker probe technique for diagnosis and treatment of stuttering in young children, revised edn. Modern Education Corporation, Tulsa

Stromsta C (1965) A spectographic study of dysfluencies labeled as stuttering by parents. Ther Vocis Loquellae 1:317–320

Stromsta C (1986) Elements of stuttering. Atsmorts, Oshtemo

Till JA, Reich A, Dickey S, Sieber J (1983) Phonatory and manual reaction times of stuttering and nonstuttering children. J Speech Hear Res 26:171–180

Travis LE (1931) Speech pathology. Appleton-Century-Crofts, New York

Van Riper C (1971) The nature of stuttering. Prentice Hall, Englewood Cliffs

Van Riper C (1973) The treatment of stuttering. Prentice Hall, Englewood Cliffs

Van Riper C (1982) The nature of stuttering 2nd edn Prentice Hall, Englewood Cliffs

Wall MJ, Myers FL (1984) Clinical management of childhood stuttering. University Park Press, Baltimore

Wingate, ME (1970) Effect on stuttering of changes in audition. J Speech Hear Res 13:861–873

Yeudall LT (1985) A neuropsychological theory of stuttering. Semin Speech Lang 6:197–223

Zimmermann G (1980) Stuttering: a disorder of movement. J Speech and Hear Res 25:122–136

Zwitman DH (1978) The disfluent child. University Park Press, Baltimore

Neurological Basis of Developmental Language Disorders

P. Tallal and S. Curtiss

In attempting to gain a better understanding of the pathogenesis of developmental dysphasia, research has focused on the associated characteristics of language-disordered children. These studies initially focused on the most obvious surface symptomatology, that is the linguistic dysfunction of these children. Subsequent studies have focused on nonlinguistic characteristics which commonly are associated with or accompany the language disorder, such as neuropsychological (perceptual, motor, memory, cognitive) dysfunction or social and emotional disturbance.

Language development requires the integration of sensory, attention, perception, cognitive, motor, linguistic, social, and emotional functions. When one or more of these fails to develop normally, language development may be delayed or disordered. Language disorder is often a symptom of other primary impairments, like mental retardation, hearing loss, autism, or brain lesion. Unfortunately, language disorders that are secondary symptoms of these disorders are often merged together with primary or specific developmental language disorders based on surface symptomatology rather than underlying etiology or common mechanisms of action. Considerable inconsistency in the research literature has resulted from a failure to adopt a uniform definition of the disorder or apply consistent diagnostic criteria. Using for a moment a medical model, advancement of knowledge in a field is often evidenced by progression from definitions based primarily on surface symptomatology to ones based on understanding of etiology and mechanism of action, with improved treatment following such advance.

In this chapter, we will focus only on studies that have included children with specific or primary developmental language disorders. These children will be referred to as developmentally dysphasic or specifically language-impaired.

Research into specific developmental language impairments has focused on investigating the associated characteristics of these children and classifying them according to these characteristics into distinct subgroups. In this regard, two main approaches have been taken: a linguistic approach and a neuropsychological approach. In reviewing the research literature, several principal findings have emerged from studies following a linguistic approach.

Although language-impaired children are delayed in comparison to normally developing children in the acquisition of phonology, morphology, semantics, and syntax, few examples of frank deviance have been reported in each of these areas. That is, these children rarely produce utterances that are not characteristic of human grammar-based systems. Detailed linguistic analyses have demonstrated that, on the whole, what may look like an aberrant and unprincipled system on the surface, can be revealed to be the output of normal rules. In this case the important distinction must be made between the acquisition of language-particular facts, of English for example, and what is a possible linguistic rule. This is tremendously important when it comes to determining whether language-impaired children are deviant in their ability to acquire a grammatically, rule-governed linguistic system or rather delayed in acquiring such a system with, perhaps, particular difficulties acquiring language specific rules (in our case, English). Research on the whole suggests that it is the latter which is the case, i.e., that language-impaired children are delayed, not deviant, in language acquisition.

However, results from our San Diego longitudinal study (Tallal and Curtiss 1979–1988) have shown that this may not entirely be the case. Whereas each component of the language (i.e., phonology, morphology, semantics, syntax, pragmatics) may follow this pattern individually, the interrelationship between these components may be deviant, especially in terms of order of acquisition of linguistic structure. That is, early acquired forms usually are replaced, in normal children, by new forms that more and more closely approximate the adult target form. As a new form emerges, the older form is deleted. This does not seem to be the case of language-impaired children. Rather, early acquired forms persist, or coexist, in free variation with later acquired forms over protracted periods of time. The acquisition of new grammatical forms presumably reflects changes in the child's grammar that should rule out the old forms. The coexistence of distinct and competing representations of the same form in language-impaired children is a disturbing finding which potentially marks the language development of these children as deviant. Free variation of competing forms may, however, reflect specific production difficulties rather than differences in grammar, per se. This area remains a critical one to investigate further.

Another consistent finding emerging from our linguistic studies of language-impaired children is that not all of the children perform similarly. Three subgroups have been widely recognized: (a) receptive impairment, (b) expressive impairment, and (c) global impairment (receptive and expressive). Key questions regarding subgroups have emerged in the literature over the past decade:

1. Do language-impaired children fall into distinct subgroups based on non-linguistic profiles?
2. Are classical subgroups meaningful linguistically?
3. Are performance profiles which define these subgroups consistent over time, i.e., are they clinically meaningful?

Attempts to define subgroups along linguistic dimensions have been quite inconsistent from study to study, possibly because few of these studies assessed a broad enough range of abilities in the same child and few were longitudinal. Thus, one of the major focuses of our longitudinal study is on questions and hypotheses pertaining to subgroupings. Some surprising results have emerged from fine-grained linguistic analyses.

Unexpectedly, the pattern of performance on formal tests both within and across language domains was the same for all language-impaired children, regardless of their subgroups. However, the receptively impaired children performed more poorly than the expressively impaired children on every linguistic measure, regardless of the structural linguistic parameter assessed. This quantitatively, but not qualitatively, different pattern of performance suggests that the impairments of these subgroups may be task dependent, that it may be in an area other than linguistic knowledge, per se. What does seem to characterize receptively vs expressively impaired children as different is the ability to access their knowledge of language under different tasks or processing demands. While receptively impaired children are more impaired using the language they know in structured tasks, and may perform much better in spontaneous speech, the opposite is true for expressively impaired children. Finally, in terms of outcome or prognosis, these classical subgroups are very robust. It is clear from the San Diego longitudinal study that, in terms of language outcomes, preschool-age children who have primarily an expressive language deficit do far better, long-term, than those with primarily receptive or receptive and expressive deficits. So these subgroups are clinically meaningful and consistent, but perhaps not in the way we have assumed they were. They demonstrate strong processing differences, differences between competence and performance in different settings, and they predict outcome.

These unexpected data, based on comprehensive linguistic analyses, implicating processing rather than linguistic deficits in language-impaired children, make it imperative that we take even more seriously hypotheses pertaining to a neuropsychological basis rather than a primarily linguistic basis for developmental language disorders.

Neuropsychological Studies of Developmental Language Disorders

Neuropsychological studies of language impairment in children have focused primarily on investigating nonverbal perceptual, memory, and motor development, and relating these aspects of development to the receptive and expressive verbal abilities of these children. It is clear from reviewing the literature that language-impaired children, as a group, are impaired on a variety of tasks which comprise temporal components (see Tallal et al. 1985a,b for review). Interestingly, impairments on neuropsychological tasks comprising timed components occur for perceptual, memory, and motor functioning, but do not occur for similar, nontimed tasks. For example, Tallal and Piercy (1973a,b, 1974) demonstrated that a well-selected group of specifically language-impaired children (without concomitant hearing loss, mental retardation, autism, or frank neurological signs, such as seizure disorder or hemiplegia) were unimpaired on a large series of nonverbal, auditory, and visual perceptual and memory tasks, provided that perceptual and memory information was presented over a protracted period of time. On the other hand, these same children were significantly impaired on the same nonverbal tasks when information to be processed and stored was presented more rapidly. In fact, language-impaired children required durations of presentation of basic sensory information orders of magnitude greater than did matched, normal control children. For example, whereas normally developing children (between the ages of 6 and 9 years) required 8 ms between two 75-ms nonverbal tones to discriminate them as the same or different 75% correctly, language-impaired children required more than 300 ms to respond at the same level of accuracy. A similar deficit was also observed in serial memory performance. Whereas these children were unable to remember a series of even three successive tones of 75-ms duration, the same children could remember correctly a series of five successive tones when the duration of presentation of the tones was increased to 250 ms.

Similar temporal deficits in nonverbal motor output also have been reported for language-impaired children. These children consistently require more time to correctly produce series of fine motor movements than do normal, matched controls (Tallal et al. 1985b). That is, on timed motor tasks, language-impaired children perform significantly more slowly than normal children. Increased processing time for language-impaired children has been reported for a wide variety of other perceptual, motor, and cognitive task. For example, Johnston and Weismer (1983) reported that, although language-impaired children did not differ from normal children on a task of mental rotation based on the number of accurate responses, they required consistently longer processing times than did normal children to respond correctly.

There is now considerable evidence that children with specific developmental language disorders are characterized by a slowed rate of sensory information processing and motor organization. What we next want to determine is what might be the outcome of such a temporal disorder on the developing language system. Tallal and colleagues have evaluated the interrelationship between nonverbal, temporal perception and production deficits and speech and language impairments in children. These studies have demonstrated a highly significant relationship between the degree of temporal processing deficits and (a) the pattern of speech perception and production deficits, and (b) the degree of receptive language impairment in language-impaired children. In addition to having difficulty discriminating, sequencing, and remembering rapidly presented, nonverbal stimuli, language-impaired children have been shown to have most difficulty discriminating between those speech sounds (such as/ba/, /da/, /ga/, /pa/, /ta/, and /ka/) that are characterized by brief temporal cues (Tallal and Piercy 1975; Tallal and Stark 1981). These same results are mirrored in speech production, with language-impaired children having most difficulty controlling brief temporal cues in speech output (Tallal et al. 1976; Stark and Tallal 1979). Importantly, this pattern of speech perception and production deficit is not compatible with a developmental delay hypothesis, as these speech sounds are among the first to be learned by normally developing children, as any proud "dada" can attest. The specificity of this result has been demonstrated using computer-synthesized speech, in which precise control over the acoustic spectra of individual speech sounds can be maintained. Tallal and Piercy (1975) demonstrated that language-impaired children who were impaired in discriminating between speech syllables such as /ba/ vs /da/, which incorporated very brief 40-ms duration acoustic transition, were unimpaired in discriminating between the same speech sounds when the duration of the critical acoustic transitions was increased to 80 ms.

In more recent studies (Tallal et al. 1985a), we have asked to what extent these basic perceptual/motor integration deficits predict the degree of language impairment in language-impaired children. In order to evaluate this question, we assessed the receptive language abilities of a large group of well-defined, language-impaired children in considerable detail. Children were rank ordered according to their degree of receptive language impairment, based on standardized language tests. Next, the same children were rank ordered, based on their scores on a series of nonverbal and speech temporal processing tests. The question we addressed was how similar were these rank orderings. That is, to what extent would language-impaired children's temporal processing deficits predict their receptive language deficits. The results demonstrated that a very similar and highly significant ($P = < 0.001$) rank ordering occurred between the two sets of variables, yielding a multiple correlation coefficient of 0.85. These results showed that 72% of the variance related to the level of receptive language dysfunction in these language-impaired children could be accounted for by their ability to discriminate and sequence nonverbal acoustic tones and speech stimuli that were characterized by brief duration temporal cues.

Finally, we have asked to what extent could specific neuropsychological abilities, alone, discriminate language-impaired children from normal children. More specifically, we set about determining whether there is a specific pattern of neuropsychological deficits that is characteristic of developmentally language-impaired children and, if so, to what extent this pattern of deficits, alone, could correctly classify children as normal or language-impaired. To address this question, a large neuropsychological battery of nonverbal and speech perception, motor, and memory tests was given to language-impaired and matched control subjects. This neuropsychological test battery incorporated auditory, visual, tactile, and cross-modal, nonverbal, and verbal stimuli. The perceptual test battery included the assessment of detection, association, temporal resolution, sequencing, rate processing, and serial memory skills. In addition, nonverbal and verbal motor tests evaluated rate of production and sequencing abilities. A comprehensive neurodevelopmental "soft sign" battery measured general motor control and coordination, balance and station, tactile sensation and perception, and laterality. Demographic and case history variables were also documented. A detailed description of these procedures has been given in previous publications (Tallal 1980; Stark and Tallal 1980; Johnston et al. 1981 and results are reported in full in a recently published book, Stark and Tallal, 1988).

A discriminant function analysis was used to examine the ability of independent variables to assign individuals appropriately to the defined outcome groups, language-impaired and normal, in the presence of all other variables. In a stepwise fashion, this analysis chooses the order in which the variables add information that improves the ability to discriminate subjects as language-impaired or normal. The result is a ranking of variables in the order in which they add to the ability to predict group membership. The results of the discriminant function analysis demonstrated that six variables assessing basic perceptual and motor abilities, when combined, correctly classified 100% of the normal children as normal and 96% of the language-impaired children as language-impaired. The six variables that entered the discriminant function equation, correctly discriminating language-impaired from normal subjects, were found to have in common the assessment of specific temporal capabilities, either in perception or production. Specifically, the variables had in common deficits in either perceiving or producing very basic sensory information presented to the nervous system quickly in time, regardless of the modality of presentation or whether information to be processed or produced was verbal or nonverbal. The six variables included rapid syllable production, two measures of two-point simultaneous tactile discrimination, discriminating between speech syllables characterized by brief format transitions, cross-modal integration at rapid rates of presentation, and sequencing the letters 'e' and 'k' when presented rapidly in succession. Importantly, the same tasks given at slower presentation rates did not enter the discriminant function equation discriminating language-impaired from normal children. Each of the variables entering the equation, which so accurately discriminated language-impaired from normal children, had in common the necessity to produce or perceive information either simultaneously or very rapidly in succession, regardless of whether the information was verbal or nonverbal.

Studies Linking Developmental Dysphasia and Dyslexia

It is of considerable interest that research on developmental dyslexia, another developmental communication disorder, has suggested a specific phonological basis for the disorder (Liberman et al. 1980). As language-impaired children also demonstrate serious phonological deficits, and many dyslexic children have a history of delayed language development, we might ask whether dyslexic children show similar deficits to those described for language-impaired children. In a series of studies with dyslexic children (Tallal 1980a, b; Tallal and Stark 1982), the temporal processing abilities of developmentally dyslexic children were evaluated. The results of these studies demonstrated that, within the dyslexic population, there are two subgroups. The larger subgroup demonstrated temporal processing deficits similar to those reported for language-impaired children, whereas the smaller subgroup did not. Interestingly, these subgroups were also differentiated in two other important ways. First, the large subgroup, which did have temporal processing deficits, also demonstrated deficits on standardized speech and language tests, whereas the smaller subgroup responded within normal limits. Secondly, the larger subgroup, with concomitant language deficits and temporal processing deficits, also demonstrated considerable deficits on phonetic decoding tests, whereas the other subgroup of dyslexic children did not. A further analysis demonstrated that there was a highly significant ($P < 0.001$) correlation ($r = 0.85$) between the degree of nonverbal acoustic temporal processing deficits and phonetic decoding deficits in dyslexic children. Thus, the basis of phonological decoding deficits in dyslexic children may be an inability to process temporal acoustic cues which are necessary for the fine-grained phonological analysis which underlie decoding skills.

Another approach to looking at similarities between developmental reading impairment (dyslexia) and language impairment (dysphasia) may be to follow language-impaired children from an early age to see what the outcomes are in terms of their academic achievement. This is one of the major purposes of the San Diego longitudinal study (Tallal and Curtiss 1979–1988). Since 1979, we have been following longitudinally the development of 100 specifically language-impaired children and 90 matched controls from ages 2 through 8 years. We are evaluating the outcomes of preschool language impairment on subsequent linguistic development, academic achievement, and social and emotional development. Preliminary results of this longitudinal study have shown that, by age 6, more than 75% of the children who were originally identified at age 4 as seriously language-impaired, could be correctly discriminated from matched control children based solely on their spelling scores. More than 85% of these children could be correctly identified from controls based solely on their reading vocabulary and reading comprehension scores. This longitudinal study demonstrates the co-occurrence of developmental language disorder and developmental reading disorder in the same child, only a different ages. Furthermore, our data implicate a similar underlying temporal processing deficit in both developmental dysphasia and developmental dyslexia, at least for some subtypes.

Thus, some developmentally language-impaired and developmentally reading-impaired children may not represent two distinct diagnostic groups. Rather, a single developmental disability affecting specific processing constraints on specific aspects of the language learning system may occur at different ages. It is possible that a similar underlying neuropsychological deficit may interfere with phonological analysis, resulting in speech and language disorders in young preschool-age children, and subsequently reading development in older school-age children.

Neurological Basis of Developmental Dysphasia/Dyslexia

The linguistic and neuropsychological research studies that have been reported for language-impaired children, together suggest a neurological basis for the disorder. However, few actual anatomical or physiological studies of developmental language or reading disorders have been reported in the literature that would relate to hypotheses pertaining to neural mechanisms for developmental language disorders. Only one child with a specific developmental language disorder has ever come to autopsy (Goldstein et al. 1958). This child had normal hearing, a performance IQ of 97, and a verbal IQ of 76. By 9 years of age, he had acquired some expressive language but still had considerable difficulty processing speech when it was spoken at a normal rate. His comprehension was reported to be improved when spoken to slowly. At autopsy old, bilateral infarctions of the Sylvian region and severe retrograde degeneration of both medial geniculate nuclei were found.

More recently Galaburda and Kemper (1979) have reported the results of autopsy studies on the brains of adults with a lifelong history of language and reading disorders. At autopsy these brains showed signs of neuroanatomical abnormalities, all confined to the left hemisphere. There was no evidence of neuron loss or gliosis. However, there was a striking area of polymicrogyria, with adjacent molecular layers of the abnormal gyri fused, no normal cortical lamination, and no cell-free layer. These abnormalities were confined to the posterior parts of Heschl's gyrus into the left planum temporale, an area corresponding roughly to the auditory association region known as Wernicke's area. Remaining auditory fields were relatively unaffected. The cause of this type of malformation is unknown, but family history reports suggest a possible genetic component. Galaburda and Kemper speculate that a familial form of localized polymicrogyria may be responsible for the language impairments of these patients.

Recently Jernigan et al. (1987) have reported preliminary results from magnetic resonance imaging studies of language-impaired and control children. Whereas all five of the control children scanned had a larger left than right posterior region of the brain, six of the ten language-impaired children had a larger right than left posterior region. These six children also had a decrease in gray matter and an increase in fluid in these brain regions.

There is a tremendous need for anatomical, physiological, and functional studies of the normally developing brain at different stages of development, especially with regard to cytoarchitectonic studies of the organization and timing of development of each area. The study of the brains of dysphasic children and adults through the use of new technological advances, such as magnetic resonance imaging, electrophysiological recording, and positron emission tomography, coupled with detailed linguistic and neuropsychological assessment, is the focus of the San Diego Center for Neurodevelopmental Studies (Tallal et al. 1985–1990). There is a continuing need for more longitudinal studies of normal and abnormal brain development. It is very difficult to answer the most pressing questions about developmental disorders based on studies that cannot assess development over time. In addition, it is increasingly important for future research to gather multidimensional, multidisciplinary data from the same child: behavioral, neurophysiological, anatomical, and biochemical, as well as complete medical, social, and genetic family histories. It may be important to develop new cross-geographical collaborations to facilitate the study of a large enough sample to justify the multivariate analysis that will be needed to analyze data from such studies.

References

Galaburda AM, Kemper TL (1979) Cytoarchitectonic abnormalities in developmental dyslexia: a case study. Ann Neurol 6:94

Goldstein R, Landau WH, Kleffner FR (1958) Neurological assessment of some deaf and aphasic children. Trans Am Otol Soc 46:122–136

Jernigan TL, Hesselink J, Tallal P (1987) Cerebral morphology on magnetic resonance imaging in developmental dysphasia. Soc Neurosci Abstr 13:1–651

Johnston J, Weismer S (1983) Mental rotation abilities in language disordered children. J Speech Hear Res 26:397–403

Johnston RB, Stark RE, Mellits ED, Tallal P (1981) Neurological status of language-impaired and normal children. Ann Neurol 10:159–163

Liberman IY, Shankweiler D, Camp L, Blachman B, Werfelman M (1980) Steps toward literacy: a linguistic approach. In: Levinson PJ, Sloan C (eds) Auditory processing and language: clinical and research perspectives. Grune and Stratton, New York, pp 189–215

Stark R, Tallal P (1979) Analysis of stop consonant production errors in developmentally dysphasic children. J Acoust Soc Am 66:1703–1712

Stark R, Tallal P (1980) Perceptual and motor deficits in language-impaired children. In: Keith RW (ed) Central auditory and language disorders. College Hill Press, Texas, pp 121–144

Stark, R, Tallal P (1988) Language speech and reading disorders in children: Neuropsychological studies, R. McCauley (ed), College hill press, Texas pp 124

Tallal P (1980a) Auditory temporal perception, phonics and reading disabilities in children. Brain Lang 9:182–198

Tallal P (1980b) Language and reading: some perceptual prerequisites. Bull Orton Soc 30:170–178

Tallal P, Piercy M (1973a) Defects of non-verbal auditory perception in children with developmental aphasia. Nature 241:469–469

Tallal P, Piercy M (1973b) Developmental aphasia: impaired rate of nonverbal processing as a function of sensory modality. Neuropsychologia 11:389–398

Tallal P, Piercy M (1974) Developmental aphasia: rate of auditory processing and selective impairment of consonant perception. Neuropsychologia 12:83–93

Tallal P, Piercy M (1975) Developmental aphasia: the perception of brief vowels and extended stop consonants. Neuropsychologia 13:69–74

Tallal P, Stark R (1981) Speech acoustic-cue discrimination abilities of normally developing and language-impaired children. J Acoust Soc Am 69:568–574

Tallal P, Stark RE (1982) Perceptual/motor profiles of reading-impaired children with or without concomitant oral language deficits. Ann Dyslexia 32:163–176

Tallal P, Stark RE, Curtiss B (1976) Relation between speech perception and speech production impairment in children with developmental dysphasia. Brain Lang 3:305–317

Tallal P, Stark R, Mellits D (1985a) The relationship between auditory temporal analysis and receptive language development: evidence from studies of developmental language disorder. Neuropsychologia 23:527–534

Tallal P, Stark R, Mellits D (1985b) Identification of language-impaired children on the basis of rapid perception and production skills. Brain Lang 25:314–322

Childhood Psychoses

Clinical, Electrophysiological, and Biochemical Markers and Monoaminergic Hypotheses in Autism

N. Bruneau, C. Barthélémy, J. Martineau, J.L. Adrien, B. Garreau, J.P. Muh, and G. Lelord

Early infantile autism, originally described by Kanner (1943), is a syndrome defined by a set of clinical features including failure to develop normal social relationships, developmental disturbances of verbal and nonverbal communication, ritualistic and compulsive behaviors, and disturbances of motility and attention.

For several years, there has been increasing evidence of brain pathology in autism. The etiology of this pathology is not known, but its manifestations, in structural brain pathology, in abnormal neurophysiology, in neurological abnormalities, and in neuropsychological and psychophysiological dysfunctions are increasingly being described.

Emphasizing the impaired sensory modulation and abnormal motility, Ornitz (1974, 1983, 1985) has postulated a disorder of subcortical neurophysiological circuitry involving brain stem and diencephalic pathways that modulate sensory input and motor output. Several authors have been of the opinion that pathophysiological influences originate in telencephalic structures, particularly the mesolimbic cortex, including the temporal lobes, and the neostriatum. This hypothesis emphasizes the autistic disturbances of language and communication and assumes an underlying specific cognitive disorder, presumably of cortical origin. Research relevant to this telencephalic hypothesis has included cognitive and linguistic studies (Hoffman and Prior 1982) but also electrophysiological studies showing either significant developmental delay of cerebral lateralization (Tanguay 1976; Tanguay et al. 1976; Cantor et al. 1986) or some "deviant" sensory information processes as evidenced with event-related potential studies (see review in Courchesne 1986). The latter disorders have been hypothesized to include malfunction in attentional mechanisms which compromises sensory information analysis and normal development (Gold and Gold 1975; Courchesne 1986).

A dysfunction in attentional processes has also been proposed by Damasio and Maurer (1978) to explain many of the clinical aspects of autism. Attention then encompasses stimulus selection but also implicates a functional system which fosters stable perception, sensorimotor integration, and adaptive behavior. For these authors, disturbances of communication do not derive from an impairment of primary linguistic processing, but rather apparently from a lack of initiative to communicate and from a lack of "orientation" towards stimuli, and are suggestive of an underlying impairment in higher motor and perceptual control.

Ritualistic and compulsive behaviors are considered to reflect inability of autistic children to adapt themselves to modified environmental contingencies, perseveration being an aspect of this inability to organize and consolidate new forms of response. Their potential for normal perception and normal memory of perceived material would be intact and their behavioral abnormalities rather indicate a defective ability to maintain optimal conditions for ongoing perceptual processes.

Behavioral observations carried out by Hoffman and Prior (1982) during neuropsychological tests agree with this view. They characterized the behavior of autistic children as follows: inability to formulate a plan to deal with a task, inability to benefit from the feedback regarding correct or incorrect responses, inflexibility and perseveration in responses, and lack of hypothesis testing. In view of the similarities between these behavioral features and those shown by patients with frontal lobe damage, as described in adult neuropsychological literature (Luria 1966; Walsh 1978), the hypothesis of a frontal lobe dysfunction in autism was proposed (Damasio and Maurer 1978; Hoffman and Prior 1982).

The role of the frontal lobes in attention, and in attention disorders, has also been implicated on the basis of biochemical studies and particularly those concerning dopaminergic systems. In addition to the two well-known dopaminergic projections originating in the substantia nigra and mesencephalic tegmentum—the nigrostriatal and mesolimbic dopaminergic systems—a mesocortical dopaminergic system has also been described (Thierry et al. 1973; Hokfelt et al. 1974). The functional role of these systems has been well studied (Le Moal et al. 1977; Glowinski 1981); it seems now that they interact to regulate attentional and intentional processes to maintain ongoing behavior (Simon et al. 1980; Simon and Le Moal 1985; Bouyer et al. 1986; Clark et al. 1987a, b). The aim of this work is to consider clinical and biological data regarding these two linked hypotheses: the ongoing attentional hypothesis and the dopaminergic hypothesis.

Clinical Evaluations

Several scales and systems have been developed for the assessment of autistic children. Five of these clinical tools have been reviewed by Parks (1983): Rimland's E2 Diagnostic Checklist (Rimland 1971), the Behavior Rating Instrument for Autistic and Atypical Children or BRIAAC (Ruttenberg et al. 1966, 1977), the Behavior Observation Scale for autism or BOS (Freeman et al. 1978), the Childhood Autism Rating Scale or CARS (Schopler et al. 1980), and the Autism Behavior Checklist or ABC (Krug et al. 1980). In our child psychiatry day care unit, two of these instruments (Rimland's E2 Diagnostic Checklist and the BOS), which were adapted in French (Leddet et al. 1986; Adrien et al. 1987), were used for the diagnosis in addition to the DSM III criteria (American Psychiatric Association 1980). Apart from these diagnosis assessment tools, a

Table 1. Behavioral Summarized Evaluation (BSE) scale

I. *Autistic withdrawal* 1. Is eager for aloneness 2. Ignores people 3. Poor social interaction 4. Abnormal eye contact	IV. *Motility disorders* 11. Stereotyped sensorimotor activity 12. Agitation, restlessness 13. Bizarre posture and gait
II. *Impairment in verbal and non-verbal communication* 5. Does not make an effort to communicate using voice and/or words 6. Lack of appropriate facial expressions and gestures 7. Stereotyped vocal and voice utterances—echolalia	V. *Inappropriate affective responses* 14. Autoaggressiveness 15. Heteroaggressiveness 16. Soft anxiety signs 17. Mood difficulties VI. *Impairment in main instinctual function* 18. Eating problems
III. *Bizarre responses to the environment* 8. Lack of initiative, poor activity 9. Inappropriate relating to inanimate objects or to doll 10. Resistance to change and to frustration	VII. *Attention and perception disturbances* 19. Unstable attention, easily distracted 20. Bizarre responses to auditory stimuli

clinical scale was developed ("Behavioral Summarized Evaluation" Scale, or BSE Scale) (Lelord et al. 1981). It represents a clinical instrument summarizing the common psychiatric evaluation performed by any clinician, intern, or nurse at the hospital or at any outpatient clinic. It consists of 20 items (Table 1) filled in by the physician and by the nurses who take care of the children. Each item is rated from 0 to 4 according to the importance of the disturbance. This behavior scale was specifically designed for the measurement of clinical modifications in children involved in educational programs (Lelord et al. 1987) and/or therapeutic studies (Lelord et al. 1981; Martineau et al. 1985; Barthélémy et al. 1989).

Validation of this clinical scale was performed by Barthélémy et al. (1990). A principal component analysis performed on results obtained with the scale in a group of 90 autistic children exhibited two main components (factors I and II) (Fig. 1).

Nine items described factor I. Most of these items fitted in with the autistic DSM III diagnostic criteria: Is eager for aloneness; Ignores people; Poor social interaction; Abnormal eye contact; Lack of appropriate facial expressions and gestures; but also Lack of initiative, poor activity, and stereotyped sensorimotor activity. A remarkable result was that attention (unstable attention, easily distracted) and perception (bizarre responses to auditory stimuli), which were not usually considered as core symptoms of autism, appeared to belong to factor I.

Five items described factor II and were considered as "associated" symptoms of autism (Resistance to change and to frustration; Agitation, restlessness; Autoaggressiveness; Heteroaggressiveness and Mood difficulties).

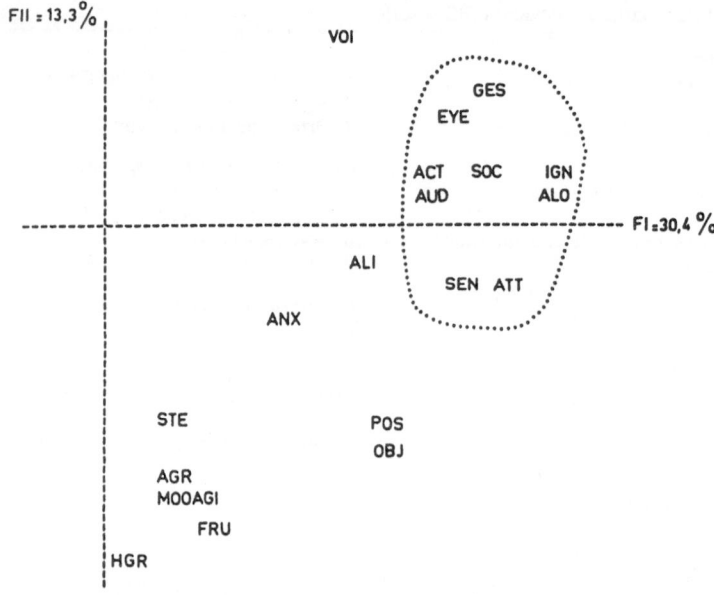

Fig. 1. Graphic representation of the results from principal component analysis performed on the 20 items of the BSE scale in 90 autistic children: Distribution of the factor loading for the 20 items on factors I and II. Nine items described factor I (see within *dotted line*): *ALO*, Is eager for aloneness; *IGN*, Ignores people; *SOC*, Poor social interaction; *EYE*, Abnormal eye contact; *GES*, Lack of appropriate facial expressions and gestures; *ACT*, Lack of initiative, poor activity; *SEN*, Stereotyped sensori-motor activity; *ATT*, Unstable attention, easily distracted; *AUD*, Bizarre responses to auditory stimuli. Five items described factor II: *FRU*, Resistance to change and to frustration; *AGI*, Agitation, restlessness; *AGR*, Autoaggressiveness; *HGR*, Heteroaggressiveness; *MOO*, Mood difficulties

So, these results emphasized that intention, attention, perception and gestures are deeply modified in autistic children.

This scale appeared a good tool for evaluating the intensity of autism since significant correlations were found between the factor I scores and the severity of autism as evaluated by experienced clinicians.

Biochemical Evaluations

There have been many attempts to define the biochemistry of autistic children. These studies have mainly centered on the metabolism of monoamines (review in Coleman and Gillberg 1985; Elliot and Ciaranello 1986; Yuwiler and Freedman 1986).

The implication of a dysfunctional serotoninergic system in the sensory and perceptual impairments characteristic of autism was proposed by Coleman (1973) and Young et al. (1982). This was supported by several studies showing that about 40% of autistic children display increased levels of serotonin in whole blood or platelets (Schain and Freeman 1961; Ritvo et al. 1970; Takahashi et al. 1976; Hanley et al. 1977), blood platelets being proposed as an indicator of serotonin storage in the presynaptic nerve terminals (Paasonen 1968; Pletcher 1968). Coleman and Gillberg (1985) underlined the great variability of results and the existence of a subgroup of autistic children with lower levels of serotonin than controls. Elevation of serotonin levels was found in the more active autistic children with low IQ (Campbell et al. 1975). It was also found that the efflux of serotonin out of the platelets was increased in autistic children (Boullin et al. 1970, 1971); however, this was not replicated by Yuwiler et al. (1975).

Results are also heterogeneous when considering the major pathway for the metabolism of serotonin resulting in the formation of 5-hydroxyindoleacetic acid (5-HIAA): results obtained in the cerebrospinal fluid in autistic children showed 5-HIAA levels to be increased (Cohen et al. 1974), decreased (Cohen et al. 1977), or similar (Gillberg et al. 1983b) when compared with controls.

From a therapeutic perspective, this biochemical hypothesis led to treatment of autistic children with fenfluramine, previously known as an antiserotoninergic drug. However, several studies showed that there is no relationship between therapeutic improvement and fenfluramine-induced serotonin reduction (Ritvo et al. 1986; Barthélémy et al. 1989).

A possible involvement of brain dopamine (DA) and norepinephrine (NE) in the production of autistic symptoms was also proposed. Both are known to be implicated in the regulation of behaviors and emotional states including motor function, cognition, attention, learning, and responses to stress: a variety of functions and behaviors which are markedly impaired in autistic children. This biochemical hypothesis received some support from pharmacological and biochemical data. Stimulants, such as amphetamine and methylphenidate— indirect DA and NE agonists—exacerbate autistic symptoms in most cases (Young et al. 1982). Neuroleptic agents—DA receptor blockers—significantly reduce behavioral symptoms such as stereotypies and withdrawal, and they facilitate learning (Campbell et al. 1982a; Anderson et al. 1984).

Changes in catecholaminergic activity are known to be reflected in the content of their major metabolites in cerebrospinal and peripheral body fluids. Results from studies of NE metabolism may appear to be somewhat heterogeneous. It was reported that plasma NE levels were elevated in autistic patients in both the lying and the standing positions (Lake et al. 1977). In contrast, a decrease in urinary free catecholamines and 3-methoxy-4-hydroxyphenyl glycol (MHPG, the principal brain metabolite of NE) was found in autistic children (Young et al. 1978, 1979). Preliminary studies of CSF and plasma MHPG levels in autistic subjects have shown that they are in the normal range (Young et al. 1981).

Results regarding DA metabolism are more homogeneous. High levels of homovanillic acid (HVA, the main DA metabolite) have been found in the CSF of severely impaired autistic children (Cohen et al. 1977; Gillberg et al. 1983b). Because of the noninvasive nature of the method, urinary assays are of interest

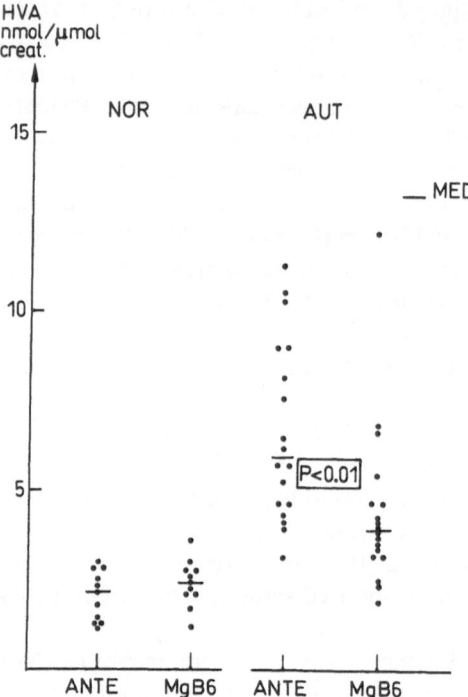

Fig. 2. Modifications of urinary HVA levels after 2 weeks of vitamin B6-magnesium (*MgB6*) treatment in normal (*NOR*) and autistic (*AUT*) children. The decrease in HVA levels in patients is concomitant with a significant improvement in autistic symptoms as rated with the BSE scale

for repeated application in child psychiatry. The work of our group has specifically dealt with research into dopaminergic clues at the urinary level. Lelord et al. (1978, 1981) and Garreau et al. (1980) first reported that urinary HVA levels are higher in autistic children than in controls. Muh et al. (1987) and Barthélémy et al. (1988) confirmed this result and showed that urinary DA amounts are significantly lower in autistic children than in controls and retarded children. Lelord et al. (1981) first reported the sensitivity of this biochemical marker to pharmacological treatment: the clinical improvement of young autistic patients under vitamin B_6–magnesium treatment was associated with a decrease in their urinary HVA excretion (Fig. 2).

Recently, Barthélémy et al. (1989) studied biochemical modifications at the urinary level during fenfluramine treatment. They showed that urinary serotonin levels decrease significantly in response to fenfluramine whatever the clinical effect obtained. However, siginificant differences were found when considering HVA according to the clinical results: HVA increased in responders whereas no modifications were found in nonresponders. Moreover, the pretreatment level of HVA was significantly lower in responders than in nonresponders and seemed to be a good indicator of fenfluramine responsiveness in children with autistic behavior.

Electrophysiological Evaluations

Clinical electrophysiological studies reported incidences of EEG abnormalities ranging from 10% to 83% in autism, but no EEG pattern unique to autism has been found (Creak and Pampiglione 1969; Tsai and Stewart 1982; Garreau et al. 1985).

Electrophysiological methods using averaged evoked potentials (AEPs) in response to sensory stimulation have been used to probe possible disorders in central stimulus information processing in autism.

The results of studies on brain stem auditory evoked responses (BAERs) have been equivocal. The majority of BAER studies either do not report abnormalities which characterize the autistic population (Student and Sohmer 1978; Ornitz et al. 1980; Novick et al. 1980; Courchesne et al. 1985a) or describe small subgroups of that population with one or another abnormal response component (Skoff et al. 1980; Tanguay et al. 1982; Taylor et al. 1982; Gillberg et al. 1983a; Garreau et al. 1984; Rumsey et al. 1984).

Event-related brain potentials (ERPs) recorded at the cortical level (mainly N1, P2, P300) have evidenced defects in mechanisms of alerting to novel stimulation (Courchesne et al. 1984, 1985b), in attentional processes (Small 1971), or in memory (Novick et al. 1979, 1980; Niwa et al. 1983) but also a more general defect in hemisphere specialization (Tanguay 1976; Tanguay et al. 1976; Dawson et al. 1989). In most of the studies, responses were found to be smaller in autistic children than in controls (Small et al. 1971; Ornitz et al. 1972; Lelord et al. 1973; Martineau et al. 1980; Novick et al. 1979, 1980; Newa et al. 1983; Courchesne et al. 1984). Moreover, responses are not elicited as frequently in autistic children as in controls. This increased wave form variability is found both for BAERs (Rosenblum et al. 1980) and for ERPs (Walter et al. 1971; Novick et al. 1979; Courchesne 1986; Martineau et al. 1987). Recently, Bruneau et al. (1989b) described a decrease in ERP variability in a subgroup of autistic children improved by administration of fenfluramine and showing dopaminergic metabolism modifications with this treatment.

Our approaches were based on studies pointing out a "faulty modulation of sensory input" and a "defect in cross-modal association" in autistic children. Disorders of individuals with infantile autism have been suggested to include defects in modulation of sensory input (Bergman and Escalona 1949; Goldfarb 1963; Ornitz and Ritvo 1968; Ornitz 1985). This faulty modulation is manifest as either underreactivity or overreactivity to sensory stimuli. Both types of abnormal reactivity can occur in the same child. While all sensory modalities are affected, these deficits are particularly evident in the auditory modality.

Previous studies of individual differences in response to auditory stimulation have suggested a model of a mechanism in the central nervous system (CNS) that modulates the intensity of incoming signals (Buchsbaum 1976). Although AEP amplitude generally increases with increasing stimulus intensity (augmenting), individual differences are found in the degree of increment at high intensity

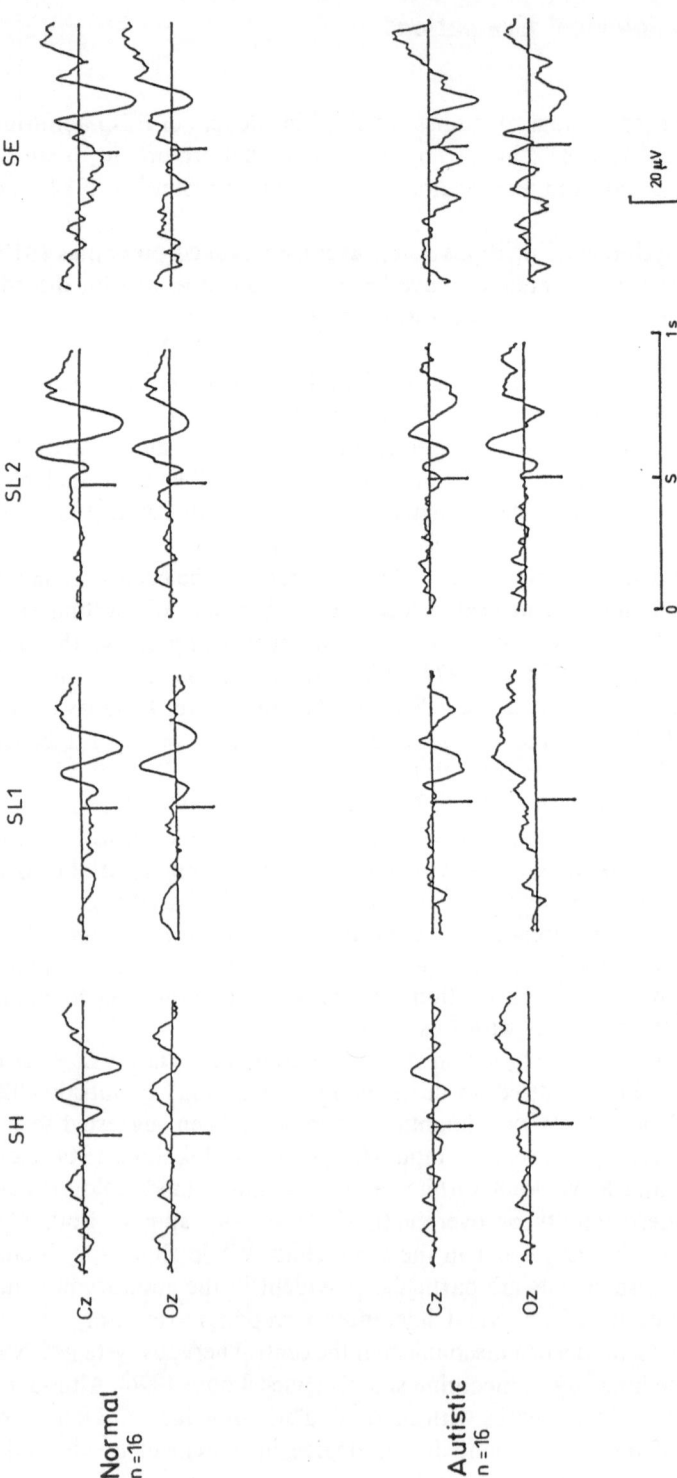

Fig. 3. Grand average responses of normal (n = 16) and autistic (n = 16) children recorded at Cz and Oz sites during sound alone series (*SH*), conditioning series of day 1(*SL1*) and day 2(*SL2*), and extinction series (*SE*)

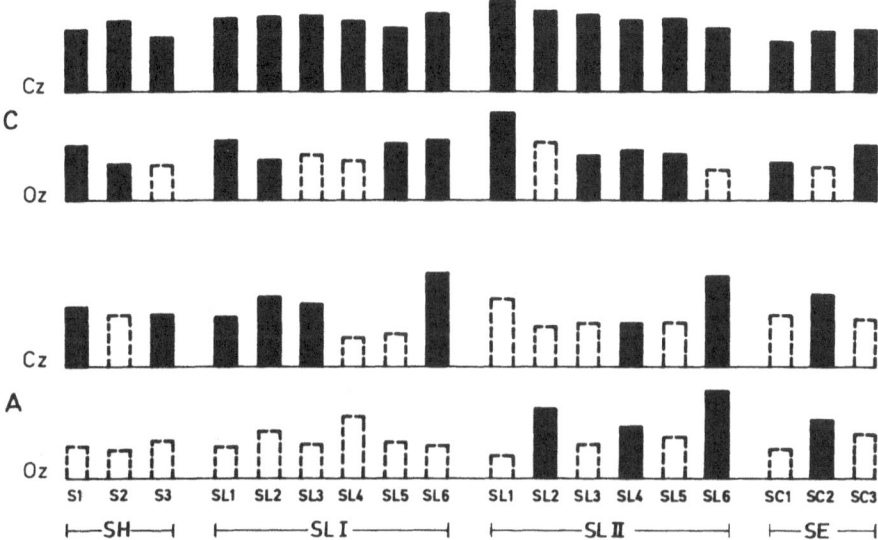

Fig. 4. Representation of presence (*black bars*) and absence (*dotted bars*) of AEPs obtained at Cz and Oz sites during a conditioning paradigm from a control child (C) and an autistic one (*A*). The criterion chosen to determine the presence of AEPs was a signal/noise ratio corresponding to the standard deviation of the averaged postsound/presound EEG (250 ms duration for each). The lower limit value of this ratio to assert the presence of each AEP was 1.3 (Roux 1981). This figure illustrates the great inconstancy of the responses in autistic children

level; many people actually show a ceiling effect or a decrease in amplitude with increasing stimulus intensity (reducing). Thus, we used this paradigm requiring minimal behavioral compliance to assess the capacity for "stimulus intensity modulation" in young autistic children. Preliminary results were detailed in Bruneau et al. (1987). Two subgroups showing different AEP reactivity were evidenced. One subgroup displayed large AEPs at both high and low intensity levels, similar to those obtained in young children. The other subgroup showed smaller AEPs than controls, particularly at a high intensity level. The children with the most hyper-reducing pattern were also those most clinically disturbed.

A deficit in the ability to form auditory—visual and auditory—tactile cross-modal associations has been observed in autism (Bryson 1970; Morton-Evans and Hensley 1978; Hetzler and Griffin 1981). Using a conditioning paradigm, Lelord et al. (1973) studied the effects on AEPs of coupling sound and light. Such an association leads to an increase in the AEP amplitude at the occipital site and a decrease in these responses at the vertex in adults and children. In this previous study, and in Martineau et al. (1980), results underlined the difficulties in evidencing such modifications in autism since responses were smaller and more variable than in controls. In a recent work, Martineau et al. (1987) replicated such a study and found some islands of associated schemes in the autistic children but emphasized the great inconstancy of these abilities (Figs. 3 and 4).

Discussion

"Behavioral Summarized Evaluation" provides a useful way of recording behavior which is needed not only for evaluation of drug responsiveness and assessment of relationships with biological data but also for the identification of clinically meaningful patterns of behavior which could be related to underlying neurophysiological dysfunctioning. Results of statistical analyses performed to validate this rating scale have underlined the fact that perceptual, attentional, and intentional disturbances as well as abnormal motor activity are involved in core symptoms of autism.

The importance of disturbances of sensory modulation in the phenomenology of infantile autism has been emphasized. Ornitz (1969, 1974, 1983, 1985) proposed that autism results from "perceptual inconstancies" such that the children affected are unable to maintain stable perception over time, which precludes the establishment of a coherent meaningful external reality. This was later defined more specifically as a disorder of sensorimotor integration, abnormal motor activity being considered as compensation for an insufficiency of sensory input. Kootz et al. (1982) also linked perceptual processes and motor abnormalities; they proposed that autistic children may develop self-stimulating stereotypies to flood sensory receptors in order to actively avoid stimuli from the external world.

The implication of attentional disturbances as basic defects in autism has been proposed (Gold and Gold 1975) and is supported by data showing abnormal sensitivity and attention to novel stimulation (James and Barry 1980; Kootz and Cohen 1981; Rutter 1979; Courchesne et al. 1984). However, as emphasized in the introduction, the attention concept can be enlarged to encompass not only stimulus selection but a more general process including readiness to accept stimuli (attentiveness), ability to identify relevant stimuli and to ignore or inhibit the response to irrelevant ones, ability to generate an appropriate motor program for response and to monitor its efficacy continuously, ability to initiate and terminate a response at the right time, and ability to shift to a stimulus with greater relevance than the one being evaluated. As detailed in the introduction, clinical observations (Damasio and Maurer 1978; Hoffman and Prior 1982) have emphasized that defects in programming, regulation, and verification of activity could explain most of the symptoms in autism, leading to the proposal of frontal lobe dysfunctioning.

Some biological data represent interesting clues with regard to this frontal lobe hypothesis. Courchesne et al. (1984) studied auditory P300 associated wih the detection of novel, surprising stimuli; this component is largest at scalp electrode sites situated midway between the superior parietal and frontal cortical areas. They found it to be much smaller in autistic than in normal subjects and pointed out that this was also found in subjects with prefrontal lobe lesions (Knight 1984).

Modulation of response amplitude according to stimulus intensity was proposed to implicate frontal lobe functioning (Petrie 1958; Bruneau et al. 1985,

1986): The hyper-reducing pattern obtained in most autistic subjects could be related to defects of frontal lobe regulation of sensory input.

An anterior cortex switch-off mechanism was proposed by Muller (1985) as interfering to different degrees with higher mental functioning, from a very transient impairment to complete arrest of mental function. Such mechanism was also proposed by Courchesne (1986) and could explain the disruptions observed at the behavioral as well as at the electrophysiological level. For example, the capacity of autistic children for cross-modal association (Martineau et al. 1987) or for detecting novelty (Courchesne 1986) is evidenced by AEPs, but the responses are occasional, as if the neural system functioning is being abnormally interfered with or hindered by some other system.

Study of regional cerebral blood flow (CBF) in autism (Sherman et al. 1984) showed abnormalities of blood flow which support frontal dysfunctioning. Habituation to testing sessions resulted in a decrease in resting CBF in normals; this failed to appear in autistic subjects, again supporting the hypothesis of frontal lobe impairment in autism.

Though Damasio and Maurer (1978) pointed out similarities in autistic children and in patients suffering from frontal damage, they extended their hypothesis toward a more general dysfunctioning of the dopaminergic system including limbic and striatal structures. Biochemical results supported this hypothesis.

It is worth emphasizing again the special interest of urinary assays in pediatric research. Because they are atraumatic, such examinations can be repeated and thus provide helpful indications for evaluation in therapeutic studies. The relationships between clinical improvement, electrophysiological "normalization", and urinary HVA modifications were very striking in all of our pharmacological studies.

The increase in urinary HVA in autism is in agreement with results obtained at the CSF level. Although urine does not seem a primary source of information about central biogenic amine metabolism, several lines of evidence indicate that urinary HVA may derive in part from CNS metabolism (Maas et al. 1979, 1980; Swann et al. 1980). This was previously discussed in detail in Bruneau et al. (1986). Compared with control values, the increase in HVA level is quite similar in autistic and retarded children. However, in the former this increase derives from the free part of the metabolite whereas in the latter it stems from the conjugated part (Garreau et al. 1988). Moreover, the decrease in the conjugated part of HVA with maturation found in controls and retarded children was not observed in autism (Martineau et al. 1987). Further research is needed to explain such data. Some experimental results obtained in animals show that central DA lesions differentially affect free and conjugated parts of the urinary HVA (Peyrin et al. 1978, 1982).

The dopamine metabolism follows two main routes, one being transformation into NE, the second, degradation into HVA. The usual degradation implies transformation into dihydroxyphenylacetic acid (DOPAC). Another less important pathway leads to HVA via methoxytyramine (MT). In a recent study,

Muh et al. (1987) found that MT levels were increased in autistic children whereas DOPAC levels were decreased (Barthelemy et al. 1988). This suggested a preferential degradation pathway of dopamine via MT.

Results concerning NE and DA are not contradictory. Certain NE systems are in direct relationship with DA neurons of the ventral tegmental area. The increased turnover of the DA system could directly result from the hyperactivity of the NE excitatory pathways.

The recent results of Courchesne et al. (1987, 1988) evidencing cerebellar maldevelopment in autism are of interest in view of the DA hypothesis. Several reports have demonstrated the existence of efferent pathways from deep cerebellar nuclei to the substantia nigra and ventral tegmental area, these pathways of cerebellar origin modulating the activity of DA neurons in the brain (Snider and Snider 1977; Nieoullon et al. 1978; Simon et al. 1979).

Rather than being a primary defect in autism, DA dysfunctioning might be considered as involved in the form of a disrupted dopamine/serotonin balance or as a result of regulatory dysfunctions involving other neuromediators. The data supporting serotonin or NE dysfunctioning in autism have been reported above. Possible DA/opioid dysregulation has also been implicated (Sandyk 1986), and data from several authors support the brain opioid hypothesis in autism (Weizman et al. 1984; Gillberg et al. 1985; Deutsch 1986; Sahley and Panksepp 1987).

These results underlined the limit of only considering the relationship between a given disorder and a single neurochemical system. Indeed, complex interactions among neuroregulators exist. It would, then, be more appropriate to study the patients' biochemical profiles (drawn from assessment of several neurotransmitters) according to their clinical and behavioral features.

Future Directions for Research

Several developments within the neurosciences are providing powerful methods for studying brain anatomy and functioning and allow some hope for the future.

New anatomical data of Courchesne et al. (1988) showing cerebellar abnormalities with MRI are very interesting and, as mentioned above, can be considered in view of the biochemical hypothesis. The other anatomical studies using computed tomographic scanning have shown no consistent structural abnormalities in the brain of autistic patients (Campbell et al. 1982b; Prior et al. 1984; Rosenbloom et al. 1984; Creasy et al. 1986).

The possibility that biological damage could be of a functional nature requires other kinds of investigation. In clinical practice, the topography of basic EEG rhythms (quantitative EEG mapping) brings a new dimension to the understanding of normal and pathologic electrogenesis in the brain. Preliminary results from studies of metabolism with positron emission tomography (PET) (Rumsey et al. 1985) and from studies of cerebral blood flow with Single Photon

Computerized Emission Tomography (SPECT) (Sherman et al. 1984) and Transcranial Doppler Ultrasonography (Bruneau et al. 1989a) are promising. These functional technologies are still in the early stages: standard methodology for isotope selection and examinations in active conditions (during sensory stimulations and during task performances) will allow new developments. The evaluation of neuroreceptor characteristics in the living brain is now possible with quantitative PET studies (Sedvall et al. 1986); it should be a suitable tool for directly elucidating the different biochemical dysfunctionings proposed as underlying infantile autism.

Other fields, including immunology and molecular biology, also look very promising for the future understanding of such pathology. However, for all of these approaches, one difficulty in attempting to identify common localized abnormalities remains the heterogeneity of the autistic population. A worthwhile research strategy would be the use of markers such as those described in the present study, or the use of differences in therapeutic responses to specific drug treatments to select more homogeneous subgroups of autistics for further explorations with powerful new techniques in order to identify different forms of functional neuropathology.

Acknowledgment. This study was supported by INSERM 4316 and by grants from INSERM (no. 859014), Fondation pour la Recherche Médicale, Fondation H. Langlois, Sécurité Sociale, and Conseil Régional de la Région Centre.

References

Adrien JL, Ornitz E, Barthélémy C, Sauvage D, Lelord G (1987) The presence or absence of certain behaviors associated with infantile autism in severely retarded autistic and nonautistic retarded children and very young normal children. J Autism Dev Disord 17:407–416

American Psychiatric Association (1980) Diagnostic and statistical manual of mental disorders (DSM III), 3rd ed. American Psychiatric Association, Washington

Anderson LT, Campbell M, Grega DM, Perry R, Small AM, Green WH (1984) Haloperidol in the treatment of infantile autism: effects on learning and behavioral symptoms. Am J Psychiatry 141:1195–1202

Barthélémy C, Adrien JL, Tanguay P, Garreau B, Fermanian J, Roux S, Sauvage D, Lelord G (1990) The Behavioral Summarized Evaluation (BSE): validity and reliability of a scale for the assessment of autistic behaviors. J Autism Dev Disord (in press)

Barthélémy C, Bruneau N, Cottet-Eymard JM, Jouve J, Garreau B, Lelord G, Muh JP, Peyrin L (1988) Urinary free and conjugated catecholamines and metabolites in autistic children. J Autism Dev Disord 18:583–590

Barthélémy C, Bruneau N, Jouve J, Martineau J, Muh JP, Lelord G (1989) Urinary dopamine metabolites as indicators of the responsiveness to fenfluramine treatment in children with autistic behavior. J Autism Dev Disord 19:241–254

Bergman P, Escalona SK (1949) Unusual sensitivities in very young children. Psychoanal Study Child 3:333–353

Boullin DJ, Coleman M, O'Brien RA (1970) Abnormalities in platelet 5-hydroxytryptamine efflux in patients with infantile autism. Nature 226:371–372

Boullin DJ, Coleman M, O'Brien RA, Rimland B (1971) Laboratory predictions of infantile autism, based on 5-hydroxytryptamine efflux from platelets, and their correlation with the Rimland E2 scores. J Autism Child Schizophr 1:63–71

Bouyer JJ, Montaron MF, Fabre-Thorpe M, Rougeul A (1986) Compulsive attentive behavior after lesion of the ventral striatum in the cat: a behavioral and electrophysiological study. Exp Neurol 92:698–712

Bruneau N, Roux S, Garreau B, Lelord G (1985) Frontal auditory evoked potentials and augmenting-reducing. Electroencephalogr Clin Neurophysiol 62:364–371

Bruneau N, Barthélémy C, Jouve J, Lelord G (1986) Frontal auditory evoked potential augmenting-reducing and urinary homovanillic acid. Neuropsychobiology 16:78–84

Bruneau N, Garreau B, Roux S, Lelord G (1987) Modulation of auditory evoked potentials with increasing stimulus intensity in autistic children. In: Johnson RJ, Parasuraman R, Rohrbaugh JW (eds) Current trends in event-related potential research (EEG Suppl 40). Elsevier, Amsterdam, pp 584–589

Bruneau N, Arbeille P, Dourneau MC, Garreau B, Pourcelot L, Lelord G (1989a) Evaluation du débit sanguin cérébral par Doppler transcrânien chez des enfants présentant un comportement autistique. C R Acad Sci Paris 308 III:255–260

Bruneau N, Barthélémy C, Roux S, Jouve J, Lelord G (1989b) Auditory evoked potential modifications according to clinical and biochemical responsiveness to fenfluramine treatment in children with autistic behavior. Neuropsychobiology 21:48–52

Bryson CQ (1970) Systematic identification of perceptual disabilities in autistic children. Percept Mot Skills 31:239–246

Buchsbaum MS (1976) Self-regulation of stimulus intensity. In: Schwartz GE, Shapiro D (eds) Consciousness and self-regulation. Plenum, New York, pp 101–135

Campbell M, Friedman E, Green WH, Collins PJ, Small AM, Breuer H (1975) Blood serotonin in schizophrenic children. A preliminary study. Int Pharmacopsychol 10:213–221

Campbell M, Anderson LT, Small AM, Perry R, Green WH, Caplan R (1982a) The effects of haloperidol on learning and behavior in autistic children. J Autism Dev Disord 12:167–175

Campbell M, Rosenbloom S, Perry, R, George AE, Kricheff II, Anderson L, Small AM, Jennings SJ (1982b) Computerized axial tomography in young autistic children. Am J Psychiatry 139:510–512

Cantor DS, Thatcher RW, Hrybyk M, Kaye H (1986) Computerized EEG analysis of autistic children. J Autism Dev Disord 16:169–187

Clark CR, Geffen GM, Geffen LB (1987a) Catecholamines and attention I: Animal and clinical studies. Neurosci Biobehav Rev 11:341–352

Clark CR, Geffen GM, Geffen LB (1987b) Catecholamines and attention II: Pharmacological studies in normal humans. Neurosci Biobehav Rev 11:353–364

Cohen DJ, Shaywitz BA, Johnson NT, Bowers MB (1974) Biogenic amines in autistic and atypical children. Cerebrospinal fluid measures of homovanillic acid and 5-hydroxyindoleacetic acid. Arch Gen Psychiatry 31:845–853

Cohen DJ, Caparulo BK, Shaywitz BA, Bowers MB (1977) Dopamine and serotonin metabolism in neuropsychiatrically disturbed children: CSF homovanillic acid and 5-hydroxyindoleacetic acid. Arch Gen Psychiatry 34:545–550

Coleman M (1973) Serotonin and central nervous system syndromes of childhood: a review. J Autism Child Schizophr 3:27–35

Coleman M, Gillberg C (1985) The biology of the autistic syndromes. Praeger, New York

Courchesne E (1986) A neurophysiological view of autism. In: Schopler E, Mesibov GB (eds) Neurobiological issues in autism. Plenum, New York, pp 285–324

Courchesne E, Kilman BA, Galambos R, Lincoln AJ (1984) Autism: processing of novel auditory information assessed by event-related brain potentials. Electroencephalogr Clin Neurophysiol 59:238–248

Courchesne E, Courchesne RY, Hicks G, Lincoln AJ (1985a) Functioning of the brainstem auditory pathway in non-retarded autistic individuals. Electroencephalogr Clin Neurophysiol 61:491–501

Courchesne E, Lincoln AJ, Kilman BA, Galambos R (1985b) Event-related brain potential correlates of the processing of novel visual and auditory information in autism. J Autism Dev Disord 15:55–76

Courchesne E, Hesselink J, Jernigan T, Yeung-Courchesne R (1987) Abnormal neuroanatomy in a non-retarded person with autism: unusual findings with magnetic resonance imaging. Arch Neurol 44:335–341

Courchesne E, Yeung-Courchesne R, Press G, Jernigan TL, Hesselink JR (1988) Hypoplasia of cerebellar vermal lobules VI and VII in infantile autism. N Engl J Med 318:1349–1354

Creak M, Pampiglione G (1969) Clinical and EEG studies on a group of 35 psychotic children. Dev Med Child Neurol 11:218–227

Creasey H, Rumsey JM, Schwartz M, Duara R, Rapoport JL, Rapoport SI (1986) Brain morphometry in autistic men as measured by volumetric computed tomography. Arch Neurol 43:669–672

Damasio AR, Maurer RG (1978) A neurological model for childhood autism. Arch Neurol 35:777–786

Dawson G, Finley C, Phillips S, Lewy A (1989) A comperieson of hemisheric asymmetries in speech-related brain potentials of autistic and dysphasic children. Brain Lang 37:26–41

Deutsch SI (1986) Rationale for the administration of opiate antagonists in treating infantile autism. Am J Ment Defic 90:631–635

Elliott GR, Ciaranello RD (1986) Neurochemical hypotheses of childhood psychoses. In: Schopler E, Mesibov S (eds) Neurobiological issues in autism. Plenum, New York, pp 245–261

Freeman BJ, Ritvo ER, Guthrie D, Schroth P, Ball J (1978) The Behavior Observation Scale for autism: initial methodology, data analysis and preliminary findings on 89 children. J Am Acad Child Psychiatry 17:576–588

Garreau B, Barthélémy C, Domenech J, Sauvage D, Muh JP, Lelord G, Callaway E (1980) Troubles du métabolisme de la dopamine chez des enfants ayant un comportement autistique. Résultats des examens cliniques et des dosages urinaires de l'acide homovanilique. Acta Psychiatr Belg 80:249–265

Garreau B, Tanguay P, Roux S, Lelord G (1984) Etude des potentiels évoqués auditifs du tronc cérébral chez l'enfant normal et chez l'enfant autistique. Rev Electroencephalogr Neurophysiol Clin 14:25–31

Garreau B, Barthélémy C, Martineau J, Bruneau N, Lelord G (1985) Aspects électrophysiologiques dans l'autisme de l'enfant. Encephale 11:145–155

Garreau B, Barthélémy C, Jouve J, Bruneau N, Muh JP, Lelord G (1988) Urinary homovanillic acid levels of autistic children. Dev Med Child Neurol 30:93–98

Gillberg C, Rosenhall H, Johansson E (1983a) Auditory brainstem responses in childhood psychosis. J Autism Dev Disord 13:181–195

Gillberg C, Svennerholm L, Hamilton-Hellberg C (1983b) Childhood psychosis and monoamine metabolites in spinal fluid. J Autism Dev Disord 13:383–396

Gillberg C, Terenius L, Lonnerholm G (1985) Endorphin activity in childhood psychosis. Arch Gen Psychiatry 42:780–783

Glowinski J (1981) Present knowledge on the properties of the mesocortico-frontal dopaminergic neurons. In: Mathysse S (ed) Psychiatry and the biology of the human brain: a symposium dedicated to Seymour HS. Kety. Elsevier, New York, pp 139–164

Gold MS, Gold JR (1975) Autism and attention: theoretical considerations and a pilot study using set reaction time. Child Psychiatry Hum Dev 6:68–80

Goldfarb W (1963) Self-awareness in schizophrenic children. Arch Gen Psychiatry 8:47–60

Hanley HG, Stahl SM, Freedman D (1977) Hyperserotonemia and amine metabolites in autistic and retarded children. Arch Gen Psychiatry 34:521–531

Hetzler BE, Griffin JL (1981) Infantile autism and the temporal lobe of the brain. J Autism Dev Disord 11:317–330

Hoffmann WL, Prior MR (1982) Neuropsychological dimensions of autism in children: a test of the hemispheric dysfunction hypothesis. J Clin Neuropsychol 4:27–41

Hokfelt T, Ljungdahl A, Fuxe K, Johansson O (1974) Dopamine nerve terminals in the rat limbic cortex: aspects of the dopamine hypothesis of schizophrenia. Science 184:177–179

James AL, Barry RJ (1980) Respiratory and vascular responses to simple visual stimuli in autistics, retardates and normals. Psychophysiology 17:541–547

Kanner L (1943) Autistic disturbances of affective contact. Nerv Child 2:217–250

Knight R (1984) Decreased response to novel stimuli after prefrontal lesions in man. Electroencephalogr Clin Neurophysiol 59:9–20

Kootz JP, Cohen DJ (1981) Modulation of sensory intake in autistic children. J Am Acad Child Psychiatry 20:692–701

Kootz JP, Marinelli B, Cohen DJ (1982) Modulation of response to environmental stimulation in autistic children. J Autism Dev Disord 12:185–193

Krug DA, Arick JR, Almond PJ (1980) Behavior checklist for identifying severely handicapped individuals with high levels of autistic behavior. J Child Psychol Psychiatry 21:221–229

Lake CR, Ziegler MG, Murphy DL (1977) Increased norepinephrine levels and decreased β-hydroxylase activity in primary autism. Arch Gen Psychiatry 34:553–556

Leddet I, Larmande C, Barthélémy C, Chalon F, Sauvage D, Lelord G (1986) Comparison of clinical

Lelord G, Laffont F, Jusseaume P, Stephant JL (1973) Comparative study of conditioning of averaged evoked responses by coupling sound and light in normal and autistic children. Psychophysiology 10:415–427

Lelord G, Callaway E, Muh JP, Arlot JC, Sauvage D, Garreau B, Domenech J (1978) L'acide homovanilique urinaire et ses modifications par ingestion de vitamine B6: Exploration fonctionnelle dans l'autisme de l'enfant? Rev Neurol 134:797–801

Lelord G, Muh JP, Barthélémy C, Martineau J, Garreau B (1981) Effects of pyridoxine and magnesium on autistic symptoms—initial observations. J Autism Dev Disord 11:219–230

Lelord G, Barthélémy C, Sauvage D, Boiron M, Adrien JL, Hameury L (1987) Thérapeutiques d'echange et de dévelopement dans l'autisme de l'enfant, Bases physiologiques. Bull Acad Natl Med (Paris) 171:137–143

Le Moal M, Stinus L, Simon H, Tassin JP, Thierry AM, Blanc G, Glowinski J, Cardo B (1977) Behavioral effects of a lesion in the ventral mesencephalic tegmentum: evidence for involvement of A10 dopaminergic neurons. In: Costa E, Gessa GL (eds) Non-striatal dopaminergic neurons. Adv Biochem Psychopharmacol 16:237–245

Luria AR (1966) Higher cortical functions in man. Basic Books, New York

Maas JW, Hattox SE, Martin DM, Landis DH (1979) A direct method for determining dopamine synthesis and output of dopamine metabolites from brain in awake animals. J Neurochem 32:839–843

Maas JW, Hattok SE, Greene NM, Landis DH (1980) Estimates of dopamine and serotonin synthesis by the awake human brain. J Neurochem 34:1547

Martineau J, Laffont F, Bruneau N, Roux S, Lelord G (1980) Event-related potentials evoked by sensory stimulation in normal, mentally retarded and autistic children. Electroencephalogr Clin Neurophysiol 48:140–153

Martineau J, Barthélémy C, Garreau B, Lelord G (1985) Vitamin B6, magnesium and combined B6-Mg: therapeutic effects in childhood autism. Biol Psychiatry 20:467–478

Martineau J, Garreau B, Roux S, Lelord G (1987) Auditory evoked responses and their modifications during conditioning paradigm in autistic children. J Autism Dev Disord 17:525–539

Morton-Evans A, Hensley R (1978) Paired associate learning in early infantile autism and receptive developmental aphasia. J Autism Child Schizophr 8:61–69

Muh JP, Barthélémy C, Jouve J, Martineau J, Mariotte N, Bruneau N (1987) Les monoamines urinaires dans l'autisme de l'enfant. In: Grémy F, Tomkiewicz S, Ferrari P, Lelord G (eds) Autisme infantile/Infantile autism. INSERM, pp 129–136 (Colloque INSERM, vol. 146)

Muller HF (1985) Prefrontal cortex dysfunction as a common factor in psychosis. Acta Psychiatr Scand 71:431–440

Niéoullon A, Cheramy A, Glowinski J (1978) Release of dopamine in both caudate nuclei and both substantiae nigrae in response to unilateral stimulation of cerebellum in the cat. Brain Res 148:143–152

Niwa S, Ohta M, Yamazaki K (1983) P300 and stimulus evaluation process in autistic subjects. J Autism Dev Disord 13:33–42

Novick B, Kurtzberg D, Vaughan HG (1979) An electrophysiologic indication of defective information storage in childhood autism. Psychiatry Res 1:101–108

Novick B, Vaughan HG, Kurtzberg D, Simson R (1980) An electrophysiologic indication of auditory processing defects in autism. Psychiatry Res 3:107–114

Ornitz EM (1969) Disorders of perception common to early infantile autism and schizophrenia. Compr Psychiatry 10:259–274

Ornitz EM (1974) The modulation of sensory input and motor output in autistic children. J Autism Child Schizophr 4:197–215

Ornitz EM (1983) The functional neuroanatomy of infantile autism. Int J Neurosci 19:85–124

Ornitz EM (1985) Neurophysiology of infantile autism. J Am Acad Child Psychiatry 24:251–262

Ornitz EM, Ritvo ER (1968) Perceptual inconstancy in early infantile autism. Arch Gen Psychiatry 18:76–98

Ornitz EM, Tanguay PE, Lee JCM, Ritvo ER, Sivertsen B, Wilson C (1972) Effects of stimulus interval on the auditory evoked response during sleep in autistic children. J Autism Child Schizophr 2:1450–1456

Ornitz EM, Mo A, Olson ST, Walter DO (1980) Influence of click sound pressure direction on brainstem responses in children. Audiology 19:245–254

Paasonen MK (1968) Platelet 5-hydroxytryptamine as a model in pharmacology. Ann Med Exp Biol Fenn 46:416

Parks SI (1983) The assessment of autistic children: a selective review of available instruments. J Autism Dev Disord 13:255–267

Petrie A (1958) Effects of chlorpromazine and of brain lesions on personality. In: Pennes HD (ed) Psychopharmacology. Harper, New York, pp 99–115

Peyrin L, Cottet-Emard JM, Javoy F, Agid Y, Herbet A, Glowinski J (1978) Long term effects of unilateral 6-hydroxydopamine destruction of the dopaminergic nigrostriatal pathway on the urinary excretion of catecholamines (dopamine, norepinephrine, epinephrine) and their metabolites in the rat. Brain Res 143:567–572

Peyrin L, Simon H, Cottet-Emard JM, Bruneau N, Le Moal M (1982) 6-Hydroxydopamine lesions of dopaminergic A10 neurons. Long-term effects on the urinary excretion of free and conjugated catecholamines and their metabolites in the rat. Brain Res 235:363–369

Pletcher A (1968) Metabolism, transfer and storage of 5HT in blood platelets. Br J Pharmacol 32:1–16

Prior MR, Tress B, Hoffman NL, Boldt D (1984) Computed tomographic study of children with classic autism. Arch Neurol 41:482–484

Rimland B (1971) The differentiation of childhood psychoses: an analysis of checklists for 2218 psychotic children. J Autism Child Schizophr 1:161–174

Ritvo ER, Yuwiler A, Geller E, Ornitz EM, Saeger K, Plotkin S (1970) Increased blood serotonin and platelets in early infantile autism. Arch Gen Psychiatry 23:566–572

Ritvo ER, Freeman BJ, Yuwiler A, Geller A, Schroth P, Yokata A, Mason-Brothers A, August GJ, Klykylo W, Leventhal B, Lewis K, Piggott L, Realmuto G, Stubbs EG, Umansky R (1986) Fenfluramine treatment of autism: UCLA collaborative study of 81 patients at nine medical centers. Psychopharmacology 22:133–140

Rosenbloom S, Campbell M, George AE, Kricheff II, Taleporos E, Anderson L, Reuben RN, Korein J (1984) High resolution CT scanning in infantile autism: a quantitative approach. J Am Acad Child Psychiatry 23:72–77

Rosenblum SM, Arick JR, Krug DA, Stubbs EG, Young NB, Pelson RO (1980) Auditory brainstem evoked responses in autistic children. J Autism Dev Disord 10:215–225

Roux S (1981) Aspects electrophysiologiques du conditionnement des potentiels évoqués par des stimulations sensorielles chez l'homme. Doctoral thesis, University of Poitiers

Rumsey JM, Grimes AM, Pikus AM, Duara R, Ismond DR (1984) Auditory brainstem responses in pervasive developmental disorders. Biol Psychiatry 19:1403–1418

Rumsey JM, Duara R, Grady C, Rapoport JL, Margolin RA, Rapoport SI, Cutler NR (1985) Brain metabolism in autism: resting cerebral glucose utilization rates as measured with positron emission tomography. Arch Gen Psychiatry 42:448–455

Ruttenberg BA, Dratman ML, Fraknoi J, Wenar C (1966) An instrument for evaluating autistic children. J Am Acad Child Psychiatry 5:453–478

Ruttenberg BA, Kalish BI, Wenar C, Wolf EG (1977) Behavior rating instrument for autistic and other atypical children. Developmental Center for Autistic Children, Philadelphia

Rutter M (1979) Language, cognition and autism. In: Katzman R (ed) Congenital and acquired cognitive disorders. Raven, New York, pp 247–264

Sahley TL, Panksepp J (1987) Brain opioids and autism: an updated analysis of possible linkages, J Autism Dev Disord 17:201–216

Sandyk R (1986) Further speculations on possible dopamine-opioid link in autism. J Autism Dev Disord 16:89–90

Schain RJ, Freedman DX (1961) Study on 5-hydroxyindole metabolism in autistic and mentally retarded children. J Pediatr 58:315–320

Schopler E, Reichler RJ, De Vellis RF, Daly K (1980) Toward objective classification of childhood autism: childhood autism rating scale (CARS). J Autism Dev Disord 10:91–103

Sedvall G, Farde L, Persson A, Wiesel FA (1986) Imaging of neurotransmitter receptors in the living human brain. Arch Gen Psychiatry 43:995–1005

Sherman M, Ruth N, Shapiro TH (1984) Regional cerebral blood flow in autism. J Autism Dev Disord 14:439–446

Simon H, Le Moal M (1985) Influences des neurones dopaminergiques du mésencéphale sur les processus d'attention et d'intention. Psychologie Médicale 17:939–945

Simon H, Le Moal M, Calas A (1979) Efferents and afferents of the ventral tegmental-A10 region studied after local injection of ^{3}H leucine and horseradish peroxidase. Brain Res 178:17–40

Simon H, Scatton B, Le Moal M (1980) Dopaminergic A10 neurons are involved in cognitive functions. Nature 286:150–151

Skoff BF, Mirsky AF, Turner D (1980) Prolonged brainstem transmission time in autism. Psychiatry Res 2:157–166

Small JG (1971) Sensory evoked responses of autistic children. In: Churchill DW, Alpern GD, De Myer MK (eds) Infantile autism. Thomas, Springfield, pp 224–242

Small JG, De Myer MK, Milstein V (1971) CNV responses of autistic and normal children. J Autism Child Schizophr 1:215–231

Snider SR, Snider RS (1977) Alterations in forebrain catecholamine metabolism produced by cerebellar lesions in the rat. J Neural Transm 40:115–128

Student M, Sohmer H (1978) Evidence from auditory nerve and brainstem evoked responses for an organic brain lesion in children with autistic traits. J Autism Dev Disord 8:13–20 (Erratum 9:309)

Swann AC, Maas JW, Hattox SE, Landis DH (1980) Catecholamine metabolites in human plasma as indices of brain function: effects of debrisoquin. Life Sci 25:1857–1862

Takahashi S, Kanai H, Miyamoto Y (1976) Reassessment of elevated serotonin levels in blood platelets in early infantile autism. J Autism Child Schizophr 6:317–326

Tanguay PE (1976) Clinical and electrophysiological research. In: Ritvo ER (ed) Autism: diagnosis, current research and management. Spectrum, New York, pp 75–84

Tanguay PE, Ornitz EM, Forsythe AB, Ritvo ER (1976) Rapid eye movement (REM) activity in normal and autistic children during REM sleep. J Autism Child Schizophr 6:275–288

Tanguay P, Edwards RM, Buchwald J, Schwafel J, Allen V (1982) Auditory brainstem evoked responses in autistic children. Arch Gen Psychiatry 39:174–180

Taylor MJ, Rosenblatt B, Linschoten L (1982) Electrophysiological study of the auditory system in autistic children. In: Rothenberger A (ed) Event-related potentials in children. Elsevier, Amsterdam, pp 379–386 (Developments in neurology, vol. 6)

Thierry AM, Blanc G, Sobe A, Stinus L, Glowinski J (1973) Dopaminergic terminals in the rat cortex. Science 182:499–501

Tsai L, Stewart MA (1982) Handedness and EEG correlation in autistic children. Biol Psychiatry 17:595–598

Walsh KW (1978) Neuropsychology: a clinical approach. Churchill Livingstone, Edinburgh

Walter WG, Aldridge VJ, Cooper R, O'Gorman G, McCallum C, Winter AL (1971) Neurophysiological correlates of apparent defects of sensorimotor integration in autistic children. In: Alpern GD, De Myer MK (eds) Infantile autism. Thomas, Springfield, pp 265–276

Weizman R, Weizman A, Tyano S, Szekely G, Weissman BA, Sarne Y (1984) Humoral-endorphin blood levels in autistic, schizophrenic, and healthy subjects. Psychopharmacology 82:368–370

Young JG, Cohen DJ, Brown SL, Caparulo BK (1978) Decreased urinary free catecholamines in childhood autism. J Am Acad Child Psychiatry 17:671–678

Young JG, Cohen DJ, Caparulo BK, Brown SL, Maas JW (1979) Decreased 24 hours urinary MHPG in childhood autism. Am J Psychiatry 136:1056–1057

Young JG, Cohen DJ, Kavanagh ME, Landis HD, Shaywitz BA, Maas JW (1981) Cerebrospinal fluid, plasma, and urinary MHPG in children. Life Sci 28:2837–2845

Young JG, Kavanagh ME, Anderson GM, Shaywitz BA, Cohen DJ (1982) Clinical neurochemistry of autism and associated disorders. J Autism Dev Disord 12:147–165

Yuwiler A, Freedman DX (1986) Neurotransmitter research in autism. In: Schopler E, Mesibov GB (eds) Neurobiological issues in autism. Plenum, New York, pp 263–284

Yuwiler A, Ritvo E, Geller E, Glousman R, Schneiderman G, Matsuno D (1975) Uptake and efflux of serotonin from platelets of autistic and nonautistic children. J Autism Dev Disord 5:83–98

Childhood-Onset Schizophrenia

P. E. Tanguay, R. Asarnow, and R. Strandberg

Although prepubertal children have been known since the turn of the present century to exhibit symptoms of "schizophrenia" (DeSanctis 1906), it has only been in the past 20 years that careful studies of childhood-onset schizophrenia have been reported. In the present chapter we will describe the acute symptoms of the disorder, review the few studies available which describe the prodromal features of the illness, and discuss what information we may learn from study of the children of schizophrenic mothers in regard to the phenotypic manifestations of schizophrenia in children. We will also review studies which attempt to identify core psychological impairments in schizophrenic children, and which examine certain information processing functions in these children. These studies indicate that schizophrenic children have deficits which parallel those seen in adult-onset schizophrenia. We will conclude with a discussion of treatment, as well as some practical suggestions for future research.

Symptoms

From studies which have relied upon DSM-III or similar diagnostic criteria (Green et al. 1984; Kolvin et al. 1971; Russell et al. 1986) it would appear that childhood-onset schizophrenia is rare (occurring less frequently than infantile autism) and cannot be diagnosed with certainty much below 8 years of age. The latter observation may .n part be a result of the difficulty in applying the adult criteria for "thought disorder" to younger children whose language and cognitive development are immature. It would appear, however, that other psychotic symptoms such as hallucinations are difficult to find in preschool children. The male-female ratio in childhood-onset schizophrenia would appear to be between 3:1 and 2:1.

All of the symptoms described in DSM III or DSM III-R for schizophrenia can be found in children under the age of 12 years. These include characteristic delusions, auditory hallucinations, and marked thought disorder. The children's affect is often flat, blunted, or inappropriate. In the sample of 35 children studied by Russell et al. (1986) auditory hallucinations were the most common schizophrenic symptom present (80%), followed by delusions (63%) and marked thought disorder (40%). Children in this group frequently exhibited several kinds of auditory hallucination. Hallucinatory experiences have occasionally been reported in nonpsychotic children (Kotsopoulos et al. 1987; Burke et al. 1985), very often following traumatic experiences or in association with severe and chronic stress. Although Kotsopoulos et al. reported that, like schizophrenic children, the nonpsychotic children who hallucinated heard voices, the voices were generally transitory in nature, and they tended to occur at bedtime. All of the children in the latter study quickly lost their hallucinations shortly after supportive treatment was begun.

Relationship of Childhood-Onset Schizophrenia to Pervasive Developmental Disorder and Infantile Autism

While there have been a number of retrospective studies of the prodromal features of adult-onset schizophrenia (Erlenmeyer-Kimling and Cornblatt 1987; Marcus et al. 1987), none have described that the subjects had anything at all resembling either pervasive developmental disorder or infantile autism in their early history. Conversely, the few follow-up studies of autistic persons (Kanner et al. 1972; Rutter 1978; Rumsey et al. 1985) have not found that autistic persons become schizophrenic. In one of the longer-term studies of this type, Rumsey et al. (1986) found that autistic adults had marked poverty of speech, poverty of content of speech, and affective flattening. While the latter symptoms are similar to those found in some schizophrenic subjects, the autistic group did not have characteristic delusions or hallucinations found in schizophrenia. An exception to the latter observations is a report by Petty et al. (1984) that described three autistic children followed from an early age who lost their symptoms of autism and began to exhibit symptoms of schizophrenia by middle childhood. It should be understood, however, that the latter subjects were from a very large population of autistic subjects seen over a period of years in a medical center known for its interest in the disorder. Schizophrenia has an estimated prevalence in the adult population of 1%, and it is possible that a similar percentage of autistic persons might be genetically at risk for schizophrenia, and eventually show symptoms of the disorder. The three cases reported by Petty et al. might represent such a subgroup.

More puzzling is the report by Watkins et al. (1988) of the prodromal symptoms of a series of children diagnosed prior to age 12 years as having childhood-onset schizophrenia. All of the children were found to have had various degrees of social, language, and emotional disturbances since early childhood. Seven of the 18 children had been described by psychologists and speech pathologists who had seen them prior to age 5 as having symptoms strongly resembling those seen in autistic children. This finding was quite surprising. It may indicate not so much that autism and schizophrenia are somehow related, but that early disturbances in language, social, and cognitive functioning can only be expressed by a limited number of specific symptoms and may antedate a number of childhood behavior disorders (including conduct disorder).

A broad view of the manifestations of schizophrenia in childhood is afforded by the large number of studies in which the offspring of schizophrenic mothers have been evaluated at various points during childhood and beyond (Érlenmeyer-Kimling and Cornblatt 1987; Marcus et al. 1987). A few findings have been consistently reported in a number of projects. A substantial percentage of the children at risk show soft neurological signs and perceptual-motor difficulties; they have also been described as having certain specific attention and information processing abnormalities. More recently, information processing abnormalities have been reported as present in children who actually have the diagnosis of childhood-onset schizophrenia (see below).

Attention and Information Processing Deficits

Psychobiological Studies of Schizophrenic Children

Do schizophrenic children have the same psychobiological impairments as schizophrenic adults? A number of studies have addressed this question and have also attempted to detail the nature and the diagnostic specificity of these impairments. These studies have been carried out by one of the authors (RA) and have been reported in several publications (Asarnow and Sherman 1984; Asarnow et al., in press).

To determine whether schizophrenic children and adults had similar attention impairments, we examined the performance of schizophrenic children on a task shown in prior research to detect impairment in adult-onset schizophrenia. This task, the forced choice, partial report span of apprehension task originally developed by Estes and Taylor (1964), indexes the efficiency of visual information processing. The task detects deficits in a variety of individuals who share the trait of vulnerability to schizophrenic disorder, including acutely disturbed and partially recovered schizophrenic adults, children at risk for schizophrenic disorder, the nonaffected parents of schizophrenic probands, and adults with no personal history of psychiatric disorder who obtain elevated scores on psychometric indices of psychosis proneness (Asarnow et al., in press).

In the first of a series of three studies (study I), schizophrenic children, mental age matched normal children, and a group of younger normal children were administered the span task (see Asarnow and Sherman 1984 for a more detailed description of these studies). Inspection of Fig. 1, which presents span data for the three groups, indicates that for all groups there was a decrement in performance as the number of letters in the array was increased. At all array sizes except the one-letter array, the schizophrenic children detected significantly fewer target stimuli than the mental age matched normals. The schizophrenic and younger normal groups did not differ from each other on any of the arrays.

The fact that schizophrenic children show the same kind of impairment on the span task as other groups of individuals vulnerable to adult-onset schizophrenia

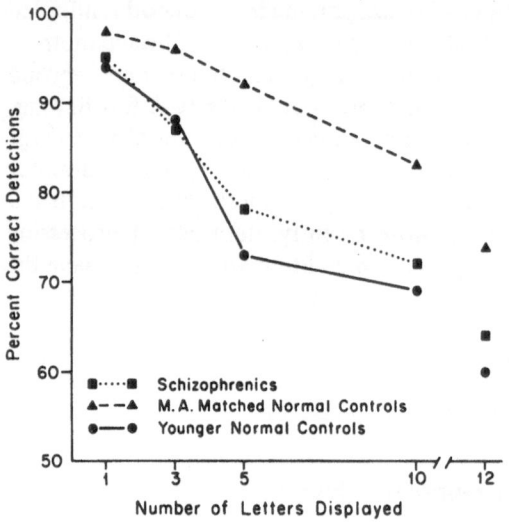

Fig. 1. Span data for the three groups of children investigated. *M.A.*, mental age

suggests that childhood-onset schizophrenia falls on the continuum of the "schizophrenia spectrum." In general, impaired attention tends to be a characteristic of adult schizophrenics with a poor premorbid history (Cromwell 1975). That schizophrenic children are also impaired suggests that attentional impairments are characteristic of the more severe forms of schizophrenia, i.e., those with the earliest onset and with the poorest prognosis.

A second study (study II) tried to define more precisely the cognitive processes underlying the impaired performance of schizophrenic children on the partial report span task. This was accomplished by manipulating task parameters to evaluate the contribution of a number of cognitive processes tapped by the span task. Four cognitive processes were examined: (a) information acquisition strategies, (b) stimulus discrimination skills, (c) general response biases, and (d) fatigue/learning effects.

We were particularly interested in examining information acquisition strategies. The role of information acquisition strategies was examined by determining the effect of the location of target stimulus within the array on the accuracy of detection for the three groups.

The results of this study suggested that the impaired performance of the schizophrenic children on the span task was *not* due to (a) a general fatigue habituation effect—they showed the same increase in accuracy level over four trial blocks as the other two groups of subjects—or (b) the differential effects of some response bias—schizophrenic children showed the same response bias as the other two groups of children, a tendency to report Ts slightly more often than Fs. The impaired span performance of the schizophrenic children was not due to poor stimulus discrimination skills. All three groups detected fewer target stimuli when the target was adjacent to five letters than when it was not immediately adjacent to any other letters or was adjacent to eight letters.

Of great interest, the schizophrenic children showed the same pattern of information acquisition as was seen in the mental age matched control group. Both the mental age matched controls and the schizophrenic children were more likely to detect target stimuli presented in the upper half of the screen than target stimuli presented in the lower half of the screen. In contrast, the younger normal children showed an equal probability of detecting the target stimuli across stimulus locations. Thus, in one important way the schizophrenic children did not function merely as the young children did. These results suggest that the schizophrenic children used the same information acquisition strategy as the older mental age matched normal control group, but less efficiently since their overall performance levels were lower than those of the mental age matched control children.

The sclizophrenic children in studies I and II showed impaired performance on two partial report span tasks relative to a group of children matched to them on general intellectual abilities. This indicates that the information processing impairment found in the schizophrenic children is a relatively specific deficit which is not attributable to a global impairment of intellectual abilities.

Adequate iconic and immediate memory are necessary for successful performance on the partial report span tasks employed in the previous two studies. Could the impaired performance of schizophrenic children be due to deficient iconic and/or immediate memory? We tested this hypothesis in a third study (study III) by presenting the children with another version of the span task (a full report version) which makes greater demands on iconic and immediate memory and less demands on early attentive processing than do partial report versions of the span. The full report version of the span task assesses immediate and iconic memory by requiring subjects to report out loud as many of the tachistoscopically presented letters as they can remember seeing. While the partial report version of the span of apprehension also involves short-term memory, the full report version places a greater demand on short-term memory since stimuli have to be held in memory while the verbal response is being executed.

The schizophrenic children did not differ from the mental age matched controls in the number of letters reported in the full report span task. This suggests that the impaired performance of the schizophrenic children on partial report span tasks was not due to deficient iconic and/or immediate memory. Perhaps more importantly, the results of the third study indicated that subjects were not performing on the partial report span by simply naming all the letters they could retrieve from iconic and/or immediate memory. The mean number of letters reported (2 to 4) on the full report version of the span was far less than the number of letters which had to be processed (6 to 9) on the partial report version to yield the detection rates obtained in studies I and II.

What strategy do children use to perform on partial report span tasks? The data from study I suggest that all groups of subjects were engaged in a serial search task when performing partial report span tasks. Serial processing is characterized by the direction of attention "serially to different locations, to integrate the features within the same spatio-temporal 'spotlight' into a unitary percept" (Treisman and Gelade 1980). One of the defining characteristics of a serial mode of processing is that there is an incremental cost (increased reaction time or errors) associated with detecting the presence of a target in displays with increasing numbers of distractors. This is exactly what happened on the partial report span task in study I. As the number of distractors increased fron 0 to 2 to 4 to 9, the probability of detecting the target decreased in all groups. Moreover, the fact that the schizophrenic children and the younger normals showed greater "cost" with increased number of distractors than the mental age matched normals suggests that their serial search was either initiated more slowly or employed less efficiently than that of the older control group.

A convergent result emerges from study II. Both the schizophrenics and the mental age matched normals were more likely to detect target stimuli located in the upper quadrants than target stimuli located in the lower quadrants. This suggests that both schizophrenics and the older normals consistently began their serial search in the upper quadrants, and that their iconic image of the stimulus display faded before they could finish processing stimuli in the lower quadrants.

From these data, we cannot determine whether the advantage of the older normals was due to their having found a better set of features to search for, or whether all groups searched for the same set of critical features but the older normals were more efficient in either the initiation or the application of their search. Cash et al. (1972) suggested that schizophrenic adults may have special difficulty in ignoring irrelevant features and therefore may have problems with tasks involving speeded search. On the other hand, there is evidence (Russell et al. 1986) suggesting that schizophrenics may be delayed in the initiation of visual search.

There is a general consensus (Pick and Frankel 1974) that much of perceptual and attentional development in children consists in a transition towards more active control of information acquisition. This is achieved, in part, through more efficient deployment of attentional capacity. It is interesting to note that it was in precisely these later developing systems that schizophrenic children showed their characteristic impairments. The results of these studies suggest that schizophrenic children, like other schizophrenic individuals (Neale and Oltmans 1980), show information processing impairments under conditions which entail controlled processing. Controlled processes (which are often distinguished from automatic processes) are momentary sequences of mental operations that are under the subject's control and which make demands on a central, limited pool of information processing capacity.

Comparative Studies of Schizophrenic and Autistic Children

The next two studies compared the cognitive functioning of autistic and schizophrenic children. These studies tested the hypothesis that schizophrenic children have impairments in controlled attentional processes using tasks which require more temporally extended processing than the partial report span task. These studies also attempted to determine the diagnostic specificity of the impairments found in schizophrenic children. Are impairments in controlled attentional processes a relatively specific characteristic of children with a schizophrenic disorder, or are they a more general characteristic of children with severe psychiatric disorders? Children with an autistic disorder are a useful comparison group because autism, like schizophrenia, is a severe, early-onset disorder in which the number of male cases exceeds the number of female cases. Even though cognitive impairments are considered to be key features of both disorders, the two studies below were among the first to compare directly the cognitive functioning of children with autistic and schizophrenic disorders.

An initial study (Asarnow et al. 1987) compared the performance of non-retarded autistic and schizophrenic children on a standardized measure of intellectual functioning, the WISC-R (Wechsler 1974). The performance of the autistic and schizophrenic children was compared on three factors, verbal comprehension, perceptual organization, and factor three (a "distractibility" factor), which have consistently been found to account for the variance on the WISC-R (Sattler 1974; Kauffman 1979). Since there are age-related changes in the factor structure of intellectual tests, the two groups were matched in age at time of testing as well as full-scale IQ.

The autistic and schizophrenic children did not differ significantly on the verbal factor, but the schizophrenic children had significantly lower scores than autistic children on both the perceptual organization and distractibility factors (Fig. 2). Of particular importance, the scores of the schizophrenic children on factor 3 were significantly lower than those of the autistic children, below the range of normal children, and significantly lower than the scores they obtained on the verbal comprehension and perceptual organization factors. The subtests included in factor 3 (arithmetic, coding, and digit span) tap a range of abilities including attention, short-term memory, visual-motor coordination, speed of responding, and mental arithmetic (Sattler 1974). While there is dispute over the general meaning of factor 3, the subtests included in this factor make extensive demands on controlled attentional processes because of their require-ments for working memory, attention, and speed of responding. These results are thus consistent with the hypothesis that schizophrenic children have impairments in controlled attentional processes. These impairments appear to be somewhat circumscribed since the autistic and schizophrenic children were matched in full-scale IQ and the schizophrenic children performed within the normal range on tasks tapping overlearned abilities (such as those loading on the verbal factor) which make less demands on controlled attentional processes.

The autistic children did not show a global deficit in language functioning, as measured by verbal IQ. The comprehension subtest was the only subtest that the autistic children scored significantly lower on than the schizophrenic children. The sensitivity of the comprehension subtest to the relatively subtle impairments found in nonretarded autistic children may result from the demands this task makes for the integration of two abilities, verbal abstraction (Rutter

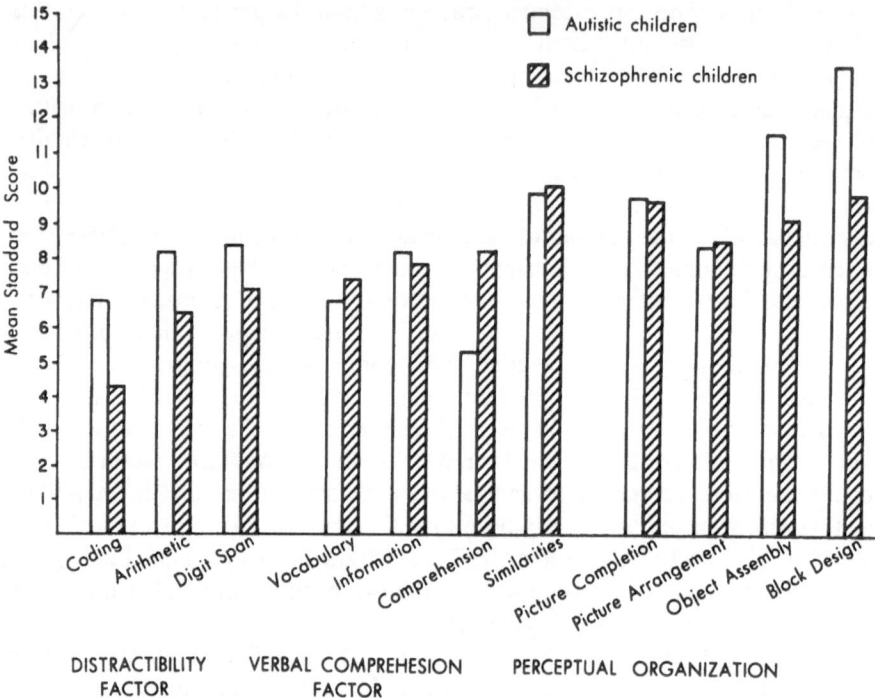

Fig. 2. Scores of schizophrenic and autistic children on three factors from the WISC-R

1983) and social cognition (Sigman et al. 1986), hypothesized to be core cognitive impairments in autism.

A second study (see Schneider and Asarnow 1987 for details) attempted to circumscribe further the cognitive impairment found in schizophrenic children by testing subjects on two neuropsychological tasks which make extensive demands on controlled attentional processes, the Wisconsin Card Sorting Test (WCST, Grant and Berg 1981) and Rey's Tangled Line Test (RTLT, Rey 1964). The WCST makes extensive demands on both verbal mediation and the subject's ability to engage in temporally extended problem solving. Thus, the WCST taps the hypothesized core deficits in both autism and schizophrenia. Autistic (Hoffman and Prior 1982) and schizophrenic (Stuss et al. 1983) individuals have shown deficits on this task. To differentiate the core cognitive impairments underlying the performance of the autistic and schizophrenic children on the WCST, subjects were explicitly taught the appropriate sorting principles halfway through this task. If autistic children are unable to use language to symbolize their experience and regulate their behavior, teaching them a verbal strategy should have little effect on their sorting performance. In contrast, previous studies (e.g., Koh 1978) have indicated that when schizophrenic individuals are taught the appropriate strategy their performance can be normalized on certain memory tasks. Consequently, it was predicted that teaching schizophrenic children the correct sorting principle would improve their WCST performance while autistic children would not improve.

The RTLT is a guided visual search task which requires the maintenance of set over time and space. Since this task requires controlled attentional processes for an extended period of time, it was predicted that schizophrenic children would show impaired performance on this task.

The above tests were administered to groups of schizophrenic, autistic, and normal control children matched in mental and chronological age. As predicted, the schizophrenic children performed significantly worse on the RTLT than the normal children. The impaired performance of the schizophrenic children on the RTLT is consistent with the hypothesis that schizophrenic children should show impairments on tasks which require temporally extended, controlled processes.

The major difference between the schizophrenic and autistic children on the WCST was that the schizophrenic children increased the number of non-perseverative errors (random responding) from the first to the second half of the WCST while autistic and normal children showed no change. Providing a sorting strategy, contrary to our predictions, did not facilitate the performance of the schizophrenic children on the WCST. Rather, their performance significantly deteriorated after they were taught the relevant sorting principles. A similar manipulation had a deleterious effect on the WCST performance of chronic schizophrenics with prefrontal leukotomies (Stuss et al. 1983). In addition, Pueschel (1980; Neuropsychological assessment and "hypothesis testing" in schizophrenic and brain damaged patients, unpublished manuscript) found that providing schizophrenic individuals with verbal directions impaired their performance on a hypothesis-testing task. The relevant information provided in this study requires the momentary integration of information from a variety of sources to direct an ongoing activity. Providing task-relevant information to the schizophrenic child, whose momentary processing capacity may be already overburdened, may further encumber processing capacity, resulting in the paradoxical effect of further impairment in performance.

To summarize, the data to date suggest that schizophrenic children show impaired performance on tasks which make extensive demands on controlled attentional processes while performing relatively normally on tasks which make less demands on controlled attentional processes. The impairments in controlled attentional processes observed in schizophrenic children may reflect abnormalities in the allocation and/or recruitment of processing capacity.

Psychophysiological Studies of Schizophrenic Children

Psychophysiological studies may provide a means of more directly examining the recruitment and allocation of processing capacity. It has been demonstrated that identified components of scalp-recorded event-related potentials (ERPs) are sensitive to variations in preparedness (CNV, contingent negative variation), selective attention (Nd, negative difference wave or PN, processing negativity), pattern discrimination (Na, early negative wave), stimulus categorization (N2, negative wave around 200 ms post stimulus), and response set or context updating (P3) (for reviews see Donchin et al. 1978; Ritter et al. 1984). The amplitude of these components can be viewed as an estimate of the resources allocated to each of these processes.

While there are few ERP studies of childhood—onset schizophrenia, diminished CNV, Nd, and P3 (positivity around 300 ms post stimulus) amplitudes have been well documented in schizophrenic adults (for a recent review see Pritchard 1986). These findings suggest that adult schizophrenics may be deficient in their processing capacity and/or their allocation strategy.

In studies carried out by one of the authors (RS) and his colleagues, EEG activity was recorded from normal and schizophrenic children during performance of the partial report version of the span task to evaluate recruitment and allocation of processing capacity in these younger individuals (mean age 11.5 years). In the initial study of 10 schizophrenic and 13 mental age matched normals (Strandberg et al. 1984), the results were similar to those previously observed in adults. In the schizophrenic children, CNV activity was smaller and slower to develop and resolve than in normals. N1 and P3 amplitudes were similarly reduced in this group. This was interpreted as evidence that the schizophrenics were impaired in their ability to regulate processes involved in mobilization and direction of attention as well as discrimination of target stimuli.

The span stimuli in this study were 250 12-letter arrays that could be grouped into four levels of difficulty. While behavioral performance was sensitive to processing demand in both groups (declining as difficulty increased), only the normal children showed an increase in N1 amplitude (negative wave around 100 ms post stimulus) with each increment in task difficulty. The normal variation in N1 with task demand appears to be the result of changes in endogenous processing negativities (Nd, PN, or Na) which partially overlap the N1 component. While N1 amplitude is largely dependent on exogenous factors (physical properties of the eliciting stimuli), the net ERP in this time region can also include endogenous activity associated with task-specific processing. For example, the Nd or PN components reflect additional CNS activity involved in selective attention (Hansen and Hillyard 1980; Näätänen and Michie 1979) and Na reflects activity associated with pattern discrimination (Ritter et al. 1982, 1983).

Evidence from both ERP studies and magnetoencephalography suggests that Nd or PN is generated in sensory cortex (see Näätänen and Picton 1987). However, Knight et al. (1981) have demonstrated that the development of Nd is dependent on regulatory inputs from frontal cortex. Thus, it is possible that the N1 results in Strandberg et al. (1984) could be explained by frontal impairments in the schizophrenic children. In fact, abnormal, asymmetric frontal activity was observed in the schizophrenic children which spanned the entire recorded epoch.

To determine whether differences in the N1 measures in the Strandberg et al. (1984) study were indeed the result of differences in endogenous processing, ERPs were recorded from 13 schizophrenic and 17 normal children utilizing the span stimulus arrays in two conditions which differ only in task instructions (Strandberg et al., to be published). The first was a reaction time task in which the children were required to press a button as quickly as possible after the onset of each visual span array. Following this, instructions for the partial report span were given, and the stimuli from the first condition were repeated. By subtracting primarily exogenous, low task demand ERP activity recorded during the RT task from the ERPs obtained during the span processing, a difference potential was obtained which reflected that portion of CNS activity associated with the additional, endogenous processing required to make the span discrimination.

Schizophrenic children produced significantly less endogenous activity in response to the span processing demands than normal children at all leads ($F = 5.96$, $P < 0.023$). While seven of the ten schizophrenic children were drug free in our first study, in the second study five were off medications, six were receiving maintenance doses of neuroleptics, and two had just begun drug trials. We compared the five drug-free children with the six children for whom optimal doses levels had been achieved, and no significant differences were obtained.

It is important to emphasize that these difference potentials measure relative activity levels between two tasks which differ in processing demand (RT and span). While N1 amplitude was greater during span processing than during the RT condition in both normal and schizophrenic children, this condition effect was significant only for the normal group. In the span condition, N1 amplitude was quite similar in the two groups, but they differed in net processing negativity (the difference potential) because in the RT condition the N1 was considerably larger in the schizophrenic than in the normal children. Thus, it is possible that the schizophrenic children did not differ in their processing capacity from the normal children, but rather were deficient in their allocation strategy, allocating excessive capacity to the simpler RT task and failing to increase this capacity adequately in response to the greater span processing demands. It should also be noted that, as expected from the purely behavioral studies discussed above, this information processing deficit in schizophrenic children was apparent in the ERP record within the first 200 ms of stimulus processing.

Treatment

Perhaps because childhood-onset schizophrenia is relatively rare, and because investigators have been reluctant to use major antipsychotic medications on children, study of the treatment of childhood-onset schizophrenia has been relatively neglected. In general, and this has been our experience on the children's psychiatric wards at UCLA, the emphasis in therapy should be to provide external structure, to teach the children cognitive coping strategies such as "thought stopping" to divert their attention from hallucinations, and to allay their fears. The family environment of schizophrenic children, like that of schizophrenic adults, is characterized by high rates of expressed emotion and communication deviance (Goldstein 1987). These family stresses may increase the risk of developing schizophrenia and appear to contribute to the poor outcomes observed in schizophrenic children (Asarnow and Goldstein 1986). Decreasing the expression of expressed emotion through family intervention has been suggested to be therapeutically effective in adult-onset schizophrenia (Anderson 1986; Falloon 1985).

Research Opportunities

There are a number of areas of study which seem particularly important in regard to childhood-onset schizophrenia. One pertains to the family genetics of the disorder: Is the incidence of schizophrenia higher in families in which there are childhood-onset forms of the disorder, as has been found for certain other genetically mediated illnesses, and are the types of attention and information processing disorders seen in schizophrenic children also found in their relatives? A second important focus would be upon imaging studies: With the development of newer imaging techniques which involve relatively small doses of radioactive materials, it will be possible to carry out studies of brain activity while schizophrenic children are engaged in tasks in which they are deficient, and in particular the attention and information processing tasks described earlier. Lastly, a search for more effective methods of treatment must be undertaken. Are psychotropic medications effective in treating some of the symptoms of schizophrenia in children, and if so, which ones? It would appear that in the last decade there has been a greater interest in childhood-onset schizophrenia; perhaps the next decade will see this interest extended to issues of treatment.

References

Anderson CM (1986) Schizophrenia and the family: a practitioner's guide to psychoeducation and management, Guilford, New York
Asarnow J, Goldstein M (1986) Schizophrenia during adolescence: a developmental perspective on risk research. Clin Psychol Rev 6:211–235
Asarnow RF, Sherman, T (1984) Studies of visual information processing in schizophrenic children. Child Dev 55:249–261
Asarnow RF, Tanguay PE, Bott L, Freeman BJ (1987) Patterns of intellectual functioning in non-retarded autistic and schizophrenic children. J Child Psychol Psychiatry 28:273–280
Asarnow RF, Granholm E, Sherman T (in press) Span of apprehension in schizophrenia. In: Steinhauer S, Gruzelier J, Zubin J (eds) Handbook of schizophrenia: neuropsychology, psychophysiology and information processing. Elsevier, Amsterdam

Burke P, Delbeccara H, McCauley E, Clark C (1985) Hallucinatious in Children. J Am Acad Child Psychiatry 24:71-75

Cash TF, Neale JM, Cromwell RL (1972) Span of apprehension in acute schizophrenia: full-report procedure. J Abnorm Psychol 3:322-326

Cromwell RL (1975) Assessment of schizophrenia. Rev Psychol 26:593-619

DeSanctis S (1906) Sopra alcure varieta della demenza precoce. Riv Sper Freniatr Med Leg Alienazioni Ment 141-165

Donchin E, Ritter W, McCallum W (1978) Cognitive psychophysiology: the endogeneous components of the ERP. In: Callaway E, Tueting P, Koslow S (eds) Event-related brain potentials in man. Academic, New York, pp 349-442

Erlenmeyer-Kimling L, Cornblatt B (1987) The New York high-risk project: a follow-up report. Schizophr Bull 13:451-461

Estes WK, Taylor HA (1965) A detection method and probabilistic models for assessing information processing from brief visual displays. Proc Natl Acad Sci USA 52:446-454

Falloon IRH (1985) Family management of schizophrenia: a study of clinical, social, family, and economic benefits. Johns Hopkins University Press, Baltimore

Goldstein M (1987) The UCLA high risk project. Schizophr Bull 13:505-514

Grant DA, Berg EA (1981) Wisconsin card sorting test. Psychological Assessment Resources, Odessa, Fl

Green WH, Campbell M, Hardesty AS, Grega DM, Padron-Gayol M, Shell J, Erlenmeyer-Kimling L (1984) A comparison of schizophrenic and autistic children. J Am Acad Child Psychiatry 23:389-409

Hansen HC, Hillyard SA (1980) Endogeneous brain potentials associated with selective auditory attention. Electroencephalogr Clin Neurophysiol 49:277-290

Hoffman WL, Prior MR (1982) Neuropsychological dimensions of autism in children: a test of the hemisphere dysfunction hypothesis. J Clin Neurol 4:513-531

Kanner L, Rodriquez A, Ashenden B (1972) How far can autistic children go in matters of social adaptation? J Autism Child Schizophr 2:9-13

Kauffman AS (1979) Intelligence testing with the WISC-R. Wiley, New York

Koh SD (1978) Remembering of verbal materials by schizophrenic young adults. In: Schwartz S (ed) Language and cognition in schizophrenia. Erlbaum, New Jersey, pp 55-99

Knight RT, Hillyard SA, Woods D, Neville HJ (1981) The effects of frontal cortex lesions on event-related potentials during auditory selective attention. Electroencephalogr Clin Neurophysiol 52:571-582

Kolvin I, Ounsted C, Humphrey M, McNay A (1971) Studies in childhood psychosis. II. The phenomenology of childhood psychoses. Br J Psychiatry 118:385-395

Kotsopoulos S, Kanigsberg J, Cote A, Fiedorowicz C (1987) Hallucinatory experiences in non-psychotic children. J Am Acad Child Adolesc Psychiatry 26:375-380

Näätänen R, Michie PT (1979) Early selective attention effects on the evoked potential. A critical review and reinterpretation. Biol Psychol 8:81-136

Näätänen R, Picton T (1987) The N1 wave of the human electric and magnetic response to sound: a review and an analysis of the component structure. Psychophysiology 24:375-425

Neale JM, Oltmans TF (1980) Schizophrenia. Wiley, New York

Marcus J, Hans SL, Nagler S, Auerbach JG, Mirsky AF, Aubrey A (1987) Review of the NIMH Israeli kibbutz-city study and the Jerusalem infant development study. Schizophr Bull 13:425-437

Petty LK, Ornitz EM, Michelman JD, Zimmerman EG (1984) Autistic children who become schizophrenic. Arch Gen Psychiatry 41:129-135

Pick AD, Frankel GW (1974) A developmental study of strategies of visual selectivity. Child Dev 45:1162-1165

Pritchard WS (1986) Cognitive event-related potential correlates of schizophrenia. Psychol Bull 100:43-66

Pueschel (1980) Neuropsychological assessment and "hypothesis testing" in schizophrenie and brain damaged patients, unpublished manuscript

Rey A (1964) L'Examen clinique en psycologie (The clinical exam in psychology). Presses Universitaires de France, Paris

Ritter W, Simson R, Vaughan HG, Macht M (1982) Manipulation of event-related potential manifestations of information processing stages. Science 218:909-911

Ritter W, Simson R, Vaughan HG (1983) Event-related potential correlates of two stages of information processing in physical and semantic discrimination tasks. Psychophysiology 20:168-179

Ritter W, Ford JM, Gaillard AWK, Harter MR, Kutas M, Näätänen R, Polich J, Renault B, Rohrbaugh J (1984) Cognition and event-related potentials. I. The relation of negative potentials and cognitive processes. In: Karrer R, Cohen J, Tueting P (eds) Brain and information: event-related potentials. Ann NY Acad Sci 25:24–38

Rumsey JM, Rapoport JL, Sceery WR (1985) Autistic children as adults: psychiatric, social, and behavioral outcomes. J Am Acad Child Psychiatry 24:465–473

Rumsey JM, Andreasen NC, Rapoport JL (1986) Thought, language, communication, and affective flattening in autistic adults. Arch Gen Psychiatry 43:771–777

Russell AT, Bott L, Sammons C (1986) The phenomenology of childhood-onset schizophrenia. Poster session. Meeting of the American Academy of Child and Adolescent Psychiatry, Los Angeles

Rutter M (1978) Diagnosis and definition of infantile autism. J Autism Child Schizophr 8:139–161

Rutter M (1983) Cognitive deficits in the pathogenesis of autism. J Child Psychol Psychiatry 24:27–41

Sattler JM (1974) Assessment of children's intelligence. Saunders, Philadelphia

Schneider SG, Asarnow RF (1987) A comparison between the cognitive/neuropsychological impairments of non-retarded autistic and schizophrenic children. J Abnorm Child Psychol 15:29–46

Sigman M, Ungerer J, Mundy P, Sherman T (1986) Cognition. In: Cohen DJ, Donnellan A (eds) Handbook of autism and pervasive developmental disorders. Wiley, New York

Strandberg RJ, Marsh JT, Brown WS, Asarnow RF, Guthrie D (1984) Event-related potential concomitants of information processing dysfunction in schizophrenic children. Electroencephalogr Clin Neurophysiol 57:236–253

Stuss DT, Benson DF, Kaplan EF, Weir WS, Naesser MA, Lieberman I, Ferrill D (1983). The involvement of orbitofrontal cerebrum in cognitive tasks. Neuropsychologia 21:235–248

Treisman AM, Gelade G (1980) A feature-integration theory of attention. Cognitive Psychol 12:97–136

Watkins JM, Asarnow R, Tanguay PE (1988) Symptom development in childhood onset schizophrenia. J Child Psychol Psychiatry 29:865–878

Wechsler D (1974) Wechsler intelligence scale for children, revised. Psychological Corporation, New York

Children at Risk for Schizophrenia

H. van Engeland

Introduction

Research into risk factors can lead to improved early detection and attempted prevention of a particular disorder. This is probably why, in the past 20 years, risk research as part of schizophrenia research has increased considerably. In this research context risk factors are variables in a person's life which predict his or her increased risk of ever developing schizophrenia. It is customary to distinguish between intrinsic and extrinsic risk factors. *Intrinsic* risk factors are individual-related variables such as genetic constitution and consequences of perinatal damage. Social variables such as prosperity level, family composition, and quality of relations within the family, but also decisive life events (death of a parent or partner, divorce, loss of a job), are described as *extrinsic* risk factors.

This paper attempts to summarize the principal results of the hitherto published research undertaken to identify *intrinsic* risk factors, and to indicate lines along which this type of research might be continued in the near future.

A wide diversity of research strategies have been used in researching intrinsic risk factors. The early studies were mostly of cross-sectional design, but these will not be mentioned here (for a review, see Garmezy and Streitman 1974; Garmezy 1974). In these studies the variables distinguishing patients from nonpatients could not be clearly identified as either causes or effects of the diseases involved, and consequently researchers switched to longitudinal approaches which initially were retrospective. Reports on prospective longitudinal studies have appeared in the past decade.

Retrospective Studies

Retrospective research mostly aims at finding peculiarities of behavior in the (early) childhood years of schizophrenic patients. Several anamnestic studies (Bleuler 1961; Langfeldt 1956; Bleuler 1972; Astrup et al. 1962) report conspicuous premorbid behavior anomalies in childhood in some 50% of all schizophrenic patients studied. In this category the study published by Bleuler (1972) is undoubtedly the most detailed. Bleuler found that in at least

50% of both a Swiss (n = 208) and a New York cohort of schizophrenics (n = 190) the social emotional behavior during primary school years could be described as schizoid. In retrospect, two-thirds of these schizoid children were believed to have shown unmistakable psychiatric disturbances. The parents described these children as having relational and emotional problems and as being difficult to fathom, easily hurt, and socially isolated from their age mates.

Retrospective anamnestic research demands much of the memory of the persons questioned. They are often asked questions about events, situations, and developments of 10 or more years ago. Data thus obtained cannot possibly be totally reliable. In an attempt to avoid this methodological problem, some investigators have performed "objective" retrospective studies. They examined only those schizophrenic patients who had had a child psychiatric examination in their childhood or on whom detailed school reports as to their social emotional behavior were still available. The results of these objective retrospective studies show that the childhood of a number of schizophrenic patients is characterized not only by relational problems and social isolation but also by emotional instability and antisocial behavior (Robins 1966; Frazee 1953; Nameche et al. 1964; Offord and Cross 1969; Roff et al. 1976; Watt et al. 1970; Watt 1972, 1978; Hartmann et al. 1984). The studies reported by Hartmann et al. (1984) and Watt (Watt et al. 1970; Watt 1972, 1978) can be described as methodologically most acceptable. Hartmann et al. (1984) studied the data collected by the Gluecks on some thousand boys from Boston; their aim was to study the development and prognosis of delinquency in this population. At the time of the Gluecks' study 500 boys already showed delinquent behavior, while the other 500 functioned inconspicuously. The age at which the boys were examined ranged from 10 to 17 years. Forty years later 24 of the boys then examined were suffering from a schizophreniform psychosis. Subsequent analysis revealed that the following characteristics had been of predictive value with regard to the development of schizophrenia:

—Unusual anxiety, at home as well as at school, and especially with regard to anything new, unexpected, and unknown
—Lack of object constancy; no or poor contacts with age mates, marked craving for attention, and tyrannizing behavior toward the parents
—Excessive daydreaming, increased distractibility, and a tendency toward increased associative thinking
—Poor scholastic achievements and a low sense of self-esteem

Watt et al. (Watt et al. 1970; Watt 1972, 1978) studied data collected at school on the social emotional behavior of children who subsequently developed schizophrenia. These school data indicated that "preschizophrenic" boys showed a type of deviant behavior which differed from that of preschizophrenic girls. Boys more frequently showed a conduct disorder with aggressiveness without social involvement, and their (school) environment described them as irritable, negativistic, easily hurt, inhibited and depressed. Girls were mostly described as extremely inhibited, vulnerable, and tending to withdraw from contact.

This objective retrospective approach is not without methodological problems either. For example, only those schizophrenic patients are studied who have a history of child psychiatric care. However, only a minority of all schizophrenic patients have a childhood history of problems or unmanageability (Bleuler 1972). Moreover, only one-third of all children with child psychiatric problems receive psychiatric/psychological examinations and treatment; boys and acting-out problems are reported to be markedly overrepresented in this respect (van Engeland 1987). The data on childhood functioning, moreover, are often collected from different points of view and with the aid of free interviews, and consequently the results of the various objective retrospective studies cannot always be compared without qualification. Because the retrospective studies are probably not representative for the total schizophrenic population, and because the studies are not always comparable, results of these studies should be interpreted with caution. Only where the results of the studies converge and in addition prove to tally with results collected by cross-sectional studies within prospectively designed longitudinal research may these results be assumed to have some validity.

Prospectively designed studies compare the behavior of children with a schizophrenic parent (children at increased risk of schizophrenia due to genetic causes) with that of children of normal parents. Five studies of this type demonstrated that—as a group—children of schizophrenic parents are characterized by emotional instability, expressed in extreme sensitivity to stress, irritability, anxiety, low frustration thresholds, and marked affective fluctuations (Janes et al. 1984; John et al. 1982; Rolf 1972; Watt et al. 1982; Weintraub et al. 1978; Parnas et al. 1982). Four other prospective studies revealed that the behavior of children of schizophrenic parents is significantly more often characterized by reticence, a tendency to withdraw, and social isolation (MacCrimmon et al. 1980; Mednick and Schulsinger 1968; Weintraub et al. 1978; Sohlberg and Yaniv 1985), and some studies also revealed more destructive and aggressive behavior in children of schizophrenic parents (Watt et al. 1982; Weintraub and Neale 1984; Weintraub et al. 1978; Rolf 1972). Unlike the retrospective studies, the prospective studies revealed no systematic sex difference.

The results of retrospective research converge with those of prospective studies. Some schizophrenic patients had a childhood history of conspicuous behavior; like some children of schizophrenic parents, they were *socially isolated, emotionally unstable,* and inclined to show *aggressive acting-out behavior.* Could this triad serve as a schizophrenia risk indicator? We do not think so! The specificity of this triad is too low. Follow-up studies (Robins 1966) have shown that certainly not all children with the above-described behavior triad develop schizophrenia at a later age; moreover, a similar syndrome has also been observed in children of depressive parents (Weintraub and Neale 1984; Weintraub et al. 1978). In addition, the *sensitivity* of the triad in relation to schizophrenic behavior is insufficient; after all, about 50% of all schizophrenic patients have no childhood history of conspicuous behavior. Schizophrenia risk research which focuses *solely* on parameters of social emotional behavior is therefore unlikely to open up further perspectives; such research would seem to be on the wrong track.

Prospective Studies

Although prospective studies also consider social emotional behavior of high-risk children (see the above), the bulk of the research effort focuses on biopsychological schizophrenia risk indices in a search for "biological markers" which are supposed to be the phenotypic expression of a genetically determined schizophrenia risk. If it is to be a valid risk indicator, such a marker should meet several requirements (Garver 1987):

—It should be reliably measurable or determinable by several investigators using noninvasive measuring methods.
—Its distribution over a schizophrenic population should differ from that over an average population.
—In schizophrenic patients it should be demonstrable not only during psychotic episodes but also during remissions: the variables should therefore above all mark a trait and not only a psychotic state.
—The marker should occur more often in offspring of a schizophrenic patient (before they develop a schizophrenic psychosis) than in a control population.
—The marker should show a skew distribution in such a high-risk population because only 10%–15% of the children of schizophrenic parents are actually carriers of a genetic risk.
—The presence of a marker in a child of schizophrenic parents should be demonstrably associated with the development of a schizophrenic disorder later in the course of that child's life.

Performing a prospective longitudinal study of schizophrenia risk factors poses a number of logistical problems:

—The lifetime schizophrenia risk approximates 1%, which implies that in a prospective study one must examine thousands of children in order to trace a few schizophrenic adults in the study population; this problem is usually avoided by focusing on children with a genetically determined increased schizophrenia risk: children with a schizophrenic parent prove to have a lifetime risk of 10%–15% (Mednick and Schulsinger 1968).
—Because most prospective studies cover a period of at least 10 years, the initially used measuring instruments are quite likely to become obsolete or no longer relevant by the time the study is closed. An example: In a number of prospective high-risk studies, schizophrenia in the parents was diagnosed on the basis of DSM-II criteria. Currently, however, DSM-III criteria are in use, which have introduced a more limited, more European definition of the

schizophrenia concept. Consequently interpretation of the older high-risk studies has become difficult. In view of this the parents have been rediagnosed according to DSM-III criteria on the basis of the available patients' records; and consequently the composition of the test groups has had to be changed.

—Not only the measuring instruments used but also the staff composition of the research team will vary during the long study period, and this may imperceptibly lead to changes in standards of evaluation.

—Finally there is the problem of the mobility of the test subjects, which may lead to a relatively high drop-out rate because some test subjects may become untraceable in the course of time.

Nevertheless, we have at this time access to the (preliminary) results of some 20 prospective high-risk studies in which some 1200 children of schizophrenic parents, 750 children of parents with other psychiatric diseases, and more than 1400 children of average parents have been intensively studied (Goldstein and Tuma 1987). The research has focused on neurological, neuropsychological, and neurophysiological risk parameters. So far as we know, neurochemical parameters have not so far been examined in prospective studies, perhaps because the measuring techniques are of a fairly invasive character. The results of these prospective studies should be regarded as preliminary results because most of the high-risk children have not yet reached the age at which a schizophreniform psychosis is most likely to develop; most of the children involved in these studies are now adolescents.

Neurological Risk Factors

Since Kraepelin, several investigators have pointed out that schizophrenic patients show minor neurological dysfunctions (MNDs) more frequently than other psychiatric patients (Tucker et al. 1975; Quitkin et al. 1976; Cox and Ludwig 1979; Luchins et al. 1980; Owens and Johnstone 1980). In a retrospective study Bender (1947) demonstrated it to be plausible that patients suffering from childhood schizophrenia had nearly all shown an exceedingly irregular development of neurological or motor functions, and in this context introduced the term "pan-developmental retardation" (PDR) or "neuro-integrative developmental disorder." One of Bender's students Fish (1957, 1975), contended that such a PDR constitutes the biological basis of "early onset, poor premorbid chronic schizophrenia," of which childhood schizophrenia is regarded as the most serious variety. This hypothesis became the starting point of the New York Infant Study (Fish 1984), in which children of schizophrenic mothers were followed from birth far into adulthood, and in which the development of neuromotor functions in high-risk children was for the first time accurately studied. Seven of the 12 high-risk children proved to show PDR, whereas in a control group of children of mothers with other serious psychiatric diseases only one out of 12 children showed PDR. Six high-risk children with PDR were found to suffer from a schizophrenic disorder at the age of 20–22 years. The only child in the control group with PDR was found at that age to be severely disturbed in social functioning but showed no psychosis. Three other prospective high-risk studies (Ragins et al. 1975; Mednick et al. 1971; Marcus et al. 1981) likewise revealed that babies and infants of schizophrenic parents showed more MNDs than control children.

Marcus et al. (1981) reported that some 50% of the high-risk children were suffering from a neurointegrative developmental disorder. On the other hand, however, there are studies with negative findings (McNeil and Kay 1984; Ragins et al. 1975; Hanson et al. 1976; Marcuse and Cornblatt 1986). In the Swedish study of McNeil and Kay (1984) babies of schizophrenic mothers were in fact found to show less MNDs than those of mothers who had suffered from some other types of psychosis.

Interpretation of neurological findings obtained in babies generally poses problems. According to Touwen (1979), MNDs cannot be reliably identified until the age of 7 years or older. In eight prospective high-risk studies (Fish 1984; Marcus 1974; Erlenmeyer-Kimling et al. 1984b; Hanson et al. 1976; Marcuse and Cornblatt 1986; Rieder and Nichols 1979; Orvaschel et al. 1979) school-age children were neurologically examined. In all these studies high-risk children proved to show more MNDs than the controls. It is doubtful, nevertheless, whether there is a specific relationship between the incidence of MNDs in childhood and subsequent development of schizophrenia. Erlenmeyer-Kimling et al. (1982) also found significantly more MNDs in children of parents suffering from affective psychoses, and in a follow-up study of children with MNDs Schaffer et al. (1984) found a correlation with anxiety disorders and affective disorders in adolescence. Children with MNDs show an increased vulnerability to subsequent psychiatric morbidity; a specific correlation with schizophrenia, however, has yet to be demonstrated convincingly.

Attention and Information Processing

As early as 1913 Emil Kraepelin wrote of patients suffering from dementia praecox (schizophrenia): "It is quite common for them to lose both inclination and ability on their own initiative to keep attention fixed for any length of time" (Kraepelin 1919). Kraepelin distinguished between "*Auffassung*" (registration of information), which in his opinion was intact in schizophrenic patients, and "*Aufmerksamkeit*" (active, focused, sustained attention), which in his opinion was always disturbed in schizophrenics (Holzman et al. 1976). With the rise of cognitive psychology in the 1960s, Chapman and McGie (1962) reactualized Kraepelin's hypothesis. In a phenomenologically oriented study of schizophrenic experiences the authors indicated it to be plausible that disorders of attention and information processing (AIP) play a central role in the pathogenesis of schizophreniform psychoses. This hypothesis has since been confirmed several times in empirically oriented experimental neuropsychological studies, which have disclosed that disorders of AIP are demonstrable during remissions as well as during psychotic episodes (Asarnow and MacCrimmon 1978; Wohlberg and Cornetsky 1973; Nuechterlein and Dawson 1984).

Attention and information processing essentially involves the individual's ability to discriminate between information that is relevant and information that is irrelevant to him/her (signal-noise discrimination). The neuropsychological test which offers the most valid parameter for that ability has not yet been identified with certainty. Prospective high-risk research often makes use of various versions of the Continuous Performance Task (CPT), Span of Apprehension Tasks (SAT), Serial Recall Tasks (SRT), and Reaction Time Tests (RTT). These tests probably measure different neuropsychological functions of processes. In this type of research, however, it is neither the nature of the processing nor the sensory modality in which one measures, but the *task* parameters that are important. Only tests with a high task load discriminate schizophrenic patients from normal controls (Erlenmeyer-Kimling and Cornblatt 1984).

Both with the CPT (Rutschmann et al. 1977; Nuechterlein 1983; Rutschmann et al. 1986) and with the SAT and SRT (Asarnow and MacCrimmon 1978, 1981; Harvey et al. 1981; Winters et al. 1981), children of schizophrenic parents were found to perform less well than normal children. In three CPT studies, high-risk children also performed less well than children of parents with nonschizophrenic psychiatric disorders (Nuechterlein 1983; Rutschmann et al. 1986; Erlenmeyer-Kimling et al. 1983). Although with the RTT schizophrenic patients showed longer reaction times than other psychiatric patients and normal controls (Nuechterlein 1977), children of schizophrenic parents did not differ from other children (Asarnow and MacCrimmon 1978; Van Dyke et al. 1975; Erlenmeyer-Kimling and Cornblatt 1978; Rutschmann et al. 1977).

If AIP dysfunctions are to be accepted as valid risk indicators, then it is not sufficient to establish group differences between high-risk children and other children: it also has to be demonstrated that only some children in the high-risk group show such dysfunctions, and that these are related to the subsequent development of schizophrenia. Four studies have indeed demonstrated that only a subgroup of high-risk children showed AIP dysfunctions, and that the number of children with AIP dysfunctions in the high-risk group significantly exceeded that in the normal group (Cornblatt and Erlenmeyer-Kimling 1984; Asarnow et al. 1978; Winters et al. 1981; Nuechterlein 1983). So far, only two prospective studies have focused on the correlation between early AIP dysfunction and subsequent psychopathology. The Waterloo High Risk Project (MacCrimmon et al. 1980) revealed that children with AIP dysfunctions scored higher on the Minnesota Multiphasic Personality Inventory (MMPI) in adolescence than children without these dysfunctions; however, the differences were not significant. In the New York High Risk Study (Erlenmeyer-Kimling et al. 1984b) a poor performance on AIP tests in childhood proved to correlate with poor social adjustment and psychiatric treatment in adolescence. This correlation was found only in high-risk children with AIP dysfunctions, and not in children with AIP dysfunctions but normal parents or parents with affective disorders. It was also found that high-risk children who showed psychopathology in adolescence had scored lower on signal-noise detection tests at school age than had other high-risk children or children of depressive or average parents (Cornblatt and Erlenmeyer-Kimling 1984; Erlenmeyer-Kimling et al. 1984a). Two other studies disclosed no difference in AIP test performance between children of schizophrenic parents and those of depressive parents (Erlenmeyer-Kimling et al. 1978; Neale et al. 1984).

To summarize: There are some indications of a correlation between AIP dysfunctions diagnosed at an early age and the development of psychopathology in adolescence. A specific correlation between early AIP dysfunctions and schizophreniform disorders later in life has not yet been demonstrated.

Psychophysiological Parameters

Some psychological functions prove to be reflected in physiological variables. These physiological variables in turn discriminate between schizophrenic patients on the one hand and normal persons as well as patients with other psychiatric disorders on the other:

—The electrodermal orientation reaction (EOR) is an index of stimulus registration by the brain. In some 50% of all chronic schizophrenics no EOR can be evoked, and such a deviant EOR pattern is probably a trait marker (Holzman 1987).
—Event-related potentials (ERPs) reflect various aspects of the cognitive stimulus assimilation process. The early components of ERPs (latencies of 50–200 ms) are indices of selective attention, while the late components (latencies of 300–500 ms) reflect above all the cognitive evaluation of the stimulus. Both components are deviant in schizophrenic patients (Pritchard 1986) and may be regarded as a trait marker (Holzman 1987).

Both indices, the EOR and ERPs, have been studied in high-risk children. Mednick and Schulsinger (1968) compared young adolescents with a schizophrenic mother and adolescents with normal parents and found that, especially in response to auditory stimuli of irritant loudness, the former showed EORs characterized by shorter latencies, larger amplitudes, very fast recovery, and resistance to habituation. Follow-up data showed that in the male (but not the female) high-risk children, such EOR features were predictive of subsequent schizophrenic symptomatology (Mednick et al. 1978). Attempts at replication of this research in other groups of high-risk children have so far yielded discrepant results. Prentky et al. (1981) and Van Dyke et al. (1974) also found larger EORs in children of schizophrenic parents but were unable to demonstrate abnormal latencies and recovery rates. Janes et al. (1978) found no difference in EOR, but they used auditory stimuli of a nonaversive character. Two other studies in fact revealed longer latencies (Erlenmeyer-Kimling et al. 1984a, 1987), while in other respects the EORs of children of schizophrenic parents did not differ from those of children of normal parents. Kugelmass et al. (1985) were also unable to replicate the findings of Mednick's study; the only difference they found was a prolonged recovery time in children of schizophrenic parents.

The above-mentioned studies vary in several respects: in severity and chronicity of the parental schizophrenia, in degree of intactness of the families of the high-risk children, in the age of the children when examined, and in the genetic contribution of the nonschizophrenic parents. Further analyses of the influence of such differences on various aspects of the EOR have failed to provide an adequate explanation of the controversial findings (Mednick et al. 1978; Erlenmeyer-Kimling et al. 1984a). Perhaps such an explanation may be found in the fact that the EOR parameters are strongly influenced by the child's IQ and the arousal level of the electrodermal system (van Engeland 1984). None of the above studies have adequately accounted for these variables.

So far five prospective high-risk studies have collected and studied ERP data in children of schizophrenic parents. In two of these studies (Itil et al. 1974; Saletu et al. 1975) the children were exposed to stimuli which were of no cognitive significance to them. Consequently the results of these studies are not readily interpretable and may be left undiscussed. Herman et al. (1977) studied a small sample of high-risk children from the Boston High Risk Study and found that the early components of their ERPs showed larger amplitudes when the children performed a CPT test. Friedman et al. (1979) were unable to replicate this finding in a visual CPT test in the New York High Risk Study. In four of the 30 high-risk children they did find that the task complexity had no effect on the late positive components of the ERPs, whereas 30 normal control children and the other high-risk children responded to increased complexity with larger late positive components.

In an auditory task, two samples of high-risk children from the New York High Risk Study showed significantly smaller late positive ERP amplitudes (Friedman et al. 1982). This finding could not be replicated when the samples were increased in size and the parental diagnoses revised in accordance with more stringent (Feighner) diagnostic criteria (Friedman et al. 1986). Very recently, Schreiber et al. (1989) in a crosssectional study, found latencies of the N2 and P3 of the auditory event-related potential in children at risk for schizophrenia, when compared with normal children. The causes of these discrepant findings may be several: the child's chronological age, his IQ, and the task complexity can all influence the ERP parameters (Courchesne 1978; Verbaten et al. 1986); none of the research so far performed has adequately accounted for these variables.

Summary and Discussion

Two decades of schizophrenia risk research have now passed. What is the outcome of all these efforts and what are the priorities for the next 10 years? To begin with, let us acknowledge that valid intrinsic risk indices are still lacking. It is true that a by no means small minority of children who later develop schizophrenia can be identified during the last primary school years and the first years of secondary education on the basis of a syndrome which consists of the triad of social isolation, emotional instability, and aggressiveness. However, the sensitivity and specificity of this triad in relation to subsequent schizophrenia are so low that we cannot expect future studies focusing solely on behavior to open up further perspectives. The same applies to neurological or motor abnormalities: MNDs predict an increased vulnerability to psychiatric disorders at a later age, but they are not a specific precursor of schizophreniform psychoses!

Yet the notion that a neurointegrative developmental disorder which manifests itself in different neuropsychological functions provides the basis of a vulnerability to schizophrenia retains its attractiveness. After all, some parameters of AIP have "marker-like" properties. They are more common in an adult schizophrenic population than in control groups, and more frequently encountered in a subgroup of high-risk children than in other children! The one thing that would make it justifiable to describe them as schizophrenia markers is still lacking: unequivocal evidence that children with AIP dysfunctions subsequently develop a schizophrenic disorder more frequently than children without such dysfunctions. A solution may offer itself in the near future, when the children in question reach the age at which schizophreniform psychoses are likely to become manifest. In anticipation of this, new research into the pathophysiology of AIP seems to merit high priority. AIP is not a simple unitary process, and it is not clear which of the many aspects of AIP is actually measured by neuropsychological tests, like the CPT. As cited earlier in this article, Kraepelin distinguished between the automatic grasping or intake of a stimulus ("*Auffassung*") and the voluntary turning of one's notice to a stimulus ("*Aufmerksamkeit*"). Both qualities imply that attention is available in limited quantities and that it can be deployed either automatically or effortfully, the former mode of information processing requiring a lesser amount of attention than the latter. Schneider and Shiffrin (1977) have characterized these two processing requirements as "*automatic*" and "*controlled*." Impaired AIP capacity can be manifested in either automatic or controlled processing or in both.

As yet in psychophysiologically oriented high-risk research only controlled processing parameters like EOR and P300 have been studied. These investigations generally report mean differences between groups, usually without any indication of effect size. Perhaps, therefore, such studies yield controversial findings. In the near future investigators should be able to classify performance as defective or normal, even though the variable under consideration is a continuous one. In physical medicine there are more or less standardized levels for judging deviation from the norm of such measures as sedimentation rate and blood pressure level. The same standards should be sought for psychophysiological measurement. In ERPs, for example, there should be a way to quantify the degree of attenuation of the P300 wave, a procedure that would permit one to judge the extent of abnormality. Perhaps a cut-off point of two standard deviations beyond the mean of a representative normal example would be helpful in order to study the functional meaning of deviance. Knowledge of how much attenuation of the P300 wave is to be considered pathological is badly needed.

Furthermore the meaning of most of the psychophysiological procedures requires refinement. Although much work has been done on ERPs (Pritchard 1986; Näätänen 1982), there is still a need to study further the meaning of the separate components of the ERPs, in as much as veteran investigators do not fully agree on the meaning of the separate components (Holzman 1987). Only for the past few years has it been possible to produce single trial ERPs instead of averaged ERPs (Woestenburg et al. 1984). The drawback of averaging procedures is that at least ten trials have to be averaged, while it has been repeatedly found that the EOR is already habituated to an asymptotic level in such a period of time. If such changes had been present in the several waves of the ERP, the averaging procedure would have washed them out. The use of single trial ERP techniques in the future will give us better insight into the dynamic aspects of information processing. Within a few years a combination of electroencephalographic and magnetoencephalographic examination will make it possible to localize accurately the generators of various ERP waves. This will enhance our understanding of the neuronal substrate of AIP and facilitate the transition to neurochemical research into these processes.

To our knowledge children of schizophrenic patients have never been tested for physiological measures indicating automatic attention and information processes like smooth pursuit eye movements, prestimulus inhibition, and startle habituation.

Eye movement dysfunctions have been shown to be reliably associated with schizophrenia as a trait, suggesting disorders of nonvoluntary attention in association with those brain areas involved in smooth pursuit and saccadic eye movements. As many of the first-degree relatives of schizophrenic patients also show this eye movement dysfunction, smooth pursuit eye movement dysfunctions can be considered to be a schizophrenia trait marker (Holzman 1987).

The sensory disturbance in schizophrenia is often described as an inability to filter out extraneous noise from meaningful sensory inputs. The neurophysiological mechanisms of this inability to filter have been the subject of numerous inquiries. One possible mechanism could involve the failure of an inhibitory gating pathway. Such pathways are known to exist in many brain areas, where their effectivity can block or "gate" the effects of a synaptic input of a target neuron. The normal role of such gating is probably to regulate the amount of sensory input into the brain. If a person needs to be aware of any sound, as in times of danger, then the gating pathways need to allow the brain to be very sensitive. If sounds have already been detected, or if the environment is a very noisy one, then the brain needs to filter more strongly to prevent distraction by irrelevant sounds.

The rate at which a startle response to strong exteroceptive stimuli habituates is a parameter of this capability to inhibit or block stimulus input (Geyer and Braff 1987). Another way to demonstrate the central inhibition capability deficit in schizophrenia lies in the so-called prepulse inhibition design. Sounds are presented in pairs while the ERPs are simultaneously registered with particular attention to the P50 wave recorded at the vertex. The sounds (clicks) are given in pairs with 0.5 s between the first and the second stimulus. Due to a central inhibitory defect schizophrenic subjects show significantly less suppression of the response to the second stimulus than do normal subjects (Freedman et al. 1987). Both deficient startle habituation and deficient prepulse inhibition seem to be schizophrenia trait markers. Recently animal models of startle habituation and prepulse inhibition have been developed. Such animal model studies allow us to draw strong inferences about the neurobiological substrate of schizophrenia. In the near future it will be possible to elucidate the influence of dopamine, norepinephrine, and serotonin overactivity in the pathophysiology of this central inhibition deficit (Geyer and Braff 1987; Freedman et al. 1987).

It may be clear that, when new prospective high-risk studies on schizophrenia are started, particular attention should be paid to smooth pursuit eye movements, startle habituation, and prepulse inhibition registration indicating automatic processing deficits.

References

Asarnow RF, MacCrimmon DJ (1978) Residual performance deficit in clinically remitted schizophrenics: a marker of schizophrenia? J Abnorm Psychol 87:597–608

Asarnow RF, MacCrimmon DJ (1981) Span of apprehension deficits during postpsychotic stages of schizophrenia. Arch Gen Psychiatry 38:1006–1011

Astrup C, Fossum A, Holmboe R (1962) Prognosis in functional psychoses: clinical, social and genetic aspects. Thomas, Springfield

Bender L (1947) Childhood schizophrenia. Am J Orthopsychiatry 17:40–56

Bleuler E (1961) Dementia praecox or the group of schizophrenia (translated from German). International University Press, New York

Bleuler M (1972) The schizophrenic disorders: long term patient and family study. Yale University Press, New Haven

Chapman J, McGhie A (1962) A comparative study of disordered attention of schizophrenia. J Ment Sci 108:487–500

Cornblatt B, Erlenmeyer-Kimling L (1984) Early attentional predictors of adolescent behavioral disturbances in children at risk for schizophrenia. In: Watt N, Anthony EJ, Wynne L, Rolf J (eds) Children at risk for schizophrenia: a longitudinal perspective. Cambridge University Press, New York, pp 198–219

Courchesne E (1978) Neurophysiological correlates of cognitive development: changes in long latency event-related potentials from childhood to adulthood. Electroencephalogr Clin Neurophysiol 45:468–482

Cox SM, Ludwig AM (1979) Neurological soft signs and psychopathology. J Nerv Ment Dis 167:161–165

Erlenmeyer-Kimling L (in press) Electrodermal recovery data on children of schizophrenic parents. Psychol Res.

Erlenmeyer-Kimling L, Cornblatt B (1978) Attentional measures in a study of children at high risk for schizophrenia. In: Wynne L, Cromwell RL, Matthyse S (eds) The nature of schizophrenia. Wiley, New York, pp 212–241

Erlenmeyer-Kimling L, Cornblatt C (1984) Biobehavioral risk factors in children of schizophrenic parents. J Autism Dev Disord 14:357–372

Erlenmeyer-Kimling L, Cornblatt B, Friedman D, Marcuse Y, Rutschmann J, Simmens S, Devi S (1982) Neurological, electrophysiological and attentional deviations in children at high risk for schizophrenia. In: Nasrallah HA, Henn F (eds) Schizophrenia as a brain disease. Oxford University Press, New York, pp 210–221

Erlenmeyer-Kimling L, Cornblatt B, Golden R (1983) Early indicators of vulnerability to schizophrenia in children at high genetic risk. In: Guze SB, Earls FJ, Barrett JE (eds) Childhood psychopathology and development. Raven, New York, pp 247–264

Erlenmeyer-Kimling L, Kestenbaum CJ, Bird H, Hilldoff U (1984a) Assessment of the New York high-risk project subjects in sample A who are now clinically deviant. In: Watt N, Anthony EJ, Wynne L, Rolf J (eds) Children at risk for schizophrenia: a longitudinal perspective. Cambridge University Press, New York, pp 227–240

Erlenmeyer-Kimling L, Marcuse Y, Cornblatt B, Friedman D, Rainer JD, Rutschmann J (1984b) The New York high-risk project. In: Watt N, Anthony EJ, Wynne L, Rolf J (eds) Children at risk for schizophrenia: a longitudinal perspective. Cambridge University Press, New York, pp 169–189

Erlenmeyer-Kimling L, Friedman D, Cornblatt B, Jacobsen R (1984c) Electrodermal recovery. The New York high-risk project. In: Watt NF, Anthony EJ, Wynne LC, Rolf JE (eds) Children at risk for schizophrenia: a longitudinal perspective. Cambridge University Press, New York, pp 169–189

Fish B (1957) The detection of schizophrenia in infancy. J Nerv Ment Dis 125:1–24

Fish B (1975) Biologic antecedents of psychosis in children. In: Freedman D (ed) Biology of the major psychoses. Raven, New York, pp 138–162

Fish B (1984) Offspring of schizophrenics from birth to adulthood. In: Watt N, Anthony EJ, Wynne L, Rolf J (eds) Children at risk for schizophrenia: a longitudinal perspective. Cambridge University Press, New York, pp 423–440

Frazee HE (1953) Children who later became schizophrenic. Smith Coll Stud Soc Work 23:125–149

Freedman R, Adler LE, Gerhardt GA, Waldo M, Baker N, Rose GM, Drebing C, Nagamoto H, Bickford-Wimen P, Franks R (1987) Neurobiological studies of sensory gating in schizophrenia. Schizophr Bull 13:669–677

Friedman D, Vaughan HG, Erlenmeyer-Kimling L (1979) Event-related potential investigations in children at high risk for schizophrenia. In: Lehmann D, Callaway E (eds) Event related potentials in man: application and problems, Plenum, New York, pp 198–217

Friedman D, Vaughan HG, Erlenmeyer-Kimling L (1982) Cognitive brain potentials in children at risk for schizophrenia: preliminary findings. Schizophr Bull 8:514–531

Friedman D, Cornblatt B, Vaughan HG, Erlenmeyer-Kimling L (1986) Event-related potentials in children at risk for schizophrenia during two versions of the continuous performance test. Psychol Res 18:161–177

Garmezy N (1974) Children at risk: the search for the antecedents of schizophrenia. Part 2. Ongoing research programs, issues and intervention. Schizophr Bull 1:55–125

Garmezy N, Streitman S (1974) Children at risk: the search for the antecedents of schizophrenia. Part I. Conceptual models and research methods. Schizophr Bull 1:14–40

Garver DL (1987) Methodological issues facing the interpretation of high-risk studies: biological heterogeneity. Schizophr Bull 13:525–529

Geyer MA, Braff DL (1987) Startle habituation and sensorimotor gating in schizophrenia and related animal models. Schizophr Bull 13:643–668

Goldstein MJ, Tuma AH (1987) High risk research (Editors' introduction). Schizophr Bull 13:369–372

Hanson DR, Gottesman II, Heston LL (1976) Some possible childhood indicators of adult schizophrenia from children of schizophrenics. Br J Psychiatry 129:142–154

Hartmann E, Milofsky E, Vaillant G, Oldfield M, Falke R, Ducey C (1984) Vulnerability to schizophrenia. Arch Gen Psychiatry 41:1050–1056

Harvey P, Winters K, Weintraub S, Neals JM (1981) Distractibility in children vulnerable in psychopathology, J Abnorm Psychol 90:298–304

Herman J, Mirsky AF, Ricks NL, Gallant D (1977) Behavioral and electroencephalographic measures of attention in children at risk for schizophrenia. J Abnorm Psychol 86:27–33

Holzman PS (1987) Recent studies of psychophysiology in schizophrenia. Schizophr Bull 13:49–75

Holzman PS, Levy DL, Proctor LR (1976) Smooth pursuit eye movements, attention, and schizophrenia. Arch Gen Psychiatry 33:1415–1420

Itil TM, Hsu W, Saletu B, Mednick S (1974) Computer EEG and auditory evoked potential investigations in children at high risk for schizophrenia. Am J Psychol 131:892–900

Janes CL, Hesselbrock V, Stern JA (1978) Parental psychopathology, age, and race as related to electrodermal activity in children. Psychophysiology 15:24–34

Janes CL, Worland J, Weeks DG, Konen PM (1984) Interrelationships among possible predictors of schizophrenia. In: Watt NF, Anthony EJ, Wynne LC, Rolf JE (eds) Children at risk for schizophrenia: a longitudinal perspective. Cambridge University Press, New York, pp 160–166

John RS, Mednick SA, Schulsinger F (1982) Teacher reports as a predictor of schizophrenia and borderline schizophrenia: a Bayesian decision analysis. J Abnorm Psychol 91:399–413

Kraepelin E (1919) Dementia praecox and paraphrenia. Krieger, Huntington, NY (Krieger, Huntington Reprint, 1971)

Kugelmass S, Marcus J, Schmueli J (1985) Psychophysiological reactivity in high-risk children. Schizophr Bull 11:66–73

Langfeldt G (1956) The prognosis in schizophrenia. Acta Psych Scand (Suppl 10)

Luchins D, Pollin W, Wyatt RJ (1980) Laterality in monozygotic schizophrenic twins: an alternative hypothesis. Biol Psychiatry 15:87–93

MacCrimmon DJ, Cleghorn JM, Asarnow RF, Steffy RA (1980) Children at risk for schizophrenia: clinical and attentional characteristics. Arch Gen Psychiatry 37:671–674

Marcus J (1974) Cerebral functioning in offspring of schizophrenics. Int J Ment Health 3:57–73

Marcus J, Auerbach J, Wilkinson L, Burack CM (1981) Infants at risk for schizophrenia: the Jerusalem infant development study. Arch Gen Psychiatry 38:708–713

Marcuse Y, Cornblatt B (1986) Children at high risk for schizophrenia: predictions from infancy to childhood functioning. In: Erlenmeyer-Kimling L, Miller N (eds) Life span research on the prediction of psychopathology. Erlbaum, Hillsdale, pp 81–117

McNeil TF, Kay L (1984) Offspring of women with nonorganic psychoses: progress report. In: Watt N, Anthony EJ, Wynne L, Rolf J (eds) Children at risk for schizophrenia: a longitudinal perspective. Cambridge University Press, New York, pp 138–154

Mednick SA, Schulsinger F (1968) Some premorbid characteristics related to breakdown in children with schizophrenic mothers. In: Rosenthal D, Kety SS (eds) Transmission of schizophrenia. Pergamon, New York, pp 267–291

Mednick SA, Mura E, Schulsinger F, Mednick B (1971) Perinatal conditions and infant development in children with schizophrenic parents. Soc Biol 18:103–113

Mednick SA, Schulsinger F, Teasdale TW, Schulsinger H, Venables PH, Rock DR (1978) Schizophrenia in high risk children: sex differences in predisposing factors. In. Serban G (ed) Cognitive defects in the development of mental illness. Brunner Mazel, New York, pp 169–197

Näätänen R (1982) Processing negativity: an evoked potential reflection of selective attention. Psychol Bull 92:605–642

Nameche G, Waring M, Richs D (1964) Early indicators of outcome in schizophrenia. J Nerv Ment Dis 139:232–240

Nuechterlein KH (1977) Reaction time and attention in schizophrenia: a critical evaluation of the data and the theories. Schizophr Bull 3:373–428

Nuechterlein KH (1983) A signal detection in vigilance tasks and behavioral attributes among offspring of schizophrenic mothers and among hyperactive children. J Abnorm Psychol 92:4–28

Nuechterlein KH, Dawson ME (1984) Information processing and attentional functioning in the developmental course of schizophrenic disorders. Schizophr Bull 10:160–203

Offord DR, Cross LA (1969) Behavioral antecedents of adult schizophrenia: review. Arch Gen Psychiatry 21:267–283

Orvaschel H, Mednick S, Schulsinger F, Rock D (1979) The children of psychiatrically disturbed parents: differences as a function of the sex of the sick parent. Arch Gen Psychiatry 36:691–695

Owens DG, Johnstone EC (1980) The disabilities of chronic schizophrenia—their nature and the factors contributing to their development. Br J Psychiatry 136:384–395

Parnas, J, Schulsinger F, Schulsinger H, Mednick SA, Teasdale TW (1982) Behavioral precursors of schizophrenia spectrum. Arch Gen Psychiatry 39:658–664

Prentky RH, Salzman LF, Klein RH (1981) Habituation and conditioning of skin conductance responses in children at risk. Schizophr Bull 7:281–291

Pritchard WS (1986) Cognitive event-related potential correlates of schizophrenia. Psychol Bull 100:43–66

Quitkin F, Rifkin A, Klein DF (1976) Neurological soft signs in schizophrenia and character disorder. Arch Gen Psychiatry 33:845–853

Ragins N, Schachter J, Elmer E, Preisman R, Bowes AE, Harway V (1975) Infants and children at risk for schizophrenia. J Child Psychiatry 14:150–177

Rieder RO, Nichols PL (1979) Offspring of schizophrenics: III. Hyperactivity and neurological soft signs. Arch Gen Psychiatry 36:665–674

Robins LN (1966) Deviant children grown up. Williams and Wilkins, Baltimore

Roff JD, Knight R, Wertheim E (1976) A factor-analytic study of childhood symptoms antecedent to schizophrenia. J Abnorm Psychol 85:543–549

Rolf JE (1972) The social and academic competence of children vulnerable to schizophrenia and other behavior pathology. J Abnorm Psychol 80:225–243

Rutschmann J, Cornblatt B, Erlenmeyer-Kimling L (1977) Sustained attention in children at risk for schizophrenia. Report on a continuous performance test. Arch Gen Psychiatry 34:571–575

Rutschmann J, Cornblatt B, Erlenmeyer-Kimling L (1986) Sustained attention in children at risk for schizophrenia: findings with two visual continuous performance tests in a new sample. J Abnorm Child Psychol 14:365–385

Saletu B, Marasa J, Mednick S, Schulsinger F (1975) Acoustic evoked potentials in offspring of schizophrenic mothers ("high risk" children for schizophrenia). Clin Electroencephalogr 6:92–102

Schneider W, Shiffrin RM (1977) Controlled and automatic human information processing. Psychol Rev 84:1–66

Schreiber H, Stolz G, Rothmeier J, Kornhuber HH, Born J (1989) Prolonged latencies of the N_2 and P_3 of the auditory event-related potential in children at risk for schizophrenia. Eur Arch Psychiatry Neurol Sci 238:185–188

Shaffer D, O'Connor PA, Shafer SQ, Prupis S (1984) Neurological soft signs: their origins and significance for behavior. In: Rutter M (ed) Developmental neuropsychiatry. Churchill Livingstone, New York, pp 162–198

Sohlberg SC, Yaniv S (1985) Social adjustment and cognitive performance of high-risk children. Schizophr Bull 11:61–65

Touwen BCL (1979) Examination of the child with minor neurological dysfunction. Spastic Int Med Publ Heinemann Medical, London

Tucker GJ, Campion EW, Silberfarb PM (1975) Sensorimotor functions and cognitive disturbance is psychiatric patients. Am J Psychiatry 132:7–12

Van Dyke JL, Rosenthal D, Rasmussen PV (1974) Electrodermal functioning in adopted-away offspring of schizophrenics. J Psychiatr Res 10:199–215

Van Dyke JL, Rosenthal D, Rasmussen PV (1975) Schizophrenia: effects of inheritance and rearing on reaction time. Can J Behav Sci 7:223–236

Van Engeland H (1984) The electrodermal response to auditive stimuli in autistic children, normal children, mentally retarded children and child psychiatric patients. J Autism Dev Dis 14:261–277

Van Engeland H (1987) Early recognition of psychiatric disorders in children (in Dutch). Ned Tijdschr Geneesk 131:1333–1336

Verbaten MN, Roclofs JW, Sjouw W, Stangen JL (1986) Different effects of uncertainty and complexity on single trial visual ERPs. Psychophysiology 23:254–262

Watt NF, Stolorow RD, Ludensky AW, McClelland DC (1970) School of adjustment and behavior of children hospitalized for schizophrenia. Am J Orthopsychiatry 40:637–657

Watt NF (1972) Longitidinal changes in the social behavior of children hospitalized for schizophrenia as adults. J Nerv Ment Dis 155:42–54

Watt NF (1978) Patterns of childhood social development in adult schizophrenia. Arch Gen Psychiatry 85:160–170

Watt NF, Grubb TW, Erlenmeyer-Kimling L (1982) Social, emotional, and intellectual behavior at school among children at high risk for schizophrenia. J Consult Clin Psychol 50:171–181

Weintraub S, Neale JM (1984) Social behavior of children at risk for schizophrenia. In: Watt NF, Anthony EJ, Wynne LC, Rolf JE (eds) Children at risk for schizophrenia: a longitudinal perspective. Cambridge University Press, New York, pp 279–285

Weintraub S, Prinz RJ, Neale JM (1978) Peer evaluations of the competence of children vulnerable to psychopathology. J Abnorm Child Psychol 6:461–473

Winters KC, Stone AA, Weintraub S, Neale JM (1981) Cognitive and attentional deficits in children vulnerable in psychopathology. J Abnorm Child Psychol 9:435–453

Woestenburg JCW, Verbaten MN, Hees HH van, Slangen JL (1984) Single trial ERP estimation in the frequency domain using OPTA, estimation of individual habituation. Biol Psychol 17:173–191

Wohlberg GW, Cornetsky C (1973) Sustained attention in remitted schizophrenia. Arch Gen Psychiatry 28:533–537

Minimal Brain Dysfunction
and Head Injury

Development in Children After Severe Head Injury

G. Lehmkuhl and W. Thoma

Introduction

As a result of empirical studies various authors seem to agree with the notion that certain specific psychopathological and neuropsychological deficiencies are likely to occur following severe craniocerebral trauma suffered during infancy or childhood (Rutter et al. 1980; Brown et al. 1981; Chadwick et al. 1981a). Head injury and psychiatric symptoms are known often to be associated, although psychopathological symptoms do not occur as a rule. Up to now, there have been no studies investigating interactions between age at injury, severity of trauma, and ensuing psychiatric symptoms, and how these issues are related to the child's cognitive capacity. Previous studies have yielded high prevalence rates of psychiatric disturbances in children who have suffered severe head injury. Psychiatric outcome seems to be directly related to severity of trauma, social and family disadvantage, pretraumatic functioning, neurological sequelae, and development of posttraumatic epilepsy (Black et al. 1969; Rune 1970; Klonoff and Paris 1974; Shaffer et al. 1975; Brown et al. 1981).

Findings by Chadwick et al. (1981a) indicated that cognitive recovery during the first year after injury might be slightly more rapid in children aged between 5 and 10 years than in those older than 10, but differences were small and well short of statistical significance. Recently Rutter et al. (1984) stated: "While little is known on age effects with respect to cognitive deficits, even less is known on their role regarding psychiatric sequelae" (p. 99). In their studies on both generalized and localized head injuries, they did not find a relationship between age at injury and risk of subsequent psychiatric disorder. They pleaded for further studies because the available data were not sufficient to rule out more subtle effects. That head injury does increase the risk of psychological disturbances is well established. Despite increased prevalence rates of emotional disturbances, however, no behavior pattern or syndrome has been demonstrated to be characteristically associated with head injury. This is in line with Ernhart et al. (1963), who stated that "there is little evidence for the generalization that there is a unique personality syndrome characteristic of the young head-injured child." Rutter (1977) found a typical mixture of emotional and conduct disorders in head-injured children (also seen in non-head-injured children) but no unitary psychiatric syndrome. In general, the course of posttraumatic psychiatric disorders seems to be more variable than the course of cognitive impairment.

Chadwick (1985) describes individual outcome variations: "There are some children showing persistent emotional and behavioral disorders after injury, others showing a pattern of recovery comparable to that observed for cognitive outcome, and still others in whom behavioral problems may only develop some time after injury."

The major purpose of our study was to procure more specific information on the effect of both age at injury and severity of injury upon psychiatric symptoms and cognitive development. Furthermore, we aimed to examine the impact of intellectual and cognitive impairment on the development of emotional and behavioral disorders, taking age at injury into account. Other questions were how adverse psychosocial conditions influenced the further development of the child and how the experience of injury affected the educational behavior of the parents.

Methods

Subjects

We examined two samples of head-injured patients differing in severity of injury and a noninjured control group. The children were aged from 12 to 14 years at assessment. A patient subsample with severe head injury consisted of 45 children who had been admitted to hospital after head injury with a posttraumatic coma lasting at least 24 h. Children with injuries to such an extent that proper psychological assessment was not possible had to be excluded (Table 1). A second subsample consisted of 45 children having suffered mild head injury with posttraumatic coma lasting less than 24 h.

Age effects were controlled in both patient samples by subdividing them into three age groups of 15 subjects each, ranging from 3 to 5, 7 to 9, and 11 to 13 years of age at injury. In the latter group (11–13 years) we made sure that there was at

Table 1. Characteristics of patients with severe cerebral trauma

	Age at injury		
	3–5 years	7–9 years	11–13 years
Patients (n)	15	15	15
Age at assessment (months, mean)	152	162	157
Age at injury (months, mean)	67	109	147
Duration of unconsciousness			
1–3 days	3	2	1
4–7 days	3	5	2
1–3 weeks	6	5	8
> 3 weeks	3	3	4
Mean (h)	347	307	360

least a 9-month interval between injury and psychological assessment, because most intellectual recovery takes place during the first 6 months after injury.

In order to avoid selective data sampling, the hospital report archives of the clinics in Aachen, Mannheim, Ludwigshafen, and Heidelberg were systematically scanned for recruiting the patients for our study. Children are admitted to these clinics after an accident. Only those patients were included in the study for whom sufficiently exact data about the trauma and posttraumatic course were available. Moreover, the children treated during a defined time span in the rehabilitation clinics of Gailingen, Neckargemünd, and Schömberg were included insofar as they met our criteria. In total less than 10% of the patients refused co-operation.

Clinical data about the injury and its treatment were extracted from the hospital reports. Children with any known preinjury head damage were excluded.

The controls were 30 "normal" children showing signs of neither pre-, peri-, and postnatal risks nor any previous central nervous diseases.

The three groups (two patient samples and controls) were matched for age at psychological assessment, sex, psychosocial adversity, and number of previous hospital admissions.

Procedure

Neuropsychological Parameters

All children were submitted to a comprehensive test battery including the following tests:

1. *Verbal abilities:*
—Token Test (De Renzi and Vignolo 1962, German adaptation by Orgass 1976a, b) to detect receptive disturbances in aphasics
—Word Comprehension Test (Freund 1980) to differentiate words of similar meaning
—Sentence Comprehension Test (Freund 1980) to measure comprehension of differing grammatical constructions
—Spelling Test (Rathenow 1979; Rathenow et al. 1981)
—Zürcher Lesetest (Linder and Grissemann 1980), i.e., a reading test which records number of reading mistakes as well as total reading time

2. *Gesture and facial recognition:*
—Gesture comprehension was tested by expressive and receptive tasks (Müller 1979; Lehmkuhl and Weniger 1981)
—Facial recognition was tested by presentation of photographs of differing difficulty for recognition referring back to the Benton-van Allen Test (Benton and van Allen 1968, 1972)
—Miming recognition was tested with the aid of a subtyping of mimic expressions after Ekman and Friesen (1975)

3. *Visuospatial Perception*:
—Visual Organization Test (Hooper 1958)
—Ambiguous Pictures Test (Elkind 1964)
—Subtest 7 "spatial perception" of the "Prüfsystem für Schul- und Bildungs-
beratung" (PSB), i.e., an intelligence test developed by Horn (1969)

4. *Memory:*
—Subtest "digit span" of the Wechsler Intelligence Scale for Children (German
version revised, HAWIK-R, by Tewes 1984) as a measure of number memory
—Paired Associate Learning Test to quantify the capacity of the verbal memory
(Lehmkuhl 1986)
—Recurring Figures Test (Kimura 1963), which provides a measure of the visual
memory

5. *Cognitive style:*
—Matching Familiar Figures Test adapted from Kagan et al. (1964) to assess
impulsivity
—Stroop Color Word Test (Stroop 1935), which provides a measure of
distractibility to irrelevant stimuli
—Children's Embedded Figures Test (Witkin 1971) to check the relative field
dependency of visual perception

6. *Intelligence and concentration*:
—German adaptation of Cattell's Culture Fair Intelligence Test (Weiß 1978) as a
measure of general intellectual functioning
—Verbal Fluency Test to assess verbal flexibility and the ability to create verbal
associations
—An adapted Bourdon concentration test (d2, Brickenkamp 1975)

Psychopathology

Psychopathological symptomatology was recorded by a highly structured parent
interview [Graham and Rutter 1968; Rutter and Graham 1968; (German
adaptation by Poustka et al. 1977 Mannheimer epidemiologisches Eltern-
interview/Kinderinterview. Unpublished manuscript)]. Proceeding from this
interview diagnostic assessment according to ICD 9 criteria was made and
severity of psychiatric disorder was rated by three independent raters using a
four-point scale (psychiatrically undisturbed, mildly disturbed, moderately
disturbed, severely disturbed). Due to low frequencies in the specific categories
a coarser subtyping with the following diagnostic groups was used: emotional
disorders (ICD 300, ICD 313), conduct disorders and hyperactive syndromes
(ICD 312, ICD 314), and postcontusional syndromes (ICD 310.2). The latter
group also included syndromes described by Brown et al. (1981) as "disinhibited
state" and defined as follows: "... disorders which had in common the one
outstanding feature of marked socially disinhibited behavior—with undue
outspokenness without regard to social convention, the frequent making of very
personal remarks or the asking of embarrassing questions, or getting undressed

in social situations in which this would usually be regarded as unacceptable behaviour.... Some of these disinhibited disorders also included forgetfulness, overtalkativeness, carelessness in personal hygiene and dress, and impulsiveness" (Brown et al. 1981, p. 74).

Furthermore the following three additional questionnaires were completed:

— The rating scale of Conners (1973) to assess hyperactivity and impulsiveness
— A rating scale of Geisel et al. (1982) to assess the presence of expansive and introversive behavior disturbances and learning disturbances
— A more specific questionnaire focusing on organic head symptoms, e.g., perseveration, hypoactivity, and mood disturbances. The 52 items of the questionnaire were directed toward symptoms described in the literature as typical sequelae of a cerebral trauma

In addition to biographical information, data about sequelae of the injury, subsequent psychosocial integration, and possible changes in the familial milieu were recorded. The Family Adversity Index of Rutter and Quinton (1977) was used to quantify adverse familial circumstances.

Clinical Parameters

A comprehensive neurological investigation was obligatory for all children (cases and controls). In addition electroencephalography (EEG) and computerized tomography (CT) recordings were analyzed by a standardized procedure (Lehmkuhl et al. 1987).

Results

Group comparisons were made in order to study the effect of the cerebral trauma upon posttraumatic neuropsychological functioning and psychopathology. Cluster analyses (Wishart 1984) including neuropsychological and behavior parameters were used to clarify whether a unitary and specific syndrome associated with cerebral trauma, the so-called organic psychosyndrome, does occur.

Neuropsychological Sequelae

The 45 mildly head-injured patients and the controls did not differ significantly in psychological test results. This applies to all the variables tested. In contrast a comparison of the 45 severely head-injured patients against the controls yielded nonsignificant differences only on the five variables word and sentence comprehension, spelling test, miming recognition, and number memory; for all the other psychometric parameters outcome differed significantly (Table 2).

Table 2. Results of neuropsychological testing

Variable	Maximum score/ scale	Age at injury			Con- trols	Signi- ficance
		3-5	7-9	11-13		
Verbal abilities						
Token Test	50 F	5.1	6.5	6.0	0.9	*
Word comprehension	10 C	7.5	7.4	7.7	8.4	
Sentence comprehension	12 C	9.1	9.8	10.7	10.5	
Spelling Test	100 ± 10	92.0	88.5	95.5	97.7	
Reading Test						
Mistakes (number)		8.8	9.4	6.0	4.4	*
Time (s)		236.2	200.3	162.3	115.4	*
Miming, gesture						
Gesture comprehension	30 C	20.7	22.1	21.7	25.2	*
Facial recognition	24 C	14.9	17.6	16.9	20.1	*
Miming recognition	24 C	15.1	16.1	17.7	18.4	
Visuospatial perception						
Visual Organization Test	30 C	19.5	21.0	22.3	25.8	*
Ambiguous Pictures Test	21 C	11.2	12.7	12.9	17.0	*
Spatial perception						
PSB, 7	5 ± 2	3.8	3.4	3.7	5.2	*
Memory						
Recurring Figures Test	56 C	27.7	24.2	26.9	35.3	*
Verbal memory	75 C	61.6	58.5	63.1	72.3	*
Number memory	10 ± 3	8.9	9.1	9.2	9.8	
Cognitive style						
Stroop Test						
Mistakes		7.2	7.8	5.2	3.6	*
Time (s)		92.7	122.9	119.9	63.1	*
Matching Familiar Figures Test						
Mistakes/items		1.0	0.8	0.7	0.5	*
Time/items		20.5	20.0	21.2	18.7	
Children's Embedded Figures						
Test	25 C	18.3	18.5	17.8	22.5	*
Intelligence and concentration						
CFT 20	100 ± 15	85.1	88.7	95.9	98.4	*
Word fluency, number of words		18.0	15.1	17.5	28.9	*
Concentration d2	100 ± 10	90.1	88.2	94.2	109.9	*

* = $P < 0.05$: control vs patient groups
C, correct; F, false

Age at injury did not have a bearing upon observed psychological deficits, either in children with mild or with severe head injury. Although the severely affected patients show poorer outcome on nearly all variables as compared with controls and the mildly affected patients, there is no evidence for a specific pattern of cognitive deficit. Memory and attention deficits seem to be more important than speech deficiency. Our data do not suggest that there is a specific deficiency in visuospatial perception but a generalized lessening of neuro-psychological functioning.

Table 3. Relationship between neurological abnormalities and cognitive impairment

	Neurological abnormalities		
Cognitive impairment	Absent	Present	P
IQ	97.1	82.5	0.006
Verbal memory	65.4	54.6	0.03
Visual memory	35.6	18.1	0.001

Although we failed to find a specific group profile of impairment in children having incurred severe head injury there is enough evidence that individual profiles do exist. These individual profiles, however, are covered by group effects. There was a significant correlation between duration of coma on the one hand and linguistic and visuomotor capacities and memory on the other. Like Chadwick and co-workers, we found neurological symptoms to be related to cognitive performance: children with neurological deficiencies scored significantly lower than neurologically nonpathological probands on the variables intelligence and verbal and visual memory (Table 3).

Psychopathological Sequelae

Marked psychiatric disorders were found in 68.9% of the severely head-injured children, in 24% of the mildly head-injured children, and in 13% of the controls. A relationship between emotional disturbances and duration of unconsciousness did exist, but it was not as pronounced as expected (the correlation coefficient amounted to $r = 0.30$). There was no evidence for a specific syndrome or for a characteristic pattern of behavior problems. Obviously many problems can be present but this does not imply that all the problems have the same associations and are part of a unitary disorder. In the head-injured children we found the same age-specific psychiatric symptoms as they are described in the ICD also for non-head-injured children. Hyperactivity was found as well as hypoactivity (Table 4).

To sum up, from our data we can conclude that age at injury cannot be regarded as a relevant causal factor in the pathogenesis of posttraumatic psychiatric disturbances. Symptoms are more expressive, though, in temporal proximity to the injury. This effect is also observed when behavioral deviations

Table 4. Psychiatric disorders (diagnosis according to ICD criteria) in children who have suffered severe cerebral trauma (n = 45)

	Psychiatric classification (ICD)					
Age at injury	n	310.2	312/314	300/313	Total	Percent
3–5 years	15	3	2	4	9	60%
7–9 years	15	4	3	1	8	53%
11–13 years	15	10	3	1	14	93%

are diagnostically classified: whereas recently injured patients mostly suffer from a postcontusional syndrome, the other children display a variety of deviations, such as emotional, neurotic, or social disturbances, without any of them being predominant. This finding can be explained by the fact that the category "postcontusional syndrome" (ICD 310.2) comprises, besides unspecific psychic symptoms, neuropsychological deficits which occur in so many different forms and combinations in head-injured patients that it would be an oversimplification to regard them as part of one unitary syndrome.

A factor analysis of the questionnaire items yielded the following factors: hypoactivity, hyperactivity, emotional lability, concentration difficulties, and aggressivity. Age at injury is not seen to have any influence on these factors, which means there are no specific psychopathological symptoms related to age at injury. In comparison with the controls, head-injured children have significantly more school problems and social problems. Parental behavior is characterized by more control and greater care of, and indulgence in, the child (Table 5).

Clustering studies based on psychopathological and/or neuropsychological parameters aimed at finding patients with similar psychiatric disturbances (as to general occurrence of psychiatric disorder or specific diagnoses). Psychopathological and neuropsychological data were submitted to separate cluster analyses; a further run was done with both data sets pooled. In each case we examined how the patients were distributed on the resulting clusters. We did not obtain any cluster specific to a diagnostic category. As to the psychiatric symptomatology, the resulting clusters were rather heterogeneous: they contained nondisturbed children as well as patients with differing diagnoses (Fig. 1).

Severity of the injury—that is, the duration of unconsciousness—was significantly correlated to psychopathological disturbances ($r = 0.30$, $p = 0.004$), i.e., the risk of subsequent psychiatric diseases is directly dependent on the severity of trauma. The only significant correlation between neuropsychological parameters and psychiatric symptoms was that between visuospatial perception and psychiatric disturbance. This effect was only found for the subsample of severe head-injured children disregarding age at injury. In a population-based study, Rutter et al. (1970) were able to demonstrate that psychosocial factors are valid predictors for psychiatric disturbances. From our study we can add

Table 5. Features of educational behavior as assessed by the parents (parents' questionnaires)

	Age at brain injury		
	3–5 years	7–9 years	11–13 years
Indulgence	12	13	13
Strong anxiety	12	12	13
Fear of failure	9	8	12
Lesser grip on the child	4	0	0
Feelings of guilt	5	4	3
Feeling closer to child	7	9	11

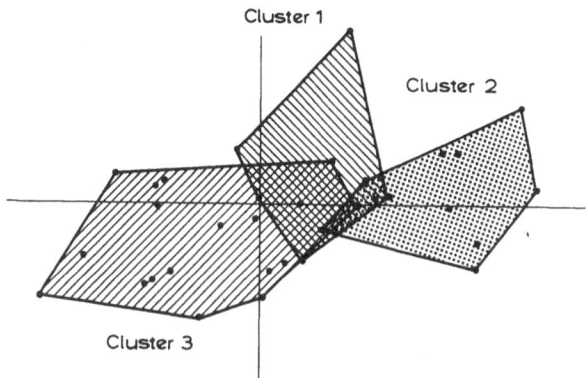

Fig. 1. Results of clustering neuropsychological and psychopathological parameters (see p. 274)

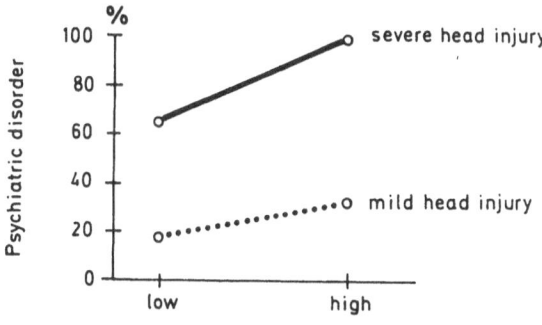

Fig. 2. Interrelationship between psychiatric disorders and psychosocial adversity

that this even holds for head-injured children (Fig. 2). Psychiatric disorders were most frequently found when severe head injury was combined with psychosocial adversity (quantified by the Family Adversity Index). The correlation between psychiatric disorder and psychosocial stress was significant for the subgroups of children having suffered head injury between ages 3–5 and 7–9, but not for those with rather recent injury (about 10 months ago). Thus we can conclude that in comparatively recent head injuries organic factors predominate over psychosocial stress and exercise the main influence on the actual behavior. The more time that has passed since the injury, the more the influence of psychosocial factors as a pathogenetic factor for evolving psychiatric symptoms is felt.

Neurological Sequelae

The results of neurological testing, EEG, and CT are summarized in Table 6. At least two-thirds of the patients show pathological findings on all three

Table 6. Results of neurological testing, EEG, and CT

	Age at injury		
	3–5	7–9	11–13
n	15	15	15
Neurological abnormalities	9	10	10
EEG			
—abnormalities	12	13	13
—focal signs	5	7	11
—hypersynchronous activity	9	9	11
CT			
—recorded	14	13	12
—abnormalities	12	12	11

Table 7. Relationship between psychiatric disorders and neurological abnormalities

Psychiatric disorders	Neurological abnormalities	
	Absent	Present
No disorder	9	2
Disorders	7	18
	$\chi^2 = 6.9$	$P = 0.009$

parameters. Changes in CT and psychiatric disorder are not significantly correlated (Lehmkuhl et al. 1987), yet there is a significant correlation between neurological abnormalities and psychiatric disorders (Table 7).

Summary and Discussion

It is generally agreed in the literature that a severe cerebral trauma suffered in childhood or adolescence increases the risk of later psychiatric disorders (Brink et al. 1970; Shaffer et al. 1975; Lezack 1978; Kleinpeter 1979; Rutter 1981; Black et al. 1981; Lange-Cosack et al. 1981; Boll 1983; Brooks 1984; Oddy 1984; Wood 1984; Rutter et al. 1984; Remschmidt 1985). The 69% disturbance rate in our study corresponds to the rates reported by Shaffer et al. (1975) and Brown et al. (1981). They found psychopathologically relevant symptoms in 62%–72% of children following severe cerebral trauma.

The sequelae of cerebral injury are above all determined by its severity. Time since injury and age at injury have no impact on neuropsychological functioning, but they do on behavioral disorders. The further emotional and cognitive development of the head-injured child is to a certain extent determined by neurological and psychosocial complications; these rank next to severity of trauma as predictors for certain sequelae.

In accordance with the literature, we found that neuropsychological deficits were more or less remedied within 9 months after injury, whereas most behavior disorders remained stable beyond that time, in some cases even for a rather long time.

The results of our study indicate that children who have suffered severe injury are not globally retarded at all levels of performance; they show not only more or less marked deficiencies in certain areas but also relative strengths. Group statistics, however, partly level these poor and strong sides of functioning.

Longitudinal studies demonstrate that $2\frac{1}{4}$ years after injury the children show impaired motor skills and poorer results on the number-symbol test (Chadwick et al. 1981a–c) as compared with controls. In our patients with severe cerebral trauma similar deficits of functioning were still observed 4–7 years after injury. Thus the suggestion of Remschmidt et al. (1980) that relatively labile and relatively stable functions/dysfunctions should be differentiated is buttressed by empirical studies. The rather increased rate of visuomotor impairment may be due to the high complexity of these tasks, which combine very different subfunctions, e.g., visual, verbal, psychomotor, and mnestic aspects (Lange-Cosack and Tepfer 1973; Michel 1983).

In our study we could find neither any relationship between age at injury and neuropsychological sequelae nor the phase specificity of cerebral injury suffered in childhood described by other authors (e.g., Kleinpeter 1971). In particular we could not find any differences in cognitive development sensu Piaget related to age at injury. Brink et al. (1970) and Lange-Cosack et al. (1979) concluded from their test psychological studies that intellectual deterioration after head injury is more pronounced in younger children. Schneider (1978) stated that the earlier the injury occurs in the ontogenesis of a child, the more likely are the sequelae to be severe. The study by Chadwick et al. (1981b, c) could not confirm this hypothesis. Our findings from a comprehensive test battery provide empirical evidence that age at injury does not have a significant effect, i.e., there is no evidence that age plays a significant role in neuropsychological deficits following head injury.

Although empirical data do not corroborate the existence of a typical persisting pattern of neuropsychological deficits following head injury, it is still possible that it exists in individual cases, that is, that certain subfunctions are more or less affected. These strong and weak points of functioning are not always the same, so that group statistics must fail to detect a specific pattern of deficits. Therefore the diagnostics should ideally proceed from the individual patient and his/her individual profile of cognitive functioning as compared with noninjured and comparably injured children. With such an approach it would be possible to assess how the individual profile of performance deviates from the age-specific norm as well as from the profile of patients with severe cerebral trauma. Thus specific patterns of cognitive deficits could be detected in the individual case which would be undetectable in group comparisons owing to the variety of affected subfunctions. Results of Shaffer et al. (1975, 1980) and Chadwick et al. (1981b, c) concerning localization of brain damage in childhood can be interpreted in this way: using comparative group statistics they found that it does not matter for neuropsychological deficits whether the lesion is localized in the left or right hemisphere.

There are no clues to indicate a uniform type of cognitive and emotional change as a consequence of traumatic head injury. Psycho-organic syndromes frequently occur immediately after the injury. This diagnosis is characterized by a coincidence of neuropsychological and psychopathological deficits, is devoid of a uniform symptomatology, and comprehends patients suffering as involving a variety of disturbances.

As to prevalence of psychiatric disorders, the groups 3–5 and 7–9 years old at injury did not differ significantly. Contrary to Rutter et al. (1984), who doubt the existence of a specific psychopathological syndrome ensuing from cerebral trauma, many child psychiatrists in the German-speaking countries still adhere to the concept of a specific postcontusional syndrome. One of the main advocates is Lempp (1972, 1973, 1980a, b). He even differentiates a specific symptomatology caused by brain damage suffered in early infancy.

Several questionnaires, especially focusing on the behavior problems of children who have suffered head injury, were used to determine the possibility of specific psychopathological deviations depending upon severity of, and age at, injury. Severely injured children were assessed significantly higher by their parents only with regard to concentration disorder/distractibility and hypo-activity. All other symptoms, e.g., aggressivity, rigidity, hyperactivity, and introversion, were equally distributed. Nor could an influence of age at injury on the parents' ratings be shown.

Rutter (1977, 1981, 1982) and Brown et al. (1981) extensively studied the influence of psychosocial adversity on the psychiatric risk following head injury. They operationalized family adversity with the Family Adversity Index (FAI, Rutter and Quinton 1977) and correlated it to posttraumatic disturbances. In this way it could be shown, in conformity with our study, that psychiatric disturbances are most pronounced in children suffering from severe head injury and exposed to high psychosocial stress. According to Rutter et al. (1984) the percent increase in psychiatric symptoms is independent of severity of trauma, i.e., the psychosocial stress bears evenly on mild and on severe injuries. This correlation between severity, psychosocial stress, and psychiatric disturbance was confirmed by our study.

When the age at injury is matched against psychiatric disturbance and psychosocial stress, the following conclusion can be drawn for severely injured patients: Children with an age at injury of 11–13 years (the injury preceding the assessment by 10 months on average) show a very high frequency of psychiatric disturbance even with low social stress, which still higher FAI scores can only increase slightly. For children aged 3–5 or 7–9 at injury, the correlation observed by Rutter et al. (1984) between psychiatric disturbance and psycho-social stress is fully confirmed, i.e., a higher stress score relates to a higher frequency of psychiatric disturbance. The findings of Brown et al. (1981) thus must be completed as follows: In close proximity to trauma, immediate cerebral sequelae dominate and lead to high psychiatric disturbance rates even when there is low psychosocial stress. For groups with less recent injury the findings of Rutter et al. (1984) are valid: psychosocial stress directly relates to psychiatric disturbance. This conclusion reflects the immediate importance of cerebral sequelae in the sense of a psycho-organic syndrome, whereas other factors come to bear only later. This is in line with the findings of Lishman (1988) evidencing that the longer postconcussional symptomatology persists after the injury, the more relevance must be attached to nonorganic factors.

Consequently, and in accordance with Rutter et al. (1984), an interaction between different factors, such as severity of trauma, psychosocial stress, family reactions, and performance, has to be assumed. Each of these factors should be regarded both in isolation and in correlation with the others.

Simplifying theories, designated as myths by Boll (1983), are revised by our findings as follows:

— There are no clues to suggest a unified pattern of cognitive and emotional deviations after head injury
— Hyperactivity and hypoactivity do not occur more frequently in severely injured patients
— The diversity of behavioral disturbances following head injury depicted by Rutter et al. (1970) and Rutter (1977) is confirmed by our study
— Psycho-organic syndromes (disinhibited state) can be observed (Arbus et al. 1969; Brown et al. 1981; Chadwick 1985) but the diagnosis is not uniform because of overlapping neuropsychological and psychopathological deviations, and because very different patients are included.

Our results show that specific sequelae of head injury do not exist, either for the neuropsychological or for the psychopathological diagnosis. This encourages us to reject the fixed concept of a "cerebral character change," in order to grasp instead the influence of the other important parameters of behavior—such as psychosocial stress, neurological symptoms, and cognitive capacities—, and thereby develop therapeutic strategies of higher efficiency, based on a dynamic view of cerebral symptoms.

Acknowledgment. This research was supported by the Stiftung Volkswagenwerk.

References

Arbus L, Moron P, Lazorthes Y, Luxey C (1969) Séquelles neuropsychiques des traumatismes craniens de l'enfant. Neurochirurgie 15:27–34

Benton AL, van Allen MW (1968) Test of facial recognition. Neurosensory Center and Departments of Neurology and Psychology, University of Iowa, Iowa

Benton AL, van Allen MW (1972) Prosopagnosia and facial discrimination. J Neurol Sci 15:167–172

Black P, Jeffries J, Blumer D, Wellner AM, Walker AE (1969) The post-traumatic syndrome in children: characteristics and incidence. In: Walker AE, Caveness WF, Critchley M (eds) The late effects of head injury. Thomas, Springfield, pp 142–149

Black P, Blumer D, Wellner AM, Shepard RH, Walker AE (1981) Head trauma in children: neurological, behavioral, and intellectual sequelae. In: Black P (ed) Brain dysfunction in children. Raven, New York, pp 171–180

Boll TJ (1983) Neuropsychological assessment of the child: myths, current status, and future prospects. In: Walker EC, Roberts MC (eds) Handbook of clinical child psychology. Wiley, New York, pp 186–208

Brickenkamp R (1975) Test d2—Aufmerksamkeitsbelastungstest. Hogrefe, Göttingen

Brink JP, Garrett AL, Hale WR, Woo-Sam J, Nickel VC (1970) Recovery of motor and intellectual function in children sustaining severe head injuries. Dev Med Child Neurol 12:545–571

Brooks N (1984) Cognitive deficits after head injury. In: Brooks N (ed) Closed head injury. Oxford University Press, Oxford, pp 44–73

Brown G, Chadwick O, Shaffer D, Rutter M, Traub M (1981) A prospective study of children with head injuries. III. Psychiatric sequelae. Psychol Med 11:63–78

Chadwick O (1985) Psychological sequelae of head injury in children. Dev Med Child Neurol 27:72–75

Chadwick O, Rutter M, Brown G, Shaffer D, Traub M (1981a) A prospective study of children with head injuries. II. Cognitive sequelae. Psychol Med 11:49–61

Chadwick O, Rutter M, Shaffer D, Shrout P (1981b) A prospective study of children with head injuries. IV. Specific cognitive deficits. J Clin Neuropsychol 3:101–120

Chadwick O, Rutter M, Thompson J, Shaffer D (1981c) Intellectual performance and reading skills after localized head injury in childhood. J Child Psychol Psychiatry 22:117–139

Conners C (1973) Rating scales for use in drug studies with children. Psychopharmacol Bull Special Issue 8:24–84

De Renzi E, Vignolo LA (1962) The token test: a sensitive test to detect receptive disturbances in aphasics. Brain 85:665–678

Ekman P, Friesen WV (1975) Unmasking the face. Prentice Hall, Englewood Cliffs

Elkind D (1964) Ambiguous pictures for study of perceptual development and learning. Child Dev 35:1391–1396

Ernhart CB, Graham FK, Eichman PL, Marshall JM, Thurston D (1963) Brain injury in the preschool child: some developmental considerations. II. Comparison of brain-injured and normal children. Psychol Monogr 77:17–33

Freund G (1980) Experimentelle Untersuchungen zum Sprachverständnis aphasischer Patienten. Doctoral thesis, University of Aachen

Geisel B, Eisert HG, Schmidt MH, Schwarzbach H (1982) Entwicklung und Erprobung eines Screening-Verfahrens für kinderpsychiatrisch auffällige Achtjährige (SKA 8). Prax Kinderpsychol Kinderpsychiatr 31:173–179

Graham P, Rutter M (1968) The reliability and validity of the psychiatric assessment of the child. Br J Psychiatry 114:581–592

Hooper HE (1958) The Hooper visual organization test. Western Psychological Services, Beverly Hills

Horn W (1969) Prüfsystem für Schul- und Bildungsberatung (PSB). Hogrefe, Göttingen

Kagan J, Rosman BL, Day D, Albert J, Phillips W (1964) Information processing in the child: significance of analytic and reflective attitudes. Psychol Monogr 78:1–37

Kimura D (1963) Right temporal lobe damage. Arch Neurol.8:264–271

Kleinpeter U (1971) Störungen der psychosomatischen Entwicklung nach Schädel-Hirn-Traumen im Kindesalter. Fischer, Jena

Kleinpeter U (1979) Folgezustände nach Schädel-Hirn-Traumen im Kindesalter und deren Begutachtung. Thieme, Leipzig

Klonoff H, Paris R (1974) Immediate, short-term, and residual effects of acute head injuries in children: neuropsychological and neurological correlates. In: Reitan RM, Davison CA (eds) Clinical neuropsychology: current status and applications. Halstead, New York, pp 179–210

Lange-Cosack H, Tepfer G (1973) Das Hirntrauma im Kindes- und Jugendalter. Springer, Berlin Heidelberg New York

Lange-Cosack H, Wider B, Schlesener HJ, Grumme T, Kubicki S (1979) Spätfolgen nach Schädelhirntraumen im Säuglings- und Kleinkindalter. Neuropädiatrie 10:105–127

Lange-Cosack H, Riebel U, Grumme T, Schlesener HJ (1981) Possibilities and limitations of rehabilitation after traumatic apallic syndrome in children and adolescents. Neuropediatrics 12:337–366

Lehmkuhl G (1986) Kognitive, neuropsychologische, psychopathologische und klinische Befunde bei 12-14 jährigen Kindern nach unterschiedlich schweren und lang zurückliegenden Schädel-Hirn-Traumen. Postdoctoral thesis, University of Heidelberg

Lehmkuhl G, Weniger D (1981) Das Gestenverständnis bei Kindern mit und ohne Hirnschädigung. 18th Meeting of the Deutsche Gesellschaft für Kinder- und Jugendpsychiatrie, Munich

Lehmkuhl G, Thoma W, Weber G (1987) Computertomographische Befunde bei Kindern nach einem schweren Schädel-Hirn-Trauma in ihrer Beziehung zu neuropsychologischen und psychopathologischen Befunden. In: Kohlmeyer K (ed) Aktuelle Probleme der Neurotraumatologie und klinischen Neuropsychologie. Regensberg and Biermann, Münster, pp 423–430

Lempp R (1972) Psychopathologie der Hirnschädigung in Kindesalter. In: Kisker KP, Meyer JE, Müller M (eds) Psychiatrie der Gegenwart, vol II/1. Springer, Berlin Heidelberg New York, pp 921–976

Lempp R (1973) Die Psychopathologie des Hirntraumas im Kindesalter. In: Müller E, Walch R (eds) Das kindliche Schädelhirntrauma. 15th Congress of the Gesellschaft für Hirntrauma, Bad Homburg

Lempp R (1980a) Die organischen Psychosyndrome. In: Harbauer H, Lempp R, Nissen G, Strunk P (eds) Lehrbuch der speziellen Kinder- und Jugendpsychiatrie, 4th edn. Springer, Berlin Heidelberg New York, pp 312–377

Lempp R (1980b) Organische Psychosyndrome. In: Bachmann D, Ewerbeck H, Joppich G, Kleihauer E, Rossi E, Stalder GR (eds) Pädiatrie in Praxis und Klinik, vol 3. Fischer, Stuttgart; Thieme, Stuttgart, pp 1970–1975

Lezak MD (1978) Living with the characterologically altered brain-injured patient. J Clin Psychiatry 39:592–598

Linder M, Grissemann H (1980) Zürcher Lesetest. Huber, Bern

Lishman WA (1988) Physiogenesis and Psychogenesis in the post-concussional syndrome. Br J Psychiatry 153:460–469

Michel M (1983) Rehabilitationsverlauf nach Schädelhirntrauma bei Kindern—eine neuropsychologische Studie. Rehabilitation 22:137–148

Müller R (1979) Untersuchung zum Gesten- und Pantomimenverständnis aphasischer und nichtaphasischer Patienten. Doctoral thesis, Free University of Berlin

Oddy M (1984) Head injury during childhood: the psychological implications. In: Brooks N (ed) Closed head injury. Oxford University Press, Oxford, pp 179–194

Orgass B (1976a) Eine Revision des Token-Tests. I. Vereinfachung der Auswertung, Itemanalyse und Einführung einer Alterskorrektur. Diagnostica 22:70–87

Orgass B (1976b) Eine Revision des Token-Tests. II. Validitätsnachweis, Normierung und Standardisierung. Diagnostica 22:141–156

Rathenow P (1979) Westermann Rechtschreibtest 4/5. Westermann, Braunschweig

Rathenow P, Laupenmühlen D, Vöge J (1981) Westermann Rechtschreibtest 6 +. Westermann, Braunschweig

Remschmidt H (1985) Psychische Störungen nach Schädel-Hirn-Traumen. In: Remschmidt H, Schmidt MH (eds) Entwicklungsstörungen, organisch bedingte Störungen, Psychosen, Begutachtung. Thieme, Stuttgart, pp 161–181

Remschmidt H, Merschmann W, Niebergall G (1980) Testpsychologische und klinische Verlaufsuntersuchungen an Kindern und Jugendlichen mit akuten Schädel-Hirn-Traumen. In: Remschmidt H, Stutte H (eds) Neuropsychiatrische Folgen nach Schädel-Hirn-Traumen bei Kindern und Jugendlichen. Huber, Bern, pp 111–145

Rune V (1970) Acute head injuries. Acta Paediatr Scand [Suppl] 209:3–112

Rutter M (1977) Brain damage syndromes in childhood: concepts and findings. J Child Psychol Psychiatry 18:1–21

Rutter M (1981) Psychological sequelae of brain damage in children. Am J Psychiatry 138:1533–1544

Rutter M (1982) Developmental neuropsychiatry: concepts, issues, and problems. J Clin Neuropsychol 4:91–115

Rutter M, Graham P (1968) The reliability and validity of the psychiatric assessment of the child. Br J Psychiatry 114:563–579

Rutter M, Quinton D (1977) Psychiatric disorder-ecological factors and concepts of causation. In: McGurk H (ed) Ecological factors in human development. North-Holland, Amsterdam, pp 173–187

Rutter M, Graham P, Yule W (1970) A neuropsychiatric study in childhood. Clinics in developmental medicine, nos 35/36. Heinemann Medical, London

Rutter M, Chadwick O, Shaffer D, Brown G (1980) A prospective study of children with head injuries.
    I. Design and methods. Psychol Med 10:633–645
Rutter M, Chadwick O, Shaffer D (1984) Head injury. In: Rutter M (ed) Developmental
    neuropsychiatry. Churchill Livingstone, Edinburgh, pp 83–111
Schneider R (1978) Hirnfunktionsstörungen im Kindesalter. Enke, Stuttgart
Shaffer D, Chadwick O, Rutter M (1975) Psychiatric outcome of localized head injury in children. In:
    Porter R, FitzSimons D (eds) Outcome of severe damage to the central nervous system. Ciba
    Found Symp 34:191–213
Shaffer D, Bijur P, Chadwick O, Rutter M (1980) Head injury and later reading disability. J Am Acad
    Child Psychiatry 19:592–610
Stroop JR (1935) Studies of interference in serial verbal reactions. J Exp Psychol 18:643–661
Tewes U (1984) Hamburg-Wechsler Intelligenztest für Kinder. Revision 1983. Huber, Bern
Weiß RH (1978) Grundintelligenztest CFT 20. Westermann, Braunschweig
Wishart D (1984) Clustan. Fischer, Stuttgart
Witkin HA (1971) A manual for the embedded figures test. Consulting Psychologists Press, Palo Alto,
    California
Wood RL (1984) Behaviour disorders following severe brain injury: their presentation and
    psychological management. In: Brooks N (ed) Closed head injury. Oxford University Press,
    Oxford, pp 195–219

What Becomes of Clumsy Children with Attention Deficits? Some Data and Reflections from a Population-Based Study in Sweden

C. Gillberg and I. C. Gillberg

Introduction

This paper deals with epidemiological and neurobiological data on a population-based group of clumsy children with attention disorders who have been followed for a 10-year period from preschool to their middle teens. In many previous publications relating to research on this group of children they have been collectively referred to as children with MBD in spite of the fact that no causal implication has been surmised by this label (Gillberg et al. 1982).

The concept of minimal brain dysfunction (MBD) has been justly criticized by many authors, among whom Rutter and his group have been most influential (Rutter 1982; Taylor 1986). We have not been able absolutely to trace the origin of this three-letter abbreviation in the literature, but Russel Schachar (1986) hold that it goes back to the promulgation of the notion of minimal brain *damage* by Alfred Tredgold (well-known mental deficiency expert and assiduous supporter of the eugenics movement) in 1908. Tredgold, and the famous pediatrician Sir George Frederic Still before him (1902), described children with a decreased capacity for sustained attention, various coordination problems, and other soft neurological signs sometimes in combination with defiant, antisocial, or outright criminal behavior. These authors believed that a "moral control defect" resulting from "organic abnormality of the higher levels of the brain" was at the root of such problems. Kahn and Cohen later (1934) coined the term "organic drivenness" to depict and explain a clinical syndrome highlighted by hyperactivity, clumsiness, and inability to remain quiet. In effect none of these authors put forward any evidence in support of their etiological speculation.

Quasiscientific empirical data were advanced by Strauss and Lehtinen in 1947 to emphasize their view that all brain lesions, regardless of localization, result in similarly disordered behavior. Their work, although hopelessly muddled and scientifically invalid, was highly influential at least up to the early 1960s, when an international study group (Bax and McKeith 1963) suggested that the minimal brain *damage* concept be discarded altogether. Since then MBD has become one of the most criticized concepts in the history of medicine. It is used less and less in the British scientific literature [though often referred to in clinical practice (Rutter 1985, personal communication)], but for various reasons remains in many other countries (e.g., Scandinavia, Germany, the People's Republic of China), where it is now taken to mean minimal brain dysfunction and to refer to, admittedly rather inconsistently, a constellation of symptoms regardless of underlying brain dysfunction or absence of brain dysfunction.

Although we are ourselves critics of the MBD concept, we would suggest that the most vociferous opponents direct some of their attacks towards equally unsound etiquettes such as cerebral palsy, mental retardation (to say nothing of mental handicap), and epilepsy. These are all labels which refer to groups of syndromes without common etiology, without common symptomatology, and with illogicalities as regards semantics. They are conceptually equally as bad as MBD, but can we do without them?

Current-day substitutes for the MBD label include "hyperkinetic syndrome" (Taylor 1986), "attention deficit disorder (ADD)"[1] (American Psychiatric Association 1980), and "attention deficit–hyperactivity disorder (ADHD)" (American Psychiatric Association 1987).

In the research we are going to report, we have particularly studied children with a combination of cross-situational ADD and motor-perceptual problems (MPD). We have not sought specifically to validate an "ADD + MPD syndrome" but rather to follow up children with this particular combination of problems. Such children are common in pediatric and child psychiatric practice and are usually those who receive a diagnosis of MBD. When we started our study in the mid-1970s, the common clinical Scandinavian notion was (and still is) that "MBD" [whatever that means (except that it is thought of as some sort of umbrella definition)] is equivalent to "ADD + MPD." Indirect data from our screening studies of the whole population of 6-year-olds in Göteborg in 1977 (Gillberg and Rasmussen 1982b and summarized in Gillberg 1987) support the notion that ADD and MPD coincide in a much stronger than chance fashion. Further evidence for the possible clinical validity of an "ADD + MPD" syndrome comes from the Maudsley studies by Taylor (1986) and Sandberg (1987), who have presented preliminary results showing cross-situational hyperactivity but not emotional disorder or classroom defiance to be combined with motor control or language problems in a majority of cases.

On balance, therefore, the results from Göteborg on children with the combination of ADD and MPD should be of interest to those who want to know more about background factors and outcome in children with problems loosely referred to as being of "MBD type."

Description of Study

Preschool Questionnaire Study

In Sweden, 95% of 6-year-old children attend public preschools. A questionnaire, designed in collaboration with preschool teachers and aimed at detecting perceptual, motor, and attentional deficits among 6-year-olds, was distributed to

---

[1] It is surprising the way the "deficit disorder" tautology has become firmly implanted with no or few objections.

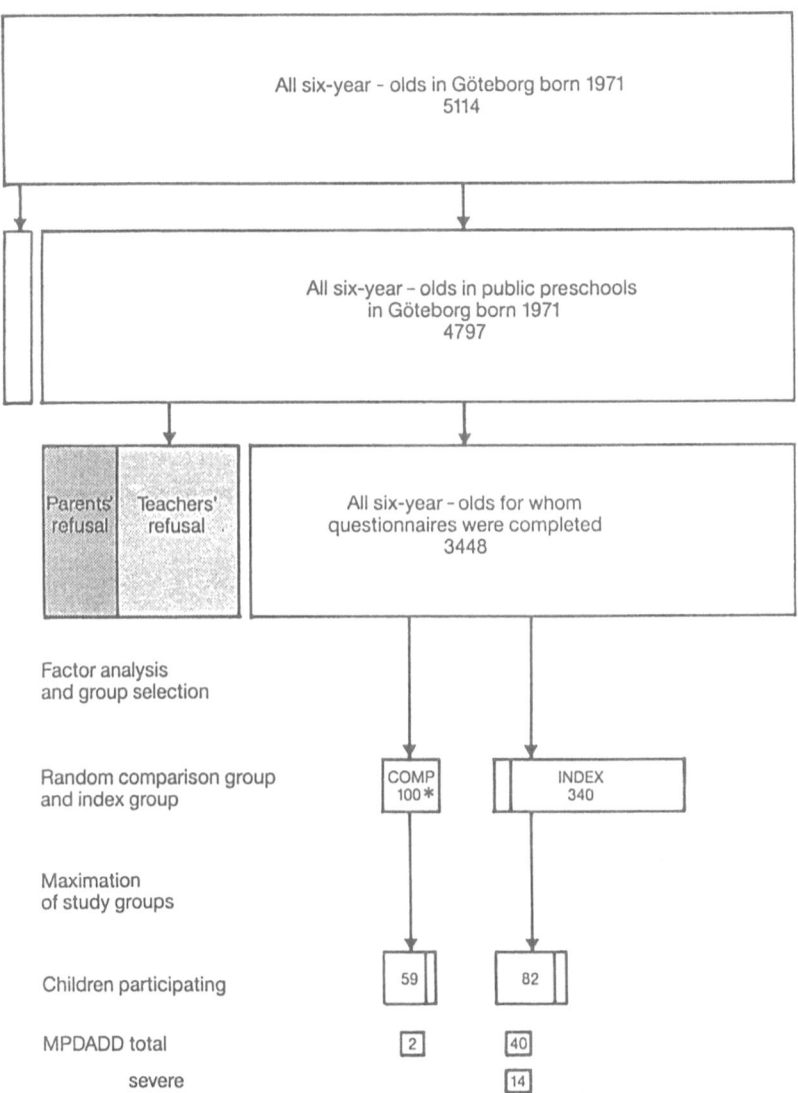

Fig. 1. Screening for MPDADD in public preschools of Göteborg

all teachers of the public preschools in Göteborg towards the end of 1977. The same questionnaire had previously been used in a large pilot study concerning 6-year-old children in Göteborg born in 1969. It contains 34 simple yes–no questions, which fall into five different groups: attention span/general behavior, speech–language, gross motor functions, fine motor functions, and perception–conceptualization. Seven of the questions refer to symptoms characteristic of the attention deficit–hyperkinetic syndrome. The teachers were requested to complete the questionnaire at the end of an observation period of at least 3 weeks, comparing each child born in 1971 with average children of the same age group. A letter of information to the parents was sent out with the questionnaire, and included a reply card for those who did not want their child to participate. There was a corresponding card for teachers who refused to participate.

The interrater reliability between two teachers rating the same child in their group has been shown to be quite acceptable, with no systematic differences and a mean Pearson r over all 34 questions of 0.91 ($n = 26$, $P < 0.001$).

The sampling procedure is shown comprehensively in Fig. 1. There were 5114 children born in 1971 living in Göteborg at the end of 1977. Of these, 317 did not attend public preschools. Many children in the group not attending preschool or school were from families who had just moved to Göteborg. There were 4797 children in public preschools at the time of the questionnaire study, and a questionnaire was completed for 72% of these—1800 boys and 1648 girls.

Reason for Nonparticipation in Preschool Study

Twenty-eight percent of the preschool children were withdrawn from the study. According to the reply cards, teachers' refusal to participate, largely due to objections in principle to investigations of this kind (which are regarded as a means for the authorities to blacklist all individuals who might later become a problem to society), was responsible for two-thirds of this failure to participate. Parents' refusal accounted for less than 9% of the total preschool population. As has been argued elsewhere in some detail (Gillberg et al. 1982), there is no obvious reason why the 72% who remained in the study should not be regarded as representative of the whole population.

Neuropsychiatric Study at Age 7 Years

One hundred and forty-one of the children from the preschool study took part in an extensive study consisting of psychiatric, neurological, and psychological examinations. Eighty-two of these children constituted the index group and 59 the comparison group. The criteria for inclusion in the index group were:

1. At least one symptom in each of the original five groups of the questionnaire
2. Abnormality in each of five factors, together found to account for 40% of the total variance in a varimax rotation factor analysis (Gillberg and Rasmussen 1982a) of the questionnaire
3. A combination of abnormality in any of the three "motor–conceptualization" factors and abnormality in either of the two "attention deficit–conduct problem" factors found in the factor analysis

Three hundred and forty children fulfilled the third criterion, and within this group 9% fulfilled either criterion 1 or 2. One hundred and twelve of the 340 cases were assigned to the study either because they fulfilled criterion 1 or 2 (all these cases were included) or because they fulfilled criterion 3 and were drawn in a one-in-four fashion from the whole of this group. For the comparison group, 100 children were sampled in accordance with a list of random numbers from the whole of the original questionnaire material. Eight of these were found in the index group. In accordance with the list of random numbers, 75 of the remaining 92 cases were contacted for participation.

All children except one were 7 years old at the time of examination.

Reason for Nonparticipation in Neuropsychiatric Study at Age 7 Years

Of the 112 index children altogether 21 (19%) had parents who refused participation. The corresponding frequency in the comparison group was 13%. Other reasons for attrition were Down's syndrome in the child (n = 2), death of parents or severe illness (n = 3), completely non-Swedish speaking family (n = 1), or moving far away (n = 6). Thus, altogether 82 index cases (64 boys, 18 girls) and 59 comparison cases (29 boys, 30 girls) participated in the first neuropsychiatric study at age 7 years.

Comprehensive Assessment of Child at Age 7 Years

All cases were examined blindly, i.e., without knowledge of which group the child belonged to, and in the same fashion by two child psychiatrists (who made home visits, examined the child in detail, and interviewed the mother for 2 h), a child psychologist [who used the performance part of the WISC-R (1974) and the Southern California Sensory Integration Tests (SCSIT) (Ayres 1972)], and a child neurologist in collaboration with an experienced child physiotherapist [who used a slightly abbreviated version of the Touwen manual (1979)].

A questionnaire concerning hereditary factors, family factors, and child's health and development was completed by the mothers.

A Swedish translation of the Carey-McDevitt questionnaire on the "behavioral style" of the child (Carey and McDevitt 1978) was completed by the mother.

Comprehensive Neurodevelopmental Diagnoses at Age 7 Years

Attention deficit disorder (ADD) was diagnosed in any child showing severe cross-situational attention deficit signs (i.e., in two or more of the following settings: child psychiatric examination, child neurological assessment, psychological evaluation, and mother's home reports). Altogether 12 children (ten boys, two girls) received this diagnosis.

Motor perception dysfunction (MPD) was diagnosed in any child showing clear dysfunction as regards gross motor, fine motor, or perception performance. These diagnoses were made on the basis of results obtained with neurological and psychological examinations. Altogether seven children (four boys, three girls) fulfilled criteria for this diagnosis.

MPD ADD was diagnosed in cases showing concomitant MPD and ADD. In several previous publications we have referred to this symptom constellation as

"MBD syndrome." *Severe MPDADD* was diagnosed in MPDADD cases with major deficits in all the following areas: attention, gross motor, fine motor, perception, and speech–language functions. Altogether 42 children (33 boys, 9 girls) were diagnosed as suffering from MPDADD, i.e., they were both "clumsy" and attention deficient.

Comprehensive Psychiatric Diagnoses at Age 7 Years (Gillberg 1983)

Six main psychiatric diagnostic categories were used for children showing marked psychiatric abnormality (i.e., a psychic condition handicapping either to the child or his environment): A few cases in this category turned out to fulfill criteria for Asperger syndrome.

1. "*Autistic type features*" were diagnosed when there were marked disturbances of social relations in combination with speech–language deficits and stereotyped or ritualistic behavior.
2. The term *depressive syndrome* was used when the child showed (and had shown for at least a month) both dysphoric mood or pervasive lack of interest or pleasure and at least three of the following: anorexia or hyperbulimia, insomnia or hypersomnia, lack of energy, psychomotor agitation or retardation, feelings of self-reproach or worthlessness, indecisiveness, or preoccupation with thoughts of death. Signs of delusions and hallucinations had to be absent. The diagnostic category of depressive syndrome thus accorded well with that described by Spitzer et al. (1975).
3. *Conduct disorder* was the diagnostic term applied in cases of socially unacceptable behavior, such as stealing, setting fire, or running away from home.
4. A diagnosis of *emotional disorder* was made in disorders where anxiety and fear without loss of sense of reality were the most incapacitating symptoms.
5. The category of *psychosomatic disorder* was used in cases of stomachache, vomiting, diarrhea, secondary enuresis, or encopresis and headache occurring in situations of psychic stress.
6. The "*other psychiatric disorder*" term was used only in cases which could not be included under one of the above headings.

Some children were diagnosed according to more than one psychiatric category. In such cases a hierarchy of diagnostic categories was used in accordance with the above order in the text.

Generalized Hyperkinesis. An operational diagnosis of generalized hyperkinesis was made when a child showed signs of attention deficit in the child psychiatric and the child neurological examination situation, as well as according to the mother.

Comparison Cases

The comparison group which will be referred to in the text hereafter consists of all 59 children in the original "population" comparison group minus those diagnosed as suffering from ADD, MPD, or MPDADD. This group consists of 51 children (24 boys and 27 girls).

Interrater Reliability of Child Psychiatric Assessments at Age 7 Years

An interrater reliability study was performed on 36 cases by the two child psychiatrists. Overall agreement was good both as regards rating of specific symptoms and global assessment (Gillberg 1983).

Familial and Hereditary Background Factors Studied

All information concerning hereditary factors refer to interview data from the mother. The information given by the mother in the questionnaire concerning hereditary factors, family factors, and child's health and development in most instances accorded well with the data obtained at direct interview, but in some instances the mother had been unwilling to put into print that which she would freely admit during the talk with the psychiatrist. Learning disorders, speech–language retardation, and left-handedness in first-degree relatives were considered nonoptimal for the development of perceptual, motor, and attention deficit (Frisk et al. 1967; Ingram 1959; Gillberg et al. 1980, 1982). Hyperactivity in first-degree male relative (Cantwell 1972, 1975) and delayed onset puberty in first- or second-degree relatives (Frisk et al. 1967) were also considered nonoptimal neurodevelopmental hereditary factors. These factors were all given scores of 1 if present and of 0 if absent and then brought together in a cluster, the maximum possible total cluster score thus being 5.

Psychiatric disorders and social maladjustment in first- and second-degree relatives were also analyzed in detail.

Neuropathogenic Background Factors Studied

Pregnancy, obstetric, and medical records were collected and examined. A reduced optimality cluster score for prenatal, intrapartal, and neonatal conditions (medical record data) was calculated on the basis of a modified version of the system suggested by Prechtl (Gillberg and Rasmussen 1982a).

"Minor neurological dysfunction" was studied as a separate background factor. Such dysfunction was rated according to the neurological examination and a total score was calculated, summing up all the scores (0, 1, and 2, where 2 indicates marked abnormality) for the individual items. This factor, as well as a minor physical anomaly (MPA) score, was calculated in accordance with the method described by Waldrop and Halverson (1971).

Psychosocial Background Factors Studied

Maternal social stress scores were calculated in accordance with a method described by Mendeloff and associates (Mendeloff et al. 1970). A cluster of nonoptimal psychosocial conditions (interview + register data) was completed in each case. A nonoptimal rearing conditions score was also computed (Gillberg 1983).

EEG Study

Electroencephalographic recordings, including rest recordings and activations by hyperventilation, intermittent photic stimulation, and sleep, were obtained in 106 of the 141 cases when the children were 8 years old (Gillberg et al. 1984).

Comprehensive Assessment of Child at Age 10 and 13 Years

A vast majority of all the 141 children in the original study were followed up when they were 10 and 13 years old (Gillberg 1987). Exactly the same methods were used on both occasions, i.e., a brief neurodevelopmental/psychiatric assessment was performed by a "new" doctor (blind to the results of the examinations performed when the child was 7 years old), and various questionnaires were completed by the children (Birleson 1981), parents (Rutter et al. 1970), and teachers (Rutter 1967; Conners 1969; Gillberg et al. 1983) in order to tap depression, other emotional and behavior problems, and school achievement difficulties.

Follow-up at Age 16 Years

A comprehensive follow-up of all the children in the original study is in progress at the moment. So far only registers concerning psychiatric and physical out- and inpatient care have been thoroughly searched for the whole group and so only results from this register search will be presented here.

Statistical Methods Used

Fisher's nonparametric permutation test (Bradley 1968) was used for univariate statistical analyses. Chi-square test was used in a few instances with 3×2 contingency tables. In some instances Pearson's coefficient of correlation was used. The Statistical Computer Library BMD P4M was used in the factor analysis of the questionnaire, using varimax rotation. A stepwise discriminant analysis using the Statistical Computer Library BMD P7M was applied in the analysis of the capacity of the questionnaire to detect MBD cases. Stepwise regression techniques, as described by Draper and Smith (1966), were utilized in the analysis of background factors. Since some of these factors were of a categorical nature, the results obtained in the regression analysis were verified by way of nonparametric partial correlation analysis (Mantel 1963).

Discussion of Major Results

Epidemiology at Age 7 Years

MPDADD

Extrapolating the results from the neuropsychiatric study at age 7 years to the general polulation, we came up with a prevalence of 1.2% severe MPDADD cases among 7-year-olds and another 3.0%–5.9% mild–moderate MPDADD cases (Gillberg et al. 1982). The result in respect of severe cases is probably relatively accurate and has been repeatedly confirmed in a number of recent Swedish studies (Ljungman 1985; Winnergård et al. 1986, unpublished manuscript, "DAMP (deficits in attention, motor control and perception) in Swedish public pre-school children. Screening methods and epidemiology."). The frequency of mild–moderate MPDADD is less firmly based, mainly because of the fact that, in spite of the screening selection of cases, we found two mild–moderate MPDADD cases in the original comparison group, consisting of random population cases *less* those who had raised suspicion of suffering from MPDADD according to preschool teachers.

The boy-to-girl population-corrected ratios were 10:1 in the severe and 1.9:1 in the mild–moderate MPDADD group. Thus "typical" cases were extremely uncommon in girls, whereas mild–moderate cases tended to be fairly common, albeit much less prevalent than in boys.

Our total maximum figure for MPDADD of 7.1% of all urban 7-year-olds is in good accord with the "MBD" prevalence reported from Beijing (Shen et al. 1985) and from inner London for "hyperkinetic syndrome" (usually with concomitant MPD) (Sandberg 1987).

As mentioned in the introduction, our study was not particularly aimed at validating the MPDADD syndrome as distinct from other developmental/neuropsychiatric problems of childhood. Rather, we accepted the notion in Scandinavia in the mid-1970s that MPDADD *is* a syndrome. Our follow-up data have shown MPDADD to have some validity as a diagnostic entity insofar as outcome is concerned. Also, indirect extrapolation data from our study suggest that MPD and ADD cluster together in much stronger than chance fashion (Gillberg et al. 1989). The findings by the Rutter group (Sandberg et al. 1978) that hyperkinetic disorders are often connected with various neuro-developmental problems yield further support for the notion of a relatively valid MPDADD or ADDMPD syndrome.

Psychiatric Problems in Children with MPDADD

The two child psychiatrists who examined the children in their homes were blind to the assessment of the psychologist and neurologist, and so were not aware of whether or not a child would be diagnosed as suffering from MPDADD or not. Therefore the child psychiatric diagnoses assigned are independent of the MPDADD label. Neither the psychiatrists nor the other examiners were informed about the results of the preschool study and therefore did not know anything about whether the child belonged in the index or comparison group.

We found autistic features in slightly more than half of the children with severe MPDADD [and not in any one child with mild–moderate problems or without such problems ($P < 0.001$)], corresponding to a population frequency of 69 per 10000 at age 7 years. Autistic features in this study constituted mild–moderate–severe problems within the range of the triad of language, social, and behavioral impairment, a term coined by Wing (Wing and Gould 1979) to describe children who show some but not all of the symptoms typical of Kanner autism or who are quantitatively as affected by problems in this realm but who qualitatively differ somewhat in one or the other fields of the triad. This prevalence of 69 per 10000 is population based and should be compared with that of 21 per 10000 found by Wing in a London screening of *handicapped* children under 15 years.

We found conduct disorders in slightly less than 10% of all the children with MPDADD. One child with mild mental retardation also had a conduct disorder, but not one child in the comparison group was similarly diagnosed ($P < 0.06$).

More than one in four of MPDADD children were diagnosed as suffering from depressive syndrome. Only one of every 25 children in the comparison group was considered to be depressed ($P < 0.001$).

Hyperkinetic Disorders in Children with MPDADD

Generalized hyperkinesis, i.e., attention deficit signs with hyperactivity in three out of three possible assessment settings, was diagnosed in 50% of children with severe MPDADD, in 20% of children with mild–moderate MPDADD, and in 0% of children without MPDADD. However, one should not draw firm conclusions on the basis of these results since attention deficits were the main part of the definition of generalized hyperkinesis and an important part of the definition of MPDADD.

Background Factors

On the basis of background data analysis and comparison of MPDADD cases with normal children we concluded (Gillberg and Rasmussen 1982a) that 33% of children with MPDADD had a severe reduction of optimality in the pre-, peri-, or postnatal period, possibly associated with increasing risk for brain damage. No child in the comparison group had a similar reduction. Further, we found 21% of children with MPDADD to have a nonoptimal hereditary background (high cluster scores for hereditary learning disorders, speech–language problems, hyperactivity, motor incoordination, delayed onset puberty, and left-handedness), not encountered in any of the comparison group cases. Finally, 17% of children with MPDADD had a combination of pre-, peri-, or postnatal problems and hereditary nonoptimal factors. Thus, altogether 71% of our group of children with MPDADD had a plausible "neuropathogenic" hereditary or environmental background factor.

It is of some interest that MPAs were much more common in the MPDADD group, the median MPA score in this group being 3 compared with 1 in the normal group ($P < 0.001$).

We then analyzed our data with regard to the influence of psychosocial factors. Social class was lower in the MPDADD group than in the comparison group, but the "neuropathogenic" background factors did not covary with social class.

Various psychosocial problems, such as maternal stress, paternal illness, one-parent family, socially disadvantaged dwelling area, and nonoptimal rearing conditions were typically associated with psychiatric problems, in particular if the child had MPDADD. Children diagnosed as suffering from MPDADD without a concomitant psychiatric diagnosis did not have psychosocial problems. On the basis of our results, I have concluded elsewhere (Gillberg 1983) that MPDADD is usually a biologically determined disorder and that if severe psychiatric problems ensue, psychosocial influences are often responsible. Also, children with MPDADD are much more prone to developing psychiatric problems under psychosocial stress than are children unaffected by MPDADD.

Follow-up at Age 10, 13, and 16 Years

There were extremely high rates of behavior, school achievement, and neuro-developmental problems among the children with MPDADD on follow-up at age 10 and 13 years. The comparison group showed significantly lower rates for all parameters studied. The rate of psychiatric and other specialist consultation at age 16 years was also extremely high in the MPDADD group as compared with the other group.

In a series of papers published in *Developmental Medicine and Child Neurology*, we have summarized the findings of the follow-up studies (e.g., Gillberg et al. 1989). Table 1 gives an overview of these findings. Behavioral abnormality was defined as scoring more than two standard deviations above the mean of the comparison group for any one of the sum or subscores of the Rutter scales, the Conners scale, and the Birleson scale. Obvious school achievement problems were diagnosed in children showing severe learning disorders according to teachers or who were at least 1 year behind in school. Neurodevelopmental problems were diagnosed in children showing a composite score for neurological dysfunction which surpassed the mean of the comparison group by more than two standard deviations.

In summary, neurodevelopmental problems, which had been present in all MPDADD cases at age 7 years, had subsided in many cases and the rates were 55% and 30% respectively at age 10 and 13 years. On the other hand, while the outcome for "clumsiness" thus seemed to be on the positive side, behavior

Table 1. Neurodevelopmental, behavior, and school achievement problems in children with MPDADD at age 10 and 13 years

	MPDADD (n = 42)		Comparison (n = 51)	
	10 yrs	13 yrs	10 yrs	13 yrs
Neurodevelopmental problems	55%	30%	4%	0%
(severe MPDADD)	(75%)	(54%)		
Behavioral abnormality	80%	64%	20%	25%
(severe MPDADD)	(93%)	(85%)		
Obvious school achievement problems	76%	67%	16%	8%
(severe MPDADD)	(100%)	(85%)		
All three types of problems	42%	22%	0%	0%
Behavior and school achievement				
problems	68%	41%	8%	3%
No problems	5%	16%	69%	70%

All differences between MPDADD and comparison groups and between severe MPDADD and comparison groups are significant at or below the 0.1% level of statistical significance (χ^2 test with Yates' correction)

Table 2. Need for specialist help in different study groups

	MPDADD			ADD	MPD	Comparison
	Severe	Mild	All			
Remedial education (Swedish, spelling, maths)	13/14** *** (93%)	18/28 *** (64%)	31/42** *** (74%)	3/12 (25%)	3/7 (43%)	12/51 (24%)
Speech therapy	7/14* *** (50%)	9/28 *** (32%)	16/42* *** (38%)	0/12	0/7	1/51 (2%)
Full-time assistant in classroom	4/14*** (29%)	1/28 (4%)	5/42* (12%)	0/12	0/7	0/51
Child psychiatric treatment	***	***	***	**		
outpatients only	10/14*** (71%)	12/28 (43%)	22/42 (52%)	4/11 (36%)	1/7 (14%)	2/51 (4%)
Inpatients	2/14 (14%)	0/28	2/42 (5%)	0/11	0/7	0/51
Treatment in hospital for a physical complaint/disorder or accident	9/14** (64%)	9/28 (28%)	18/42* (43%)	3/12 (25%)	3/7 (43%)	11/51 (22%)
Number of hospital admissions because of accidents	5/14** (36%)	5/28 (18%)	10/42* (24%)	2/12 (17%)	0/7	3/51 (6%)

***$P < 0.001$ MPDADD vs comparison; **$P < 0.01$ MPDADD vs comparison; *MPDADD vs comparison
***$P < 0.001$ MPDADD vs ADD; **$P < 0.01$ MPDADD vs ADD; *MPDADD vs ADD
———, significant differences for boys compared with boys
- - - - -, significant differences for girls compared with girls

and/or school achievement problems occurred in two-thirds or more of the cases throughout the follow-up period from preschool to the early teens.

Table 2 shows the rates of specialist consultation before age 16 years in the MPDADD and comparison groups. Almost two-thirds of children with MPDADD had consulted a child psychiatrist. Of these, more than half had been in contact with child psychiatric services before age 7 years. In the comparison group the rate was 6% ($P < 0.001$). In our original study we had not informed the parents of diagnostic speculations and had not advised them concerning need for specialist treatment except in two very problematic cases. Therefore, participation in the study should not have produced atypical results, particularly since the rate of consultation is so low in the blindly examined comparison group.

Changing Panorama of Behavior Problems

A fairly distinct pattern of behavior problems were seen at age 7 and 10 years, but the picture had become much more muddled at age 13 years. At age 7 years, the children with severe MPDADD often showed autistic-type features or had some kind of major conduct problem. Many of the children with mild–moderate MPDADD were depressed. At the age of 10, almost half of the MPDADD cases had a conduct problem and one-fifth appeared depressed. The children who had shown autistic features were still those with the highest rate of behavior problems, but many now appeared to be depressed and isolated rather than having autistic-type features. At age 13 years, a whole range of behavior and emotional problems were prevalent in the MPDADD group. "Daydreaming" was the most common type of problem and neither conduct disorder nor depressive symptoms stood out as particularly typical. A small group of cases with severe MPDADD fulfilled criteria for Asperger syndrome throughout the follow-up period.

Hyperkinesis in Particular

The 18 children diagnosed in the original study as suffering from extremes of hyperkinetic behavior disturbance ("generalized hyperkinesis") were analyzed separately. Somewhat unexpectedly they did not show a very high rate of conduct disorder either at age 10 or 13 years. To be sure, a few children were clearly conduct disordered at both ages, but the majority of severely hyperkinetic children (many of whom remained hyperkinetic throughout the follow-up period) were not.

EEG and Complex Reaction Time Measurements

A majority of the children with MPDADD and in the comparison group underwent EEG examinations at age 8 years (Gillberg et al. 1984). At age 10 and 13 years measurements of complex reaction times were performed (using

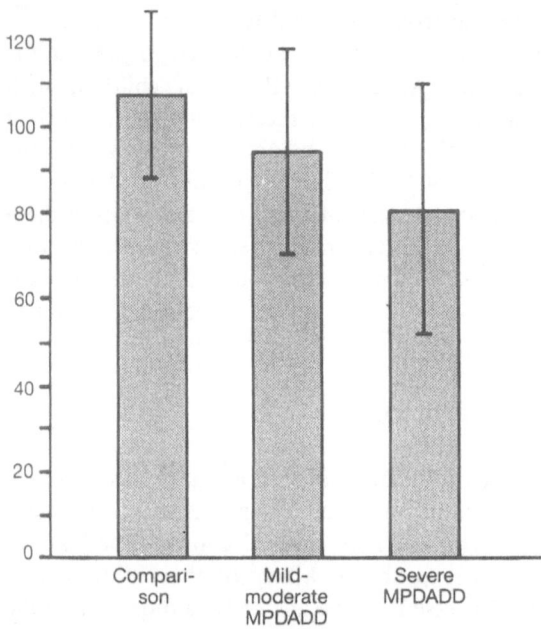

Fig. 2. "EEG normality ratio" in comparison children and children with mild–moderate MPDADD and severe MPDADD. (Product–moment correlation coefficient = 0.39, P < 0.01 for the average ratio over all four pairs of EEG derivations used in th spectral analysis)

a small computer, the size of a typewriter with flashes of colored lights and correspondingly colored buttons for the child to press, yielding an automatically calculated mean complex reaction time averaged over 10 min and 150 exposures) (Gillberg et al. 1981, 1988; Gillberg 1985).

A moderate increase in low-frequency activity was seen in 30% of children with MPDADD and in none of the children in the comparison group (P < 0.001). A totally normal EEG was seen in 25% of MPDADD cases and in 70% of the comparison group (P < 0.001). Left-sided predominance of abnormalities was more common than right-sided abnormalities and deviance was particularly pronounced in parieto-occipital derivations.

Among the group with severe MPDADD, about a quarter showed paroxysmal activity on activation. All of these children also showed autistic traits.

Frequency spectral analysis of the EEG showed a gradual decrease in the so-called normality ratio with increase in pathology on the comparison/mild–moderate MPDADD/severe MPDADD scale (Fig. 2). This ratio can be interpreted as reflecting the degree of low-frequency activity. It approaches 100 in individuals with age-appropriate frequencies and falls with increasing amounts of low-frequency activity.

Mean complex reaction time was significantly longer in the MPDADD group than in the comparison group at both 10 and 13 years. The mean time was 0.89 s (SD = 0.22) and 0.74 s (SD = 0.12) in the two groups at age 10. At age 13 the comparison group had a mean time of 0.60 s (SD = 0.11) whereas the MPDADD group now had a mean time very close to that of the comparison group 3 years earlier.

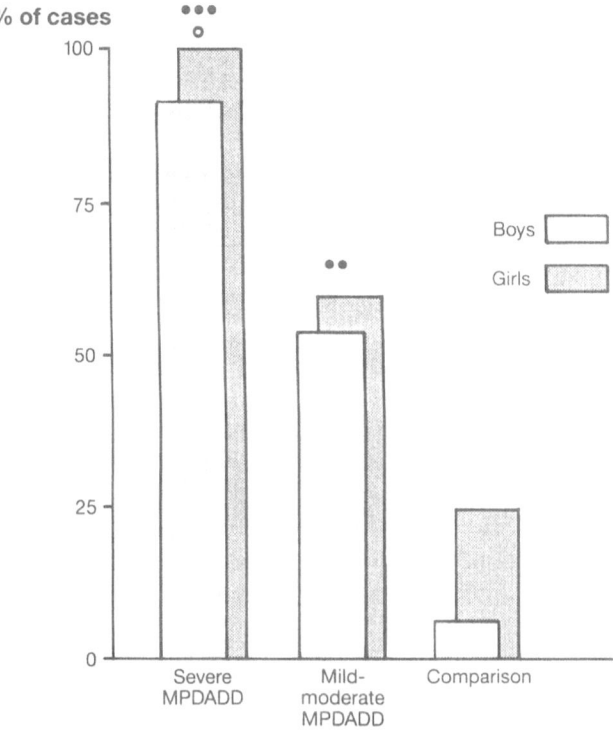

Fig. 3. Frequencies of children in MPDADD and comparison groups showing complex reaction times exceeding 0.65 s. Subdivision according to sex. Total number of cases = 78. ●●● $P < 0.001$ vs comparison; ●● $P < 0.01$ vs comparison; ○ $P < 0.05$ vs mild–moderate MPDADD

Figure 3 shows the great difference between the groups in respect of definitely prolonged reaction time (> 0.65 s) at age 13 years. The difference was particularly pronounced in boys, who could almost blindly have been allocated to the correct diagnostic group on the result of this measurement alone (at least the severely affected and the normal cases).

Indirect Implications of Epidemiological Findings for Child Psychiatry and "Developmental Medicine"

On the basis of our results we infer that one-fourth to one-half of all child psychiatric patients in a Swedish urban area show or have shown MPDADD. The relative rate depends on whether in- or outpatients are examined. The rate would be highest in the inpatient group. We found that 7.1% of the whole population of 6- to 7-year-olds in Göteborg had MPDADD. Of these, 60%

(4.3% of the whole population) had consulted a child psychiatrist before age 16 years. In this age group 10%–18% of all children in Swedish communities have consulted a child psychiatrist (Karlsson and Sonesson 1984; Gillberg 1984; von Knorring et al. 1987). A rough estimate on the basis of these figures would be that 24%–43% of all children in our community who have consulted child psychiatric services before age 16 years have or have had MPDADD. This, of course, has implications as regards the specialist education of child psychiatrists who—at least in Sweden—are generally unaware both of the scope of these problems, how they should be interpreted, and what to do in the way of treatment.

Another important implication of the findings as regards "growing-out-of" motor-perceptual problems is that a teenager presenting with "simple" psychiatric problems may well have had moderate to severe motor-perceptual problems in childhood, problems which can no longer be detected on neuro-logical–neurodevelopmental examination. This means that one must listen carefully to the history provided by the patient, parent, and teacher concerning motor clumsiness and learning disorders and not dismiss the possibility of biological influences on the psychiatric clinical picture of the patient, just because unequivocal proof of neurological–neurodevelopmental impairment can no longer be demonstrated.

Further, extrapolation of our results to the general population leads us to the following tentative conclusions. At age 10 years, 25% of children with school achievement problems have shown MPDADD at age 7 years. This relative frequency rises to 40% at age 13 years. On the other hand, of children showing severe behavior problems at age 10, 25% have an MPDADD background. This frequency drops to 17% at age 13 years. Thus, the probability that school achievement problems should be interpreted in terms of an MPDADD connection increases over time but the reverse is true in respect of behavior problems.

Our data yield indirect support for the notion that a child who shows overall motor clumsiness in the early teens always has an MPDADD background or another neurological problem of some sort.

Finally, from the very limited data we have on ADD without MPD and MPD without ADD, it appears that such problems between them constitute about as large a group as that with the combination of the two and that "simple" ADD as well as "simple" MPD has an outcome closer to—although not quite as bright as—that of the comparison group.

Conclusions

The combination of pervasive overactivity/inattention and motor-perceptual dysfunction is probably a fairly valid diagnostic concept because pervasive overactivity tends to cluster with motor/perception problems (Sandberg et al.

1978; Sandberg 1987; Gillberg 1987) and vice versa, because there are sufficiently clear indications of associated neurobiological background factors (Gillberg and Rasmussen 1982a), and also because of the particular prognostic implications (Gillberg 1987). MPDADD is a common childhood problem. Even in the severe form it affects more than 1% of all Swedish 7-year-olds and mild–moderate problems are present in at least one child in every primary school class.

Children with MPDADD are at extreme risk for the development of psychiatric problems and school achievement difficulties. The MPD problems, at least as reflected in the neurodevelopmental examination performance, tend to subside in two-thirds of cases from preschool to the early teens.

Electroencephalographic examination cannot help in general diagnostic considerations, but may still be a useful tool in research and in the individual clinical case. Measurement of so-called complex reaction times proved to be the best single instrument when screening for MPDADD-type problems among young teenagers.

Further longitudinal long-term follow-up of the children in this research is currently being undertaken. It is our assumption that psychiatric and personality disorders will turn out to be prevalent in young adults who were clumsy and attention deficient in childhood. Before firm conclusions can be drawn we must also await results from other researchers who have studied population-based groups of children with the combination of pervasive hyperactivity and motor-perceptual problems as well as of children with only one of these two major deficits.

References

American Psychiatric Association (1980) DSM III. Diagnostic and statistical manual of mental disorders, 3rd edn. American Psychiatric Association, Washington

American Psychiatric Association (1987) DSM III-R. Diagnostic and statistical manual of mental disorders. American Psychiatric Association, Washington

Ayres AJ (1972) Southern California sensory integration tests. Western Psychological Services, Los Angeles

Bax M, Mac Keith R (1963) Minimal cerebral dysfunction. Little Club clinics in developmental medicine, no 10. Spastics Society with Heinemann Medical, London

Birleson P (1981) The validity of depressive disorder in childhood and the development of a self-rating scale: a research report. J Child Psychol Psychiatry 22:73–88

Bradley JW (1968) Distribution-free statistical tests. Prentice Hall, Englewood Cliffs, pp 68–86

Cantwell D (1972) Psychiatric illness in the families of hyperactive children. Arch Gen Psychiatry 27:414–417

Cantwell D (1975) Genetic studies of hyperactive children: psychiatric illness in biologic and adopting parents. In: Fieve R, Rosenthal D, Brill H (eds) Genetic research in psychiatry. Johns Hopkins University Press, Baltimore, pp 273–280

Carey WB, McDevitt SC (1978) The measurement of temperament in 3-7 year old children. J Child Psychol Psychiatry 19:245–253

Conners CK (1969) A teacher rating scale for use in drug studies with children. Am J Psychiatry 126:884–888

Draper NR, Smith H (1966) Applied regression analysis. Wiley, New York

Frisk M, Wegelius B, Tenhunen T, Widholm O, Hortling H (1967) The problem of dyslexia in teenage. Acta Paediatr Scand 56:333–343

Gillberg C (1983) Perceptual, motor and attentional deficits in Swedish primary school-children. Some child psychiatric aspects. J Child Psychol Psychiatry 24:377–403

Gillberg C (1984) Great need for child psychiatric services in cities. (In Swedish) Läkartidningen 81:1618

Gillberg C, Rasmussen P (1982a) Perceptual, motor and attentional deficits in seven-year-old children. Background factors. Dev Med Child Neurol 24:752–770

Gillberg C, Rasmussen P (1982b) Perceptual, motor and attentional deficits in six-year-old children. Screening procedure in pre-school. Acta Paediatr Scand 71:121–129

Gillberg C, Rasmussen P, Wahlström J (1980) Neuropsychiatric problems among children born to older mothers. Acta Paedopsychiatr 46:57–65

Gillberg C, Frisk M, Carlström C, Rasmussen P (1981) "Complex reaction times" in so called minimal brain dysfunction. Acta Paedopsychiatr 47:245–252

Gillberg C, Rasmussen P, Carlström G, Svenson B, Waldenström E (1982) Perceptual, motor and attentional deficits in six-year-old children. Epidemiological aspects. J Child Psychol Psychiatry 23:131–144

Gillberg C, Matousek M, Petersén I, Rasmussen P (1984) Perceptual, motor and attentional deficits in seven-year-old children: electroencephalographic aspects. Acta Paedopsychiatr 50:243–253

Gillberg IC (1985) Children with minor neurodevelopmental disorders. Neurological and neuro-developmental problems at age 10. Dev Med Child Neurol 27:3–16

Gillberg IC (1987) Deficits in attention, motor control and perception. Follow-up from pre-school to the early teens. Thesis, Uppsala University

Gillberg IC, Gillberg C, Rasmussen P (1983) Three-year follow-up at age 10 of children with minor neurodevelopmental disorders. II. School achievement problems. Dev Med Child Neurol 25:566–573

Gillberg IC, Groth J, Gillberg C (1989) Children who had 'DAMP' (deficits in attention, motor control and perception) at age 7. Neurological and neurodevelopmental profile at age 13. Dev Med Child Neurol 31:14–24

Ingram TS (1959) Specific developmental disorders of speech in childhood. Brain 82:450–467

Kahn E, Cohen LH (1934) Organic drivenness: a brainstem syndrome and an experience with case reports. N Engl J Med 210:748–756

Karlsson A, Sonesson L (1984) How extensive is the need for child psychiatric services? (In Swedish) Läkartidningen 81:439–440

Ljungman P-G (1985) Evaluation of diagnostic methods intended for screening of 6-year-old children with presumptive reading and writing problems and MBD-related symptoms. Report No. I. Data collection during 1983-1984 among 377 6-year-olds. Värnamo local education authority Information series. No. 5 (in Swedish)

Mantel N (1963) Chi-square test with one degree of freedom. J Am Stat Assoc 58:690–700

Mendeloff AI, Monk M, Siegel CI, Linjenfeld A (1970) Illness experience and life stresses in patients with irritable colon and ulcerative colitis. N Engl J Med 282:14–17

Rutter M (1967) A children's behaviour questionnaire for completion by teachers: preliminary findings. J Child Psychol Psychiatry 8:1–11

Rutter M (1982) Syndromes attributed to minimal brain dysfunction in childhood. Am J Psychiatry 139:21–33

Rutter M, Graham P, Yule W (1970) A neuropsychiatric study in childhood. Clinics in developmental medicine nos 35/36. Spastics Society with Heinemann Medical, London

Sandberg S (1987) Epidemiological study on hyperactivity, attention deficits and conduct disorder. Paper read at the European Brain and Behavior Society workshop on attention deficit disorder and hyperkinetic syndrome, Oslo, June

Sandberg ST, Rutter M, Taylor E (1978) Hyperkinetic disorder in psychiatric clinic attenders. Dev Med Child Neurol 20:279–299

Schachar RJ (1986) Hyperkinetic syndrome: historical development of the concept. In: Taylor E (ed) The Overactive Child. Clinics in Developmental Medicine, No 97, London, pp 19–40

Shen Y-C, Wang Y-F, Yang X-L (1985) An epidemiological investigation of minimal brain dysfunction in six elementary schools in Beijing. J Child Psychol Psychiatry 26:777–787

Spitzer RL, Endicott J, Robins E, Kuriansky J, Gurland B (1975) Preliminary report on the reliability of research diagnostic criteria. In: Sudicovsky A, Gerson S, Beer B (eds) Predictability in psychopharmacology: pre-clinical and clinical correlations. Raven, New York

Still GF (1902) The Coulstonian lectures on some abnormal psychical conditions in children. Lancet 1:1008–1012, 1077–1082, 1163–1168

Strauss AA, Lehtinen LE (1947) Psychopathology and education of the brain injured child. Grune and Stratton, New York

Taylor E (1986) Childhood hyperactivity. Br J Psychiatry 149:562–573

Touwen BC (1979) Examination of the child with minor neurological dysfunction. Clinics in developmental medicine, no 71. Spastics Society with Heinemann Medical, London

Tredgold AF (1908) Mental deficiency (amentia). Wood, New York

Von Knorring AL, Andersson O, Magnusson D (1987) Psychiatric care and course of psychiatric disorders from childhood to early adulthood in a representative sample. J Child Psychol Psychiatry 28:329–347

Waldrop MF, Halverson C (1971) Minor physical anomalies and hyperactive behaviour in young children. In: Helmuth J (ed) Exceptional infants. Brunner Mazel, New York, pp 343–380

Wechsler D (1974) Manual of the Wechsler Intelligence Scale for children. (Revised edn.) Psychological Corporation, New York

Wing L, Gould J (1979) Severe impairments of social interaction and associated abnormalities in children: epidemiology and classification. J Autism Dev Disord 9:11–29

The Significance and Course of Minimal Brain Dysfunction

G. Esser and M.H. Schmidt

Introduction

Minimal brain dysfunction (MBD) and minimal brain damage can be regarded as typical examples of concepts which emphasize the interplay between biology and behavior[1]. Indeed there are many aspects of MBD which seem to have a biological basis, and most authors have discussed this point (Black 1981; Ochroch 1981; Small 1982; Gillberg and Rasmussen 1982; Lempp 1978; Strunk and Faust 1967; Sieber 1978; Gwerder 1976; Ruf-Bächtiger 1987; Rutter 1977, 1982; Rie and Rie 1980; Shaffer 1977; Nichols and Chen 1981). But, if we do not want to reduce the biological basis of MBD to the general assumption that all human behavior is biologically determined and therefore link all deviant, abnormal, or disordered behavior to a biological basis, we must find more specific relations between biology and behavior. Concerning MBD the following relations have to be tested:

— The connection of MBD with clearly defined brain injury suffered early in life
— The association between more biologically and more behaviorally determined levels of MBD case definition
— The association between MBD and psychiatric disorders, examining the question of whether a uniform psychopathology can be observed in cases of MBD
— The course of MBD with respect to the above-mentioned relations

Sample

In the past, a considerable number of related studies have employed the questionable procedure of gathering data from clinical samples and subsequently generalizing the observed results by applying conclusions to unselected popula-

---

[1] Attention deficit disorder, hyperkinetic syndrome, and learning disability represent concepts which are sometimes used synonymously with MBD. Examination of these concepts is not the subject of the present work.

tions. In an attempt to avoid such methodological shortcomings in the present Mannheim study, a purely epidemiological approach was chosen.

Out of a total population of 1444 8-year-old German children, born between March and September 1970 and living in Mannheim (FRG) on 1 March 1978, 361 (i.e., 25%) were randomly selected and asked to participate in our investigation. Of these, 129 (36%) refused to take part in the study and 16 were excluded due to their low IQ (below 70), chronic diseases, or severe handicaps, or because they had moved away in the meantime. The remaining 216 children formed our random sample.

Each of the 1444 families of our population was asked to fill out a Behavior Questionnaire (an adapted version of the Conners scale which had previously been evaluated in a pilot study on 391 children in Ludwigshafen) and to grant permission to their child's teacher to fill out the same screening instrument. After the random sample had been selected and separated, screening data for a total of 733 children were available. To increase the number of subjects with behavior problems in our investigation, 25% of the children with the highest scores in the teacher and parent questionnaire were selected. Together with the 216 subjects of the random sample, they formed our total field sample of 399 children.

In view of the rather high rate of noncompliance in the first wave, nonparticipants were asked to take part in a brief version of the interview to supplement the available information and to investigate whether they differed systematically from the rest of the sample. Of the original refusers, only 18% also refused this brief interview, so that the hard core of refusers could be estimated as 18% of 36%, i.e., 6.5%. The main results showed that nonparticipating children did not have more behavior problems, but tended to show somewhat poorer school performance. Overall, no significant noncompliance effects could be expected to bias the representativity of the final sample and the estimation of prevalence rates.

For the establishment of case definition criteria, 79 children were randomly selected from the combined field sample. This group was used to define cerebral dysfunction (item selection process, factor-analytic combination of the variables). Case definition criteria were subsequently applied to the remaining 320 children of the identification sample.

Results reported for the field sample stem from 268 subjects who remained in the identification sample after 28 subjects were excluded due to missing values and 24, because of their lower IQ scores (between 70 and 85). The latter 24 children with subnormal intellectual abilities were analyzed separately.

Out of the 399 children who participated in the first wave of the investigation, 356 (89%) took part in the first follow-up study at age 13. Of these, 22 had missing data in respect of at least one of the variables relevant to MBD. Therefore 334 subjects remained. Since the case definition procedure was done with a new definition sample of 85 13-year-old subjects, all 334 subjects could be included in the identification sample. Of these 334 subjects, 177 came from the random sample, 157 from the screening sample.

Selection of Variables and Case Definition

To ensure comprehensive assessment of MBD indicators, a wide range of potentially relevant variables were assigned to one of three diagnostic levels:

—Neurophysiological parameters (EEG and neurological examination)
—Neuropsychological measures
—Behavioral parameters (i.e., specific skills and child psychiatric symptoms)

At the behavioral level of case definition, psychopathological symptoms had to be discarded because it was intended to test this particular association between MBD and psychiatric disorders.

To achieve an approximate equivalence among the remaining three diagnostic levels, variables were subjected to a strict sequential selection process, which for EEG and neurology included four steps: plausibility control, item difficulty

Table 1. MBD indicators included in three separate factor analyses in order of their height of factor score coefficient sums

---

I. *Neurophysiological level (two factors explaining 33.5% of the total variance)*
   Increased theta activity
   Walking backwards on heels
   Abnormal movements and posture while walking on tiptoe
   Walking backwards along a straight line
   Increased delta activity
   Finger opposition test
   Absence of alpha attenuation when opening eyes
   Choreoathetotic movements
   Amplitude of the background activity
   Hopping on one leg
   Substantial side differences
   Other EEG abnormalities (i.e., beta activity, abnormal rhythms)
   Paroxysmal activity
   Mirror movements during diadochokinesia
   Spooning of the wrists and hands (arms extended, supinated)

II. *Neuropsychological level (two factors explaining 50.5% of the total variance)*
   Children's Motor Coordination Test (KTK; Kiphard and Schilling 1974): sum of scores 1
      (balancing) and 2 (high jumping on one leg)
   Choice reaction task: average reaction time
   Choice reaction task: number of false reactions
   Hand tremor (Steadiness, Motor Performance Series (MLS); Schoppe)
   Continuous performance at a multistimulus multireaction task (Wiener-Determinationsgerät;
      Schuhfried)

III. *Specific skills level (three factors explaining 66.9% of the overall variance)*
   Sound blending—meaningful words (ITPA/PET)
   Culture Fair Test (CFT 1): sum of scores 3, 4, and 5 (reasoning tasks)
   Culture Fair Test (CFT 1; Cattell et al. 1977): sum of scores 1 and 2 (perceptual tasks)
   Number of errors in the modified Matching Familiar Figures Test (adapted from Kagan)
   Number of errors in the revised Bender-Gestalt Test (GFT; Schlange et al. 1972)
   Digit span memory [Illinois Test of Psycholinguistic Abilities, German version (PET);
      Angermaier 1974]
   Performance in an adapted Bourdon concentration test
   Sound blending—nonsense words (ITPA/PET)

(not > 0.20), item–total correlation (over 0.25), and multiple correlation (item variance should not be explained to more than 50% by all other items of the same data level). For the variables of the neuropsychological and specific skills levels, the steps item difficulty and item–total correlation were dropped.

Since it was postulated that MBD could manifest itself at each of the three levels, three separate factor analyses were performed. In contrast to the Nichols and Chen (1981) procedure, variables in our study were not entered into one joint factor analysis. From a methodological point of view, it seems highly questionable to process jointly variables with such different scale levels and representing totally different levels of complexity. Furthermore, the independence of the resulting factors determined by orthogonal rotation (as described by Nichols and Chen 1981) is hardly surprising.

In our own study, individual factor scores were next calculated for each subject. Appropriateness of factor solutions was determined according to conventional statistical criteria ("Little Jiffy" criteria; Glass and Taylor 1966). In addition, robustness of the chosen solutions was demonstrated by repeated replications in split-sample analyses.

Being statistically independent due to orthogonal rotation, individual scores on the extracted factors within each diagnostic level were then added up (see Table 1 for an overview of the best MBD indicators in order of their height of factor score coefficient sums). Subjects were classified as cases of MBD if their summed factor scores on at least one of the three data levels were more than two standard deviations below the relevant mean value observed in the random sample.

Results

Connection of MBD with Clearly Defined Brain Injury Suffered Early in Life

As stated above, excluding subjects with missing values and those with an IQ below 85, 268 of the original 320 children remained in the identification sample. Of these, 31 met the definition criteria.

To verify this relation, anamnestic data reported by parents were used to determine an anamnestic risk score, indicating pregnancy and delivery complications (similar to the index described by Littman and Parmelee 1978). Comparing subjects with and without MBD, we were unable to find differences in their mean anamnestic risk scores. For both groups, the average risk index observed in our field sample amounted to 3.8 points out of a possible score of 25. Although these results must be viewed with caution due to the questionable validity of data collected retrospectively (cf. Everbeck 1970; Cox and Rutter 1977), it should be pointed out that, in clinical practice, data from this unreliable source often enough contribute to the diagnosis of MBD.

Analyses of objective pregnancy and birth records, which were available only for a smaller subgroup, yielded very similar results (Esser and Schmidt 1985). Thus, it must be concluded that MBD is not systematically associated with early brain injury (also cf. Fritsch and Haidvogl 1982).

Association Between More Biologically and More Behaviorally Determined Levels of MBD Case Definition

The rank order of the MBD levels used could be described according to their biological determination as follows: neurophysiological parameters are more biological than neuropsychological parameters, and the latter are more biological than specific skills.

Thirteen of the 31 MBD cases met the case definition at the neurophysiological level, 6 at the neuropsychological level, and 15 at the specific skills level. The total number of 34 deviant level scores shows that there was no relevant association between the three MBD levels. Only one child showed disorders at all three levels, another at two levels. In addition the correlation coefficients between the summed factor score coefficients of each data level were low (maximum 0.20). Thus the relation between biology and behavior, which should be "typical" for MBD, could not be proved.

Association Between MBD and Psychiatric Disorders

Data collected for each subject in our field sample included detailed parent interviews lasting approximately 2 h and covering a broad range of child psychiatric symptoms. Each of the 28 symptoms was rated as absent (0), mild (1), or severe (2 points). An overall psychiatric rating on a four-point scale followed, in which each child was classified as psychiatrically undisturbed (0), mildly disturbed (1), moderately disturbed (2), or severely disturbed (3). These ratings were subsequently dichotomized to indicate presence (moderately or severely disturbed) or absence (undisturbed or mildly disturbed) of psychiatric disorders. Children with psychiatric disorders were also assigned to four different diagnostic categories: emotional disorders (ICD 300 and 313), conduct disorders (ICD 312), special symptoms and syndromes (ICD 307), and hyperkinetic syndromes (ICD 314). The diagnostic categories were supplemented by diagnostic symptom sums, which allowed a more differentiated measurement.

Of 31 subjects with MBD, 12 (about 40%) were also psychiatrically disordered. Compared with a rate of 15.2% in the group without MBD this represents a significantly enhanced association between biology and behavior. A closer look at the intersection between psychiatric disorders and the three diagnostic levels of MBD reveals a tendency towards more overlapping with neuropsychological deficiencies and learning disabilities than with neurophysiological abnormalities. This suggests that enhancement of psychiatric risks is mainly associated with the impairment of more complex functions which to a large extent depend on environmental conditions and learning processes (i.e., performance deficiencies in a general sense), rather than with organic or constitutional factors. Hence, learning disabilities and related deficiencies seem to represent a key factor determining child psychiatric morbidity. Corresponding results were obtained in other studies examining reciprocal influences of performance disorders and behavioral disturbances in preschool children (Esser 1980), or associations between reading retardation and antisocial behavior (Sturge 1982). The above-mentioned association between biology and behavior therefore has to be qualified. Biologically determined parameters in a narrower sense, like EEG and/or neurological abnormalities, do not increase the probability of psychiatric disorders. Learning disabilities and, as could be demonstrated by separate analyses, unfavorable family and social circumstances are more likely to lead to higher rates of psychiatric disorders.

To test the assumption of a specific psychopathological association with MBD for each of the above-mentioned diagnostic categories, symptom sums were computed. MBD children showed higher scores in each of the four categories, not only for the hyperkinetic symptoms.

Comparing the three data levels of our MBD case definition, it was found that these overall MBD effects were caused by the 15 MBD children identified at the specific skills level. The 13 MBD subjects identified at the neurophysiological level failed to show any enhanced symptom scores, while the six MBD cases at the neuropsychological level scored between these two groups, without showing statistically significant differences from the non-MBD group.

Looking for Mistakes

In summary, our results confirmed that MBD has a certain relevance for psychiatry. However, the failure to find associations between the more biological levels of MBD and the more behaviorally determined levels, or to find an association between anamnestic indicators of brain damage and MBD, seems to disprove the view that MBD can be regarded as a nosological unit and as an example of interplay of biology and behavior.

The reported results contradict the opinion and findings of a lot of authors (Black 1981; Ochroch 1981; Small 1982; Gillberg and Rasmussen 1982; Lempp 1978; Sieber 1978; Gwerder 1976; Ruf-Bächtiger 1987). This might be due to the fact that the results are based on case definition criteria different from those normally applied in clinical settings. Furthermore, the study was an epidemiological one, which could explain significant differences from those studies which were done on clinical samples.

Besides being necessary in order to meet the specific aims of the project, these differences should overcome the methodological shortcomings of ordinary clinical procedures. However, it could be argued that the application of such different criteria only served to disprove a theoretical concept which, in this particular form, never existed anyway. To investigate this possibility, the obtained results were compared with those acquired using conventional clinical definitions, as well as a number of further alternative procedures. In addition a study on 83 8-year-old child psychiatric outpatients was conducted.

Alternative Case Definitions

In general three aspects of case definition can be distinguished:

—Selection of relevant variables
—Combination and weighting of such variables
—Determination of cut-off values to separate pathological scores from the normal range

Concerning the selection of variables, especially the role of general intelligence was tested. Among the alternative combinations and weightings of the variables the most relevant was a joint factor analysis including all variables of the three data levels (cf. Nichols and Chen 1981). To obtain a more restrictive case definition the cut-off points were placed from two to three standard deviations below the respective means. A comparison with the conventional clinical definition of MBD was performed using the following five items, of which two must be present: pre- and perinatal risks, EEG abnormalities, a substantial number of neurological soft signs, impaired performance in the Bender test, and diagnosis of hyperkinetic syndrome.

None of the alternative procedures substantially altered the presented results. Nonsignificant trends even showed the chosen case definition to confirm the basic assumption of the concept of MBD (i.e., that strong associations exist between biological and behavioral symptoms) better than any of the alternatives.

Comparison with Results from a Clinical Sample

Ninety-six consecutive 8-year-old outpatients were examined with the same instruments as the field sample. 83 of the subjects had an IQ above 85. In the clinical sample, too, there was no association between MBD and pre- and perinatal risks. There was no specific psychopathology but there was a significantly higher proportion of MBD cases with impairments at more than one data level. Although the majority of cases were only defined by deviant scores at one data level and there was not a single case with abnormal scores at all three levels, the association between biology and behavior seemed to be a little stronger than in the field sample. However, the number of cases showing such a stronger association was limited to 1% in unselected samples and reached 7% in clinical samples if the criteria of association were reduced to impairment at two data levels.

Course of MBD

Prognostic Value of MBD

The prognostic value of MBD should be tested by means of the following questions:

—What proportion of children with MBD at age 8 show child psychiatric disorders at age 13?
—Do the MBD cases show a specific psychopathology at age 13?

The latter is a meaningful question despite the fact that there was no specific psychopathology at age 8. It might be that MBD moderates the course of psychiatric disorders and that the unspecific psychopathology at age 8 could therefore lead to a specific psychopathology at age 13.

Psychiatric disorders at age 13 were determined by a highly structured 2-h parent interview and a highly structured interview with the adolescent which lasted $1\frac{1}{2}$ h. The procedure for symptom definition and global rating of psychiatric severity grade and diagnosis was just the same as for the 8-year-olds.

Of 43 children with MBD (reunification of definition and identification sample) at age 8, 13 (30%) showed psychiatric disorders at age 13. In comparison with the rate of psychiatric disorders (62 of 291 = 21.3%) in the group of children without MBD at age 8, this represented no significant difference. A decrease in psychiatric disorders during the course of MBD (possibly caused by maturation) is therefore more probable than an increase, which would mean secondary disorders following MBD. At age 13 former MBD children showed a significantly ($\chi^2 = 5.78$; $P < 0.05$) higher proportion of expansive disorders (predominantly dissocial) than children without the diagnosis of MBD at age 8. This apparently specific psychopathology was caused by two facts: (a) Children with MBD at age 8 showed higher rates of psychiatric disorders in general. (b) The course of psychiatric disorders depended on the diagnosis. Those with emotional disorders recovered, a development which was the same for non-MBD cases. The dissocial disorders were stable. Furthermore, dissocial disorders are the diagnostic category with the greatest increase in prevalence from 8 to 13.

Looking at the predictive value of the different MBD levels, a similar result to that at age 8 was observed. Deficiencies in specific skills and neuropsychological impairments tended to be more relevant than neurological and neurophysiological disorders.

A comparison with other risk factors (learning disabilities, adverse family factors, life events) proved MBD to be the one with lowest significance.

Stability of MBD

As far as we know, in the past nobody has really tested the stability of MBD. MBD is always regarded as a constant characteristic. Gillberg and co-workers (1982) merely investigated the later consequences of MBD instead of testing whether the earlier diagnosis could be confirmed 4 years later. Such a procedure may be caused by confounding trauma (brain damage) and its consequences (brain dysfunction). Brain dysfunction itself can recover or even manifest itself at a very late point in time, for it has to be seen in the context of age-appropriate tasks. From a clinical point of view, it must be possible to discover MBD at the latest by the time of school enrolment.

The clinical diagnostic procedure, however, nolens volens makes even very late manifestations possible. With the exception of pre- and perinatal risks, all characteristics used for the definition of MBD (EEG background activity, neurological soft signs, specific test failure, hyperactive behavior) have to be regarded as deviations from age-related norms. Therefore changes in the diagnosis of MBD are possible and probable. Changes in the sense of a diminished form of characteristics are accepted and even expected (Lempp 1978). Problematic cases would be those fulfilling the criteria of MBD at age 13 for the first time.

Case Definition at Age 13

To achieve an equivalent case definition at age 13, as far as possible identical instruments and constructs should be chosen. This proved to be no problem in respect of EEG and neurological parameters or neuropsychological parameters—only marginal changes had to be made. At the level of specific skills, however, this constancy of measurement could not be maintained:

—Four identical measurements were used [sound blending and digit span memory—both from the Illinois Test of Psycholinguistic Abilities (ITPA)—, the revised Bender Gestalt Test, and the Matching Familiar Figures Test]
—Three constructs could be covered by very equivalent measurements [reasoning tasks of the Culture Fair Test 1 (CFT 1) were replaced by reasoning tasks of the Prüfsystem für Schul- und Bildungsberatung (PSB), written spelling tests for the first and second grades by that for the fifth grade, Bourdon test for concentration by subtests concentration of the PSB]
—Different aspects of the construct space were covered by the perceptual tasks of CFT 1 and the space factor of the PSB
—Two additional constructs were investigated [verbal ability by PSB 1, 2, 5, and 6, and creativity by the Test für Divergentes Denken und Kreativität (TDK) 4–6]

The same sequential selection of variables as at age 8 was applied (see p. 304). as was the same factor analytic procedure with test of stability of factor solution by split sample methods (see p. 305).

At the *neurophysiological* level we obtained a three-factor solution explaining 40.8% of the total variance. The factors were named as (1) quality of EEG background activity, (2) quality of motor functions, and (3) pathological EEG deviations, At the *neuropsychological* level we obtained a two-factor solution explaining 40.2% of the total variance. The factors were named as (1) motor coordination, and (2) reaction control. At the level of *specific skills* we obtained a two-factor solution explaining 59.8% of the total variance. Factor (1) was named as verbal perception, factor (2) as cognitive organization and attention.

At each level the factor scores were summed. Again, a child was defined as an MBD case if his/her summed factor scores on at least one of the three data levels lay more than two standard deviations below the respective mean value observed in the random sample.

Results

With the above definition 42 MBD cases were found at age 13, out of the sample of 334. Twenty-two of them had already been defined as such at age 8, while 20 were new cases. On the other hand, 22 of the 43 children defined as MBD cases at age 8 kept their diagnosis, while 21 recovered. The remarkably high number of new cases was somewhat unpleasant but methodologically not unexpected.

A meticulous inspection of each single child defined as having MBD at age 8 or 13 was performed. This inspection showed that stable MBD cases were diagnosed on the basis of comparable characteristics at both time points. Of the 21 recovered cases, 12 were fully improved, 6 were only partially improved, and 3 were unchanged but just missed the cut-off point. Within the group of new cases, there were many children who had become a case on the basis of a changed EEG. Mostly, the EEG parameters did not deviate from the norm at age 8 but did so at 13. Half of the new cases could be attributed to methodological factors (new variables or new weights of the variables), the other half to changes in the characteristics of the subjects.

Alternative Case Definitions

Case definition at age 13 differed from that at age 8 in two respects:

1. Differences in the variables used
2. Differences in the weights of the variables (even for the unchanged ones)

These methodological differences were accepted in order to obtain a comprehensive set of variables relevant for 13-year-olds. As is known, characteristics represented by certain tests change their meaning in the course of development. Therefore the same tests do not necessarily have the same meaning in different age groups. Not only the meaning but also the weight of certain tests will change over time. This is especially true for our weights, which are based on factor analytic procedures. Therefore we had to test whether and to what extent the stability of MBD was influenced by these methodological factors. Two alternative case definitions were elaborated, one in which the variables and the weights were kept constant from 8 to 13, and another in which only the variables were kept constant and the weights were calculated separately for each age group. The percentages of stable cases and new cases should serve as an indication of quality.

With the first alternative the percentage of stable cases was 37% [preferred model (p.m.): 51%] and the percentage of new cases, 59% (p.m.: 48%). With the second alternative the percentage of stable cases was 49% (p.m.: 51%) and the percentage of new cases, 55% (p.m.: 48%). Although there were no statistically significant differences, in absolute numbers the preferred model always showed the best results, which means that the high number of new cases cannot be reduced by these methodological changes.

Lastly the stability was calculated for the conventional clinical model. The figures were 45% stable cases and 68% new cases. Despite the fact that in the clinical model one of the five criteria (pre- and perinatal risks) was absolutely constant, the model itself led to an absolutely weaker stability.

Conclusions

The present study evaluates the significance of the concept of minimal brain dysfunction (MBD), especially the connections between behavioral and biological characteristics. Taking into account the known objections to the concept, it was necessary to use an epidemiological approach. For validation the results of the field study were compared with those of a clinical study and those of a follow-up study. The field sample consisted of 399 8-year-old children, the clinical sample of 96 8-year-old psychiatric patients. One of the basic clinical assumptions pertaining to the concept of MBD is that strong associations exist among designated biological and behavioral symptoms, thus suggesting the notion of a homogeneous MBD syndrome. The epidemiological field investigation of 399 8-year-old children residing in Mannheim was unable to support this basic assumption. Using a multilevel procedure for case definition, potentially relevant variables were assigned to one of three diagnostic levels. Within each level, redundancy was reduced by applying strict selection criteria and by a factor-analytic data aggregation of selected indicators. Results demonstrated nearly complete independence among the three diagnostic levels of "neurophysiology", "neuropsychology," and "specific skills." A uniform psychopathology, elsewhere claimed to be associated with the MBD syndrome, was not confirmed by the present study. Anamnestic data failed to reveal higher rates of perinatal brain damage in children labeled as MBD cases. The occurrence of psychiatric disorders was increased (39% vs 15% in the rest of the sample) in children with MBD. The major portion of observed overlappings with psychiatric disorders occurred at the specific skills level, indicating the necessity to consider the importance of learning disabilities in child psychiatric morbidity. The results obtained with the original classification model were compared with those obtained with alternative strategies. Two variations of the original procedure tested the influence of intelligence measures; in one model IQ measures were omitted completely, in another the original model was applied to a sample of children with IQs between 70 and 84. In another model, all variables assigned to the three diagnostic levels were jointly processed in one common factor analysis. A further approach moved the established cut-off points in order to yield lower frequencies of (more severely affected) cases. Finally, classical clinical criteria for defining MBD were applied. Compared to the findings with the original model, none of the alternative procedures for case definition provided stronger support for the underlying assumptions.

Major conclusions drawn from the field data were replicated in the clinical sample of 96 8-year-old patients. Neither study was able to support empirically elementary suppositions of the clinical concept of MBD. Comparing these two investigations, it becomes clear that the diagnosis of MBD is predominantly based on unjustified extrapolations from isolated single cases. The prevalence of these cases is lower than 1% in unselected samples.

In half of the MBD cases the symptomatology was still present at the time of the follow-up study at age 13; the other half had improved. There was a significant number of new MBD cases. Alternative procedures for case definition were neither able to improve the stability rate nor to reduce the number of new MBD cases. Thus the presumed specific biological basis of MBD could not be confirmed. Many authors (Gearheart 1977; Lempp 1978; Sieber 1978; Gwerder 1976; Berger 1981; Small 1982) are of the opinion that specific skill deficiencies, learning disabilities, or hyperkinetic behavior are already by themselves the expression of brain dysfunctions and therefore do not need any other support from biologically determined parameters. This assumption is the result of neuromythology: there is no need to gather all these learning problems under the umbrella of MBD (Rutter 1982).

References

Angermaier M (1974) Psycholinguistischer Entwicklungstest (PET). Beltz, Weinheim

Berger E (1981) Modellvorstellungen zum Problem der hirnfunktionellen Bedingungen von Perzeptions- und Teilleistungsstörungen. In: Frostig M, Müller H (eds) Teilleistungsstörungen — Ihre Erkennung und Behandlung bei Kindern. Urban and Schwarzenberg, Munich, pp 189–200

Black P (ed) (1981) Brain dysfunction in children. Etiology, diagnosis and management. Raven, New York

Cattell RB, Weiss H, Osterland J (1977) Grundintelligenz-Test CFT 1. Westerland, Braunschweig

Cox A, Rutter M (1977) Diagnostic appraisal and interviewing. In: Rutter M, Hersov L (eds) Child psychiatry. Blackwell Scientific, Oxford, pp 271–305

Esser G (1980) Über den Zusammenhang von Verhaltens- und Leistungsstörungen im Vorschulalter (und Grundschulalter). Doctoral thesis, University of Mannheim

Esser G, Schmidt MH (1985) Die Bedeutung von Schwangerschafts- und Geburtskomplikationen für die Genese von Hirnfunktionsstörungen. Geburtshilfe Frauenheilkd 45:161–166

Ewerbeck H (1970) Frühkindliche Hirnschäden — Pathogenese, Früherkennung, Differentialdiagnose. In: Stutte H, Koch K (eds) Charakteropathien nach frühkindlichen Hirnschäden. Springer, Berlin Heidelberg New York, pp 10–14

Fritsch G, Haidvogl M (1982) Pre- and perinatal risk factors in the etiology of infantile cerebral palsy. Int J Rehabil Res 5:19–25

Gearheart BR (1977) Learning disabilities. Educational strategies. Mosby, St Louis

Gillberg C, Rasmussen P (1982) Perceptual, motor and attentional deficits in seven-year-old children: background factors. Dev Med Child Neurol 24:752–770

Gillberg C, Rasmussen P, Carlström G, Svenson B, Waldenström E (1982) Perceptual, motor and attentional deficits in six-year-old children: epidemiological aspects. J Child Psychol Psychiatry 23:131–144

Glass GV, Taylor PA (1966) Factor analytic methodology. Rev Ed Res 36:566–587

Gwerder F (1976) Das Syndrom der leichten frühkindlichen Hirnschädigung. Huber, Bern

Kiphard EJ, Schilling F (1974) Körper-Koordinationstest für Kinder (KTK). Beltz, Weinheim

Lempp R (1978) Frühkindliche Hirnschädigung und Neurose, 3rd edn Huber, Bern

Littman B, Parmelee AH (1978) Medical correlates of infant development. Pediatrics 61:470–474

Nichols PL, Chen TC (1981) Minimal brain dysfunction—a prospective study. Erlbaum, Hillsdale

Ochroch R (ed) (1981) The diagnosis and treatment of minimal brain dysfunction in children. A clinical approach. Human Science, New York

Rie HE, Rie ED (1980) Handbook of minimal brain dysfunctions. Wiley, New York

Ruf-Bächtiger L (1987) Das frühkindliche psycho-organische Syndrom. Minimale Cerebrale Dysfunktion, Diagnostik und Therapie. Thieme, Stuttgart

Rutter M (1977) Brain damage syndromes in childhood: concepts and findings. J Child Psychol Psychiatry 18:1–21

Rutter M (1982) Syndromes attributed to "minimal brain dysfunction" in childhood. Am J Psychiatry 139:21–33

Schlange H, Stein B, von Boetticher I, Taneli S (1972) Göttinger Formreproduktionstest (GFT). Hogrefe, Göttingen

Shaffer D (1977) Brain injury. In: Rutter M, Hersov L (eds) Child psychiatry. Blackwell, Oxford, pp 185–215

Sieber M (1978) Das leicht hirngeschädigte und das psychoreaktiv gestörte Kind. Huber, Bern

Small L (1982) The minimal brain dysfunctions. Free Press, New York

Strunk P, Faust VS (1967) Die Bewertung hirnorganischer Befunde bei Verhaltensstörungen im Kindesalter. Arch Psychiatr Nervenkr 210:152–160

Sturge C (1982) Reading retardation and antisocial behavior. J Child Psychol Psychiatry 23:21–31

Sleep

Sleep Disturbances in Children:
From the Physiological to the Clinical

B. Garreau, C. Barthélémy, N. Bruneau, J. Martineau, and A. Rothenberger

Introduction

During childhood, more than one child in two has a sleep problem, ranging from a simple transitory episode such as nocturnal awakening to a genuine sleep disorder such as narcolepsy. Although there may exist a familial or other predisposition to sleep disorders, conditions affecting child development such as material, familial, and emotional factors and general upbringing may favor the persistence and even the emergence of sleep disorders. The sleep of children is more vulnerable than that of the adult, since this complex function—like other functions—undergoes a slow maturation. A developmental problem of sleep function, whether the cause is exogenous or endogenous, not only may result in sleep disorders during childhood and extending into adult life, but also may be the cause of certain behavioral disorders.

This is why we will first review the basics of adult sleep and then consider the main steps in the development of this function before describing sleep disorders in the child.

Physiology and Maturation of Sleep

Sleep in the Adult

Before analyzing the maturation of the states of consciousness during development, it is first necessary to review briefly the basic physiological notions of circadian rhythms, the organization of sleep, and the nerve center involvement in the regulation of consciousness.

Circadian Rhythms

More than 100 human physiological variables are known to have a circadian rhythm (*circa*: around; *die*: day). Although their existence has been suspected since antiquity, Moore-Ede et al. (1982) stated that circadian rhythms were described for the first time by Ogle in 1866, concerning the regulation of human body temperature. The first experimental data showing that these rhythms persist

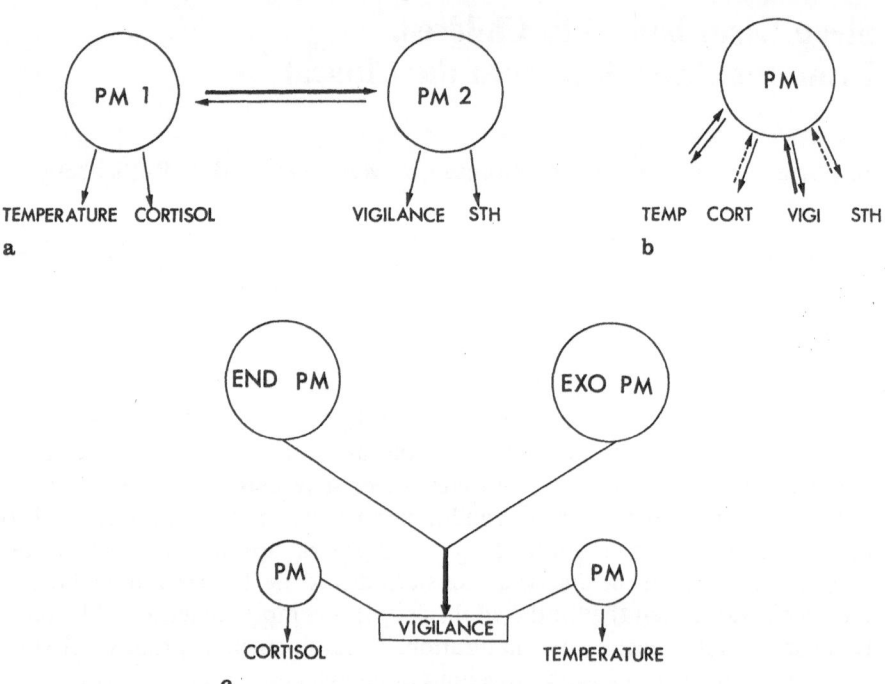

Fig. 1a-c. The three proposed models of circadian rhythm operation. *PM*, pacemaker; *END*, endogenous; *Exo*, exogenous; *Temp*, temperature; *CORT*, cortisol; *VIGI*, vigilance; *STH*, somatotropic (growth) hormone

in the absence of any external stimulation, however, are only several decades old. Thus, in 1962, Aschoff and Wever proved the endogenous nature of circadian rhythms in man (cited in Aschoff and Wever 1976).

From a physiological point of view, circadian rhythms depend on pacemakers, veritable clocks in the organism, which can measure time in the absence of external stimuli. However, these pacemakers are informed of changes in the environment by a receptor system and command the target structures in question. In man, three models of circadian rhythm operation are currently proposed (Wever 1972) (Fig. 1):

1. A multipacemaker system (Fig. 1a). Each pacemaker would command several physiological variables and the most powerful pacemaker would be the least sensitive to environmental changes
2. A single pacemaker system (Fig. 1b). A single pacemaker would control all physiological variables, with feedback control loops, some of which would be more powerful than others
3. A self-regulation system (Fig. 1c). This system would operate according to the resultant of two processes, one having an endogenous circadian rhythm, the other being dependent on environmental variations

Thus the consciousness–sleep cycle in man is not a passive fluctuation between two states of consciousness, but a part of an active circadian rhythm system.

Even though the intimate mechanism of circadian rhythm function remains to be elucidated, it has been shown in humans that relationships exist between the circadian rhythms of body temperature, consciousness, and secretion of growth hormone. The sensation of sleepiness is felt around the body temperature minimum, while awakening occurs during the ascending part of the temperature curve in 80% of cases (Zulley and Wever 1982). However, the circadian rhythm of body temperature is apparently independent of environmental conditions, and that of sleep progresses on its own when subjects are placed in conditions of sensory deprivation. The anatomical structures of these pacemakers have not yet been determined in man. In monkeys, bilateral lesions of the suprachiasmatic nuclei of the hypothalamus abolish the rhythmic nature of consciousness–sleep cycles. Lesions in the centromedian regions of the hypothalamus, on the other hand, suppress the rhythmicity of body temperature (Krieger 1980). The physiopathological determination of circadian rhythms has therapeutic repercussions, e.g., in treating depression (Wehr and Wirz-Justice 1982).

The Organization of Sleep

The Stages of Sleep
The electroencephalogram (EEG) was the first tool enabling the physiology of sleep to be studied. This was followed by polygraphy, i.e., the simultaneous recording of the EEG, eye movements, and muscle tone, with the possibility of recording other parameters such as ECG, respiratory movements, etc., which led to the definition of the stages of sleep.

An international codification (Rechtschaffen and Kales 1968) defines, in addition to consciousness, five stages of sleep, numbered from 1 to 4 with 4 being deep sleep, plus rapid eye movement (REM) sleep. The criteria defining each stage are summarized in Table 1. During stage 2, spindles and K complexes are present. Many authors now speak of slow wave sleep (SWS) and REM sleep, SWS being the combination of stages 3 and 4.

Distribution of the Stages During the Night
In a normal adult the different stages of sleep are distributed during the night according to a particular chronology. Stages 1 and 2 appear after a certain latency period, followed by SWS and then a first episode of REM sleep. This succession determines one cycle of sleep. One cycle is the time between the beginning of one REM episode and the onset of the next. The mean duration of one cycle is 90 min and one night includes four to five cycles. SWS stages predominate at the beginning of the night and REM is longer at the end (Fig. 2). Total REM sleep accounts for 20%–25% of total sleep time. Finally, the characteristics of sleep are very stable from one night to another in the same subject placed in identical physiological and environmental conditions (Benoit et al. 1981).

Table 1. Main characteristics of each sleep stage in the normal adult

	Stage 1	Stage 2	SWS	REM stage
EEG	Low voltage	Theta frequencies	High amplitude	Low voltage
	Alpha activity	Sleep spindles	Delta waves	Fast frequency
	Theta frequencies	K complexes		
EM	Slow	Slow	Absent or oscillating	Rapid
RM	Normal	Normal	Normal	Suppressed
ECG	Normal	Normal	Slow	Fast and irregular
BP	Normal	Normal	Low	Irregular
% of TST	5	50	25	20

SWS, slow wave sleep: sleep stages 3 and 4; REM stage, rapid eye movement sleep stage; EEG electroencephalogram; EM, eye movements; RM, resting muscle; ECG, electrocardiogram; BP, blood pressure; TST, total sleep time

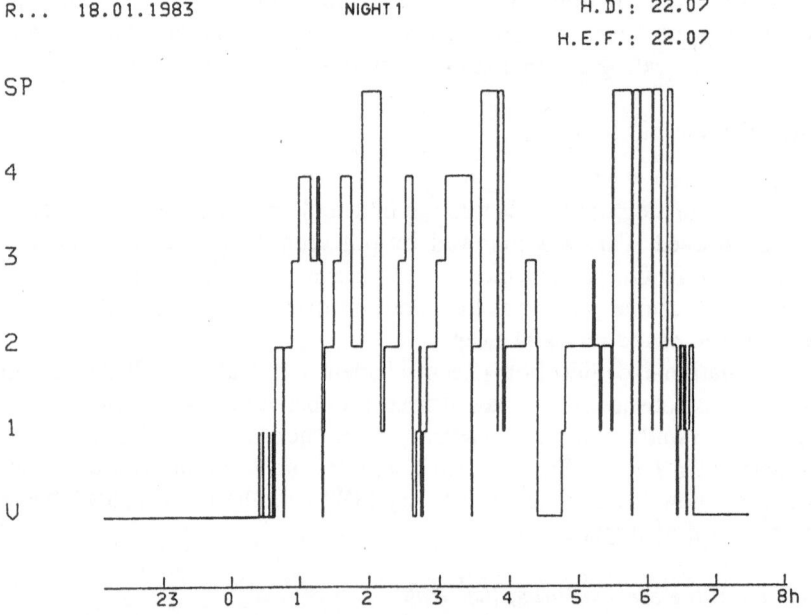

Fig. 2. Distribution of sleep stages in a normal adult. *SP*, rapid eye movement stage; *V*, awake state; *1–4*, stages of sleep; *HD*, beginning of recording; *HEF*, time of turning off the light

The Mechanism of Sleep

We will successively analyze the mechanisms of SWS and REM sleep.

Slow Wave Sleep

The role of serotonin in the onset of SWS has been shown in studies involving the creation of lesions and the administration of pharmacological agents. The destruction of the rostral raphe system containing serotoninergic neuron cell bodies causes the total disappearance of SWS without affecting REM sleep. Similarly, when serotonin synthesis is blocked by p-chlorophenylamine, total insomnia appears within 48 h, whereas the intravenous injection of serotonin precursors causes the return of SWS (Jouvet 1969; Petitjean et al. 1978). However, the activity of serotoninergic neurons, measured with minielectrodes, is higher during consciousness than during sleep (McGinty and Harper 1976), which means that serotonin is not the direct hypnogenic neurotransmitter. It in fact acts by inducing the synthesis of hypnogenic substances, probably of peptide origin, in structures containing serotoninergic neuron endings. Among these structures, we find the central hypothalamus cited above as a possible candidate for endogeneous pacemaker of circadian rhythms.

Thus, during consciousness, neurons whose cell bodies are located in the central raphe system and whose endings are located in central hypothalamic structures would synthesize and release serotonin. This release would cause the synthesis of peptide substances. Once the latter are synthesized and released, they would stimulate the effector systems of SWS (Jouvet 1984).

Rapid Eye Movement Sleep

Rapid eye movement sleep is characterized by the triad of EEG activity of consciousness, pontogeniculooccipital activity, and muscle atonia. In the normal adult, this always occurs after a period of SWS.

After ablation of the pons in animals, REM sleep disappears in about 1 week. In pons-ablated animals in which a hypothalamo-hypophyseal islet is conserved, REM sleep may be maintained indefinitely. Furthermore, selective lesions of the brain stem may suppress one index of REM sleep. For example, bilateral lesions of the alpha locus coeruleus lead to the disappearance of muscle atonia (Sakai 1981).

The biochemical factor responsible for REM sleep remains to be discovered, but a peptide origin is suspected, since certain peptidases suppress it and it seems to be independent of serotonin. Thus, a hypothalamo-hypophyseal factor is indispensable for the appearance of REM sleep. This factor is responsible for the activation of the centers responsible for expressing indices of REM sleep. Each REM sleep index has a specific center and specific neurotransmission system (Jouvet 1984).

Maturation of States of Consciousness

In order to retain a degree of unity with the section devoted to the study of sleep in the adult, we will first analyze the maturation of circadian rhythms, followed by that of the organization of sleep; finally we will briefly summarize current knowledge on the maturation of the centers involved.

Maturation of Circadian Rhythms

Figure 3 summarizes the development of the circadian rhythms of sleep from birth to adolescence. During the neonatal period, i.e., from birth until several weeks, the neonate lives according to an ultradian rhythm of 4 h, based on feeding times, although these cycles are not totally bound to alimentation, since they persist in infused newborns (Salzarulo et al. 1980). The duration of diurnal sleep periods is not less than that of nocturnal sleep periods.

At the age of about 3 weeks, a phase called consolidation begins (Coons, 1987), which lasts until the age of about 6 months. During this phase, periods of sleep become progressively fewer, their duration increases, and above all the period of nocturnal sleep becomes predominant, with the baby remaining awake during the day. Coons (1987) stated that the longest period of sleep increased from 23% of total sleep per 24 h at 3 weeks to 48% at 6 months. At that age, the longest period of sleep is nocturnal and is followed by the longest period of consciousness (Coons 1987).

In the period between 6 months and about 4–5 years, the circadian rhythm of sleep becomes progressively organized. There are fewer periods of diurnal sleep and they shorten progressively, disappearing totally at around 4 years of age. After this time, the distribution of sleep is exclusively nocturnal and the duration per 24 h decreases from 16 h at 1 year to 10 h at 6 years of age. From 6 to 12 years of age, only the total duration of sleep decreases, from 10 to about 8 h (Coble et al. 1987).

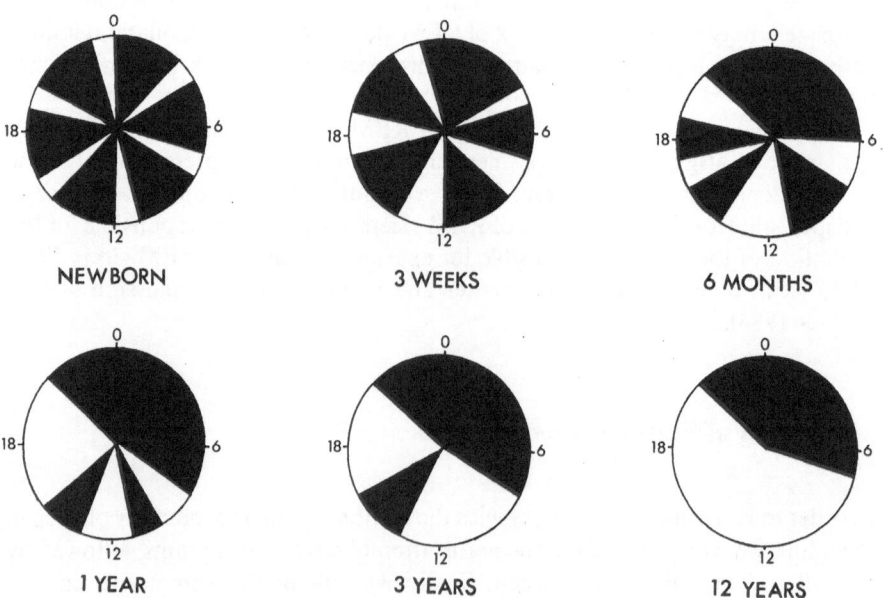

Fig. 3. Evolution of circadian rhythms of sleep from neonatal to prepubertal periods

Thus, in the neonate environmental factors have little effect on biological rhythms, which are controlled by endogeneous pacemakers generating an ultradian rhythm of 4 h. Subsequently, external factors progressively take on more and more influence, and the ultradian rhythms disappear, being supplanted by a circadian rhythm.

There are two possible explanations for this passage to circadian rhythm (a) maturation of the endogeneous pacemakers combined with the influence of external factors and (b) the endogenous pacemakers continue to generate a 4-h ultradian rhythm, even in adults, but their circadian expression is due to the predominant effect of external factors, e.g., day–night alternation and diurnal activity. Currently it is not known which explanation is correct.

Maturation of the Organization of Sleep

Although the cyclic structure of sleep exists at birth, the organization of cycles undergoes a slow maturation until the age of about 1 year. At this time, the main features of adult sleep are already encountered.

In the neonate, Monod and Pajot (1965) and then Anders et al. (1971) defined four behavioral stages based on behavioral and polygraphic criteria: consciousness, active sleep, quiet sleep, and transitional sleep. These criteria are enumerated in Table 2. After the beginning of active sleep, the newborn presents a period of quiet sleep, then returns to a stage of active sleep, with a variable interval of transitional sleep between each stage (Anders and Keener 1985). The neonate sleeps about 16–18 h per 24 h, of which 33% is active sleep, which Coons (1987) and Schulz et al. (1983) equate with REM sleep. In addition, there are numerous periods of consciousness during active sleep in the neonatal period, such that a coefficient of efficacy is often calculated by some authors (Coons and

Table 2. Main characteristics of each sleep stage in the newborn (Monod and Pajot 1965)

Sleep stage	Characteristics
Quiet sleep	Eyes closed Regular breathing pattern Absence of body movements Absence of ocular movements
Active sleep	Eyes closed Irregular breathing pattern Presence of body movements Rapid eye movements Erection
Consciousness	Quiet or active Eyes open Exploratory eye movements Vocalizations
Transitional sleep	State transitions from stage to stage

Guilleminault 1982). It is about 85% for the period of quiet sleep and 70% for that of agitated sleep.

Important changes will occur between 3 weeks and 6 months, i.e., during the period qualified as consolidation by Coons (1987). Thus, up to about 3 months, the baby falls asleep in active sleep, while after this age the process will progressively occur in quiet sleep.

The duration of sleep is close to 16 h at 3 weeks and is only 12 h at 6 months. This decrease is at the expense of REM sleep; at birth REM sleep accounts for 33% of total sleep, i.e., more than 5 h, while at 6 months, it accounts for only 25%, or 3 h. SWS remains approximately constant during this period, while the decrease in REM sleep is proportional to the increased duration of consciousness. In addition, the distribution of REM sleep also changes during this period. Thus, there is a substantial decrease in REM sleep at the onset of sleep, but there is no preferential distribution of REM sleep over the entire 24 h period. SWS progressively differentiates into stages 2–3 and stage 4 and is redistributed over the 24 h period with a considerable increase in stages 3 and 4 during sleep periods at the beginning of the night, while stage 2 predominates at the end of the night. Finally, periods of consciousness during sleep decrease notably, such that the coefficient of efficacy increases from 70% to 86% for REM sleep and from 85% to 95% for SWS.

From 6 to 8 months, the organization of sleep remains stable. Between 8 months and 1 year, the change is toward a pattern which is very close to that of the adult. The changes bear on the distribution of REM sleep which—as in the adult—follows a period of SWS and becomes longer during nocturnal sleep. Its dominance at the end of the night, however, appears only around 4 years of age. Between 1 and 2 years of age stage 2 diminishes at the end of the night. Stages 3 and 4 progressively differentiate and they are distributed preferentially at the beginning of the night.

The changes in sleep between 6 and 12 years of age are moderate. Total sleep time decreases from 10 to 7.5–8 h, with a coefficient of efficacy of 95% for all age groups (Coble et al. 1987). Stage 4 decreases somewhat, from 18% to 14%, while stage 2 increases in parallel. REM sleep remains stable, oscillating between 20% and 22% of total sleep time, with a clear-cut dominance during the cycles at the end of the night (Carskadon et al. 1987).

Finally, we may note that Williams et al. (1974) observed that beginning at 9 years of age, stage 3 is longer in boys than in girls. They advanced the hypothesis that this difference is due to greater physical activity by boys. In this context, Browman (1980) showed that REM sleep is greater after physical exercise.

Maturation of the Centers Involved

We will purposely remain schematic on this aspect of the maturation of sleep, since many unknowns remain, especially concerning the maturation of neurochemical systems, and above all the systems of regulation. We will briefly review the ages at which the development of centers involved in the processes of sleep are considered to be mature.

As already discussed, the hypothalamus is an indispensable center for both SWS and the appearance of REM sleep and is undoubtedly of fundamental importance for the regulation of circadian rhythms. The other structures concerned are localized in the brain stem: the reticular formation, the raphe system, and the locus coeruleus. For this aspect of cerebral maturation, we will refer primarily to the work of DeKaban (1970).

At birth, only certain fibers originating from the hypothalamic nuclei begin their myelinization. The nerve cells themselves have a large nucleus with depleted cytoplasm. Cytoplasmic inclusions are typically nonexistent or nonfunctional. There are few presynaptic membranes. At 3 months, some hypothalamic fibers exhibit advanced myelinization and the posthypothalamic fibers are at a very early stage of myelinization. The postmedullary fibers, on the other hand, are at an advanced stage of myelinization. The cytoplasm of nerve cells develops but cytomegalic inclusions are relatively nonfunctional. At 6 months, the myelinization of fibers arising from centers involved with sleep is well advanced, but the nerve cells have not yet acquired their functional specificity and synapses remain few in number. It is not before the age of 1 year that the main nerve structures with an important role in the physiology of sleep will have acquired an advanced stage of maturation. After 1 year, maturation involves primarily the cerebral cortex and several subcortical structures.

Conclusion

It has been shown that the hypothalamus can play an important role in the regulation of circadian rhythms and that certain hypothalamic nuclei have a predominant role in the appearance of SWS and REM sleep. In addition, certain structures in the brain stem also play a major role in the genesis of SWS and the expression of REM sleep. There are three very important dates in the maturation of circadian rhythms and the organization of sleep: birth, 3 months, and 6 months. At birth the baby sleeps a good part of the 24-h period, with a 4-h ultradian rhythm, according to an organization in stages which is very different from that in the adult and which is distributed more or less identically over the 24-h period. At 6 months, the baby sleeps primarily at night, REM sleep remains important during the day, and SWS is differentiated (especially stages 3 and 4 are primarily nocturnal). Finally at 1 year, even though episodes of diurnal sleep persist, sleep is primarily nocturnal and the organization of sleep during the cycles is similar to what it will be in the adult. It is interesting to compare the behavioral and polygraphic findings to the three phases of neuronal myelinization and cell maturation in the centers playing an important role in the regulation of sleep.

We will now discuss several major sleep disorders important in child psychiatry and will attempt—as far as is possible—to compare pathophysiological hypotheses with the findings already summarized.

Sleep Disorders in the Child

As we saw above, the sleep characteristics of the child are of a dynamic nature which depends on neurobiological maturation as well as on intellectual and emotional development. Although sleep disturbances are most often harmless and related to simple causes, they may occasionally reflect a more chronic if not frankly pathological character. Other more severe and above all atypical sleep disorders may occur in the context of a developmental disorder. Finally, preexistent pathologies may be affected by sleep disorders.

Specific Sleep Disorders

Incidents in Daily Life

We will first describe difficulties due to individual changes in sleep during development, and then discuss several hypotheses concerning nightmares, pavor nocturnus, sleepwalking, and rhythmic movements.

Difficulties Related to Individual Variations in Sleep During Development
Very early, often as soon as 3 months, parents start qualifying the sleep patterns of their baby: some are "big sleepers," others are "big nappers," and still others "don't sleep much." Starting at 2 years of age, however, certain sleep difficulties become frequent. This is the time at which the child begins to resist going to sleep, to fear the dark, and to adopt rituals for falling asleep. These harmless difficulties may take on a chronic nature and cause family conflicts.

Other types of difficulty appear when the child reaches school age, including drowsiness or even falling asleep during the daytime. These anomalies may be responsible for scholastic problems and even minor behavioral problems. This type of problem is also frequent in the teenager.

Some authors, including Ferber (1987a), believe these minor sleep disorders to be due to individual variations in the circadian rhythm. Thus, although the pacemakers for ultradian rhythms and then for circadian rhythms which control the consciousness–sleep alternation are functional at birth (Anders 1978) and become autonomous between 6 and 10 weeks (Ferber and Boyle 1983), it is difficult to explain the great interindividual variations in circadian rhythms.

Horne and Ostberg (1979) described adults who are "evening" people or "morning" people. These different types have also been observed in children as young as 2 years by Moore-Ede et al. (1982). When these tendencies to morning or evening activity become substantial, the circadian rhythms may advance or retard their phase.

In the child with a phase advance, all daily activities are shifted: he goes to bed early, even asks to go to bed, gets up very early, and is immediately active. Meals and the afternoon nap are also advanced. There is no problem if the circadian rhythm of the parents is of the same type, but early-rising children are not appreciated by some parents.

A phase delay is much more frequent and may be the cause of insomnia at any age (Weitzman et al. 1981). These children fear and resist going to bed and adopt rituals associated with falling asleep. The child cannot fall asleep early, which is often the cause of family conflict. Awakening in this case is difficult, since it is usually caused by school hours. Loss of sleep is compensated by drowsiness or even falling asleep during school activities. If this state persists, the child will become irritable, hyperactive, and may even present behavioral difficulties. Finally, scholastic performance drops markedly and there is a high rate of absenteeism.

Ferber (1987a) proposed progressive adjustments of the circadian rhythm for these children and stressed the screening of such anomalies. He recommended that parents scrupulously observe the circadian rhythms of the child outside of school periods and determine the constancy or the variability day after day.

Parasomnias

Parasomnias include nightmares, pavor nocturnus, sleepwalking, and rhythmic movements while sleeping. Some authors include enuresis but this subject will be discussed in the section on disorders influenced by sleep.

Nightmares. Nightmares, or anguished dreams, may be observed as early as 15 months of age. The child awakens screaming and crying, often has difficulty in calming down, and may not describe his dream. There is often anxiety before going to bed which lasts several days. He may even refuse to sleep in his room for a certain time (Hameury 1986). Some authors, e.g., Roffwarg et al. (1973), advanced the hypothesis that REM sleep produces intense stimulations favouring the maturation of the central nervous system. At certain stages in the maturation of the cerebral cortex, there would exist a physiological state of facilitation. Thus, during the dream the reminiscence of recent sensorial experiences could be amplified and experienced as a nightmare. Table 3 lists factors differentiating nightmares from pavor nocturnus.

Table 3. Differential diagnosis of pavor nocturnus and nightmares (adapted from Martinius 1986)

	Pavor nocturnus	Nightmares
Time	Before midnight (1–4 h after falling asleep)	Variable but more often after midnight
Initial scream	+ +	(+)
Recall	None	Vivid
Sleep stage	Non-REM	REM
EEG	Rhythmic activity	Desynchronization
Behavior	Disappears upon waking	Persists while awake
Type of reaction	Avoids body contact	Searches body contact
Falling back to sleep	Rapid	Delayed

Nocturnal Terror. Pavor nocturnus and sleepwalking are two entities on a continuum of behavior anomalies during sleep. In contrast to nightmares, they have basically the same pathophysiological background (Rothenberger 1989).

Jacobson et al. (1969) evaluated the rate of occurrence of pavor nocturnus as being between 1% and 3%. Episodes begin between 4 and 12 years and are more frequent in boys. Most authors stress the existence of a familial predisposition. Hallstrom (1972) believes that dominant transmission is responsible. Started by a scream of terror, pavor nocturnus involves repeated episodes of abrupt awakening. After screaming, the child sits up, exhibits intense anxiety, is agitated, and appears to be terrified.

There are numerous autonomic signs accompanying pavor nocturnus: intense sweating, dilation of the pupils, piloerection, and acceleration of heart rate and respiration. The episode lasts from 1 to 10 min, during which the child is inconsolable. He then relaxes and remembers fragments of dreams. Amnesia is total the next day.

Pavor nocturnus is favored by asthenia and stress. It always occurs during the first third of the night, during SWS and often during the first cycle of sleep. The episode is preceded and followed by generalized and symmetrical hypersynchronic delta activity, which is different from classic delta activity of SWS. This delta activity is concomitant with the change in heart and respiration rates (Guilleminault 1987).

There is generally no intercurrent psychopathology, but other signs may be observed. Thus, for example, in 25 children presenting parasomnias, Guilleminault and Silvestri (1982) detected five cases of motor incoordination, four cases of language retardation, and five cases in which there were slow waves, isolated spikes, or spike-waves on the standard EEG, localized in the temporal or parietotemporal regions.

The pathophysiology of this syndrome remains undefined, but it is to be noted that pavor nocturnus is associated with a certain developmental stage.

Sleepwalking. Kales et al. (1980) reported that 15% of children between 5 and 12 years of age have had at least one episode of sleepwalking. Frequently associated with enuresis, sleepwalking occurs more often in boys, and the familial predisposition described for pavor nocturnus is encountered in sleepwalking as well.

Sleepwalking involves repeated episodes of complex behavioral sequences. In the typical form, the child sits up and carries out stereotypical movements followed by motor activity which may cause him to leave his bed and even to get dressed and go out. With the exception of the empty stare and a certain lack of coordination, there are no other signs. As in the case of pavor nocturnus, these episodes occur during a stage of SWS in the first third of the night and often during the first cycle of sleep. The EEG recorded before the episode is identical to that described for pavor nocturnus, whereas during the episode there are paroxysmal outbreaks. This type of sudden onset is normally observed in 85% of children between 6 and 11 months, and only in 3% of children between

7 and 9 years. This led Kales et al. (1968) to state that sleepwalking is due to retarded maturation, but this hypothesis is unlikely, since there are no concomitant symptoms. Benoit et al. (1978) believe that sleepwalking is a disorder affecting the organization of sleep. In a study involving 23 children between 5 and 12 years of age and presenting parasomnias, these authors found EEG anomalies during the first 3 h of sleep, in the absence of any sleepwalking episode. Stages 3 and 4 were shorter, the first SWS stage was delayed, and there was a stage of atypical transitional sleep between the two. The results were confirmed by Guilleminault (1987).

Pathophysiological Mechanisms in Sleepwalking and Pavor Nocturnus. With respect to both sleepwalking and pavor nocturnus (Rothenberger 1989), children seem to show at the clinical as well as the electroencephalographic level a dissociated state which combines features of both wakefulness and non-REM sleep. The pattern of electrical activity is understood as a trigger of motor functions under incomplete conditions of alertness since sleepwalking can be experimentally elicited in healthy children as well as in those tending to develop sleepwalking phenomena by holding up the children during the non-REM sleep stage and setting them on their feet.

The more striking form of partial awakening in remote consciousness, pavor nocturnus, is viewed with its vegetative and motor reactions as the expression of a disinhibition of the functions that are automatically maintained constant in deep sleep. To what extent the observed behavior of disinhibitory reactions during partial awakening is associated with the mechanisms underlying REM sleep remains an open question. There is evidence supporting the suggestion that the central nervous system attempts (even though in vain) to develop a fully formed REM stage during partial awakening. In adults, such an REM stage occurs after the first cycle of sleep.

After completion of the first sleep stage, one is able to observe, in children, a short period within which the EEG shows a configuration of waves similar to that found in REM sleep, the electromyographic activity also being slightly decreased. It returns very promptly to non-REM sleep and spindle patterns, with a subsequent rapid decline in SWS. It is possible that we are dealing here with an REM arousal mechanism which interrupts only incompletely non-REM sleep and immediately afterwards develops a kind of undertow to non-REM activity, the latter characterizing the somnogram of the first half of the night.

Furthermore, it was suggested by Hobson (1985) that parasomnia represents a functional dysregulation in regard to temporal tuning between the central nervous oscillators responsible for sleep cycles and the central and peripheral neuronal motor systems. More specifically, it implies that the central motor systems have already been activated by the time the REM sleep takes place. This activation is thus completed before the full inhibition of spinal motor neurons by central sleep oscillators is reached. The interplay of noradrenergic and cholinergic activity leads to a cholinergic hyperactivity.

Biological Rhythms and Psychological Status in Sleepwalking and Pavor Nocturnus.
The question of the underlying cause of sleepwalking and pavor nocturnus has
yet to be elucidated. Is it the case that partial awakening activates subjective
experience and behavioral anomalies? Or is it that frightful thoughts, ideas, and
representations are increasingly intruding into consciousness during stages of
deep sleep, when the emotional defense of the child is reduced? Will REM stages
be initiated by these hypothetical processes of thought without, however, being
properly structured by the brain of the child? Does the motor behavior reflect
fragments of dream experience?

Most likely, the awakening activity develops first (Fig. 4) due to an underlying
sleep disorder or due to external stimuli; such activity can be weak or intense.
This explains best why the children, unlike those who have experienced nightmares, cannot report any definable thoughts and why tachycardia and other
vegetative dysregulations return quickly to normal values when the event is
over, i.e., as soon as the child is no longer terrified and does not show signs of
general fear. This does not signify that emotional factors are irrelevant. They
are in fact extremely important. However, instead of acting directly as a trigger,
they exert their influence indirectly.

Thus, emotional factors do not elicit the awakening activity in the sense
mentioned above, but rather influence or modulate (as do neuronal predisposition and age) the process through which the child reacts to the awakening
activities occurring at night. The effective trigger of these awakening activities
is attributed to an underlying biological system which controls the time course
of the sleep stages. This also provides an explanation for the occurrence of
events at the exact time when the end of the first or second sleep cycle is
normally expected. Within the context of sudden partial awakening, behavioral
features will thus be determined in favor of sleepwalking or pavor nocturnus
by the intensity of the awakening activity and the nature of the psychological
state of the child before going to bed. This psychological state defines the

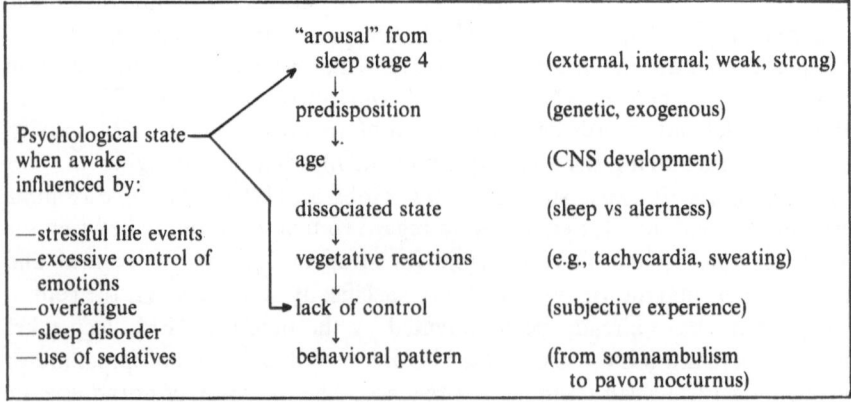

Fig. 4. Parasomnia: cascade of pathogenetic factors

response of the child when he finds himself in a dissociated internal situation characterized by a lack of behavioral control and of orienting capacity as well as by a vegetative uneasiness.

Rhythmic Movements During Sleep. This aspect of parasomnias will be dealt with briefly, since rhythmic movements are not alarming and most often regress spontaneously. Lacombe (1980) reported that 4% of children between 1 and 3 years of age carry out rhythmic movements during sleep. This agitation is more common in boys; it lasts several minutes, is repeated several times during the night. It may involve only one segment of the body, usually not perceived, or the entire body, in this case being bothersome because of the noise it creates. The movements always occur in stages of light sleep following a stage of REM sleep. The diurnal and nocturnal EEGs are strictly normal. These rhythmic movements sometimes occur concomitantly with general developmental disorders or with behavioral disorders. Regardless of this, the course is towards spontaneous disappearance before the age of 5.

Pathology of Sleep

In this section, we will summarize current knowledge of hypersomnias and will touch on the sudden death syndrome, or cot death, which is a particular syndrome deserving of a special chapter.

Hypersomnias

Without wishing to underestimate the importance and frequency of hypersomnias due to respiratory, metabolic, or neurological causes, we will simply summarize published data on narcolepsy, and briefly discuss the Kleine-Levin syndrome.

Narcolepsy. Guilleminault (1987) evaluated the frequency of narcolepsy in the entire population at between 0.04% and 0.09%. In 1957, Yoss and Daly defined the criteria of narcolepsy by a combination of four symptoms:

1. Sudden and repeated onsets of brief sleep, called "sleep attacks"
2. Episodes of cataplexy, i.e., abrupt disappearance of muscle tone with no change in the state of consciousness
3. Momentary loss of the capacity to produce voluntary movements during the sleep attacks, which lasts several minutes
4. Hallucinatory phenomena involving all the senses, occurring primarily when falling asleep, explaining their name of hypnagogic hallucinations

Episodes of microsleep can be added to the above. These episodes last 5–15 s, during which time there is a veritable suspension of activity; the subject may keep his eyes open, but perception is absent. Finally, in about half the cases there is a family history.

Narcolepsy classically appears after the age of 15, with diagnosis usually being made between 25 and 30. Navelet et al. (1976) performed a retrospective survey on 160 narcoleptic adults. The results suggested that the disorder is present much earlier in childhood, but early signs remain poorly defined. In this study, 49% of the subjects presented at least one of the four symptomatic signs before the age of 15 and 5% presented all four symptoms, although no subject was diagnosed as narcoleptic before the age of 15. Even more subtle signs are frequently found, however, since many subjects in the study had been considered as needing a lot of sleep in their childhood: 30% slept more than 10 h a night at 10 years of age, naps persisted for a long time and when they were stopped, the children manifested a desire to sleep during the day.

Hypnagogic hallucinations and parasomnias are often reported in the course of development. In addition, many of these children are considered hyperactive. Polygraphic recordings of the narcoleptic subject show a major disorder in the organization of sleep: the total time of REM sleep is normal, but the subject falls asleep in REM sleep and this REM sleep is fragmented. The duration of SWS is reduced (Fig. 5).

From a neurochemical standpoint, studies in dogs have shown that there is an overall increase in cerebral catecholamines, especially in the amygdala (Mefford et al. 1983), with no change in serotonin levels. Kilduff et al. (1986) observed a change in the level of cholinergic receptors in the brain stem. The most promising result is that of Honda et al. (1985), who found a link between narcolepsy and the DR2 antigen of the HLA system.

In conclusion, narcolepsy is a disorder which is not rare in adults and is undoubtedly much more frequent in children than is believed, but its discreet symptoms explain why it is rarely diagnosed before the age of 20.

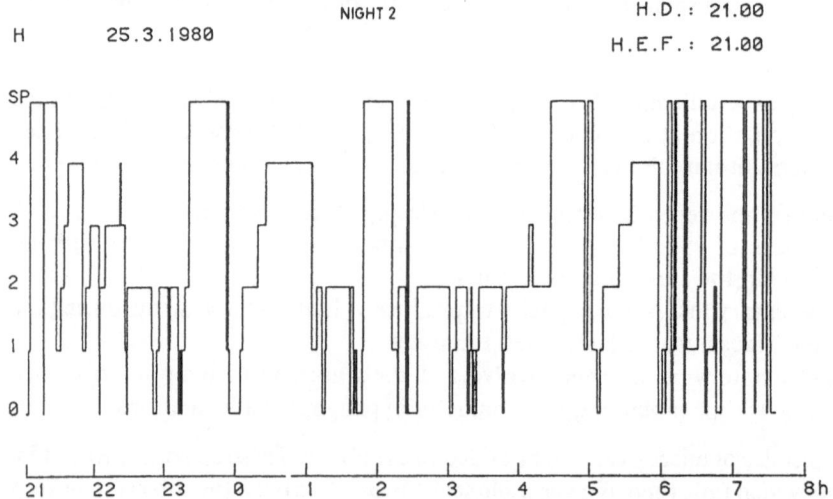

Fig. 5. Distribution of sleep stages in a 12-year-old narcoleptic child. See Fig. 2 for explanation of abbreviations. *0*, awake state

The Kleine-Levin Syndrome. The Kleine-Levin syndrome occurs in adolescence and involves hypersomnia, hyperphagia, and abnormal behavior. This syndrome begins around the age of 12 and is more frequent in boys. It is manifested by sudden outbreaks, often preceded by prodromes marked by fever, vomiting, photophobia, and irritability. The outbreak lasts between 1 and 2 weeks and stops just as suddenly as it begins. The examination is usually negative; in particular the polygraphic recording reveals nothing unusual. The course is always to spontaneous cure at between 18 and 20 years of age.

Near Miss for Sudden Infant Death Syndrome
We will mention this etiology only in the context of sleep disorders of the child. It is in fact difficult to summarize the multiple aspects of this problem. It should nonetheless be remembered that the near miss of sudden infant death syndrome concerns infants found inanimate but who regain consciousness after intensive stimulation. Guilleminault (1987) reported that 6% of these infants eventually die from sudden death. This author advanced the hypothesis of retarded maturation as the cause of death. Polygraphic recordings of these infants show a sleep organization suggesting subtle and perhaps temporarily retarded maturation.

Sleep Disorders and Psychopathology

Sleep problems are often encountered in psychopathology and are often not very serious, providing that therapeutic measures are taken without delay. Sleep difficulties also exist in developmental disorders but are more easily relegated to a secondary level as the behavioral symptomatology is more severe. In this type of pathology it is above all anomalies of sleep that are among the early signs which should alert the attending physician.

Sleep Disorders and Psychopathology of Daily Life

During the first year of life, sleep difficulties related to problems of upbringing or the psychoemotional environment are frequent. The problems most often encountered, and which may subsequently take on different clinical forms, are difficulties in falling asleep and nocturnal awakening.

Ferber (1987b) reported that as early as at the age of 1 month, about 30% of infants cry in the evening before falling asleep or wake up during the night. This is a normal phase of development: the consolidation phase, described in part one of this chapter, and is not a sleep disorder. It should therefore be respected. Carey (1974) reported than 25% of infants between 6 and 12 months of age wake up during the night. Blurton-Jones et al. (1978) reported the corresponding figure to be 23% at 2 years, while Klackenberg (1982) reported a figure of 34% at 4 years. At this age, however, only 3% of children present a true sleep disorder. During this consolidation period certain factors pertaining to upbringing or of a material, emotional, or organic nature favor the persistence or emergence of sleep problems.

The most frequent attitude is to feed the child when he wakes up. Ragins and Schachter (1971) stated that more than 30% of 3-year-old children always receive a bottle at night. Not only does this not help, but it also results in a bad habit (Moore and Ucko 1957). Ferber (1987b) observed that nighttime awakening decreased spontaneously more slowly in breast-fed children. Another factor favoring the persistence of nighttime awakening is the sharing of a bedroom. Even though almost 10% of 10-year-old children prefer sleeping several in a room, nocturnal awakening is reinforced when the awakened child is taken into the bed of the parents. Anxiety of the parents—or indifference—also favors the persistence of nocturnal awakening.

Finally, the following organic problems should also always be considered: infantile colic, infectious diseases, respiratory difficulties, otitis, dental problems, and gastroesophageal reflux. Organic problems are sometimes ignored when they are the cause of difficulty in falling asleep or reinforcing nocturnal awakening.

Regardless of the cause, the infant presents sleep stage 1 at the beginning and middle of the night, during which he acts as a conscious child but asks for no assistance. However, the child is awakened by the slightest stimulus. Earlier in this chapter we saw that this stage of consciousness is normal, with awakening yielding to a well-structured REM sleep at around 6 months. This developmental phase should thus be respected in order to preclude the emergence of disorders in the organization of sleep.

Sleep Disorders and Disturbances in the Psychoemotional Environment

Sleep disorders appearing in a disturbed psychoemotional or upbringing environment will be more severe and chronic as the pathogenic factor is durable and appears early in the life of the child. In fact the causes may be multiple and severe familial disorders are suggested, e.g., neuroses, psychoses, alcoholism, severe emotional lack, drug addiction, and violence in the family. We should not ignore certain momentary events with longlasting consequences, such as mourning, illness, absence or separation of parents, moving, birth of another child, and starting school.

A special rare form deserves mention: psychosocial dwarfism. These children exist in a highly disturbed upbringing and emotional context. Disorders of sleep are combined with retarded growth which may be substantial. This combination is interesting from a pathophysiological point of view. As discussed earlier, the circadian rhythm pacemakers of consciousness and growth hormone are the same. In a disturbed psychosocial context, these centers mature abnormally.

Sleep Anomalies in Developmental Disorders

Sleep anomalies have been reported at an early age in certain global developmental disorders as well as in specific disorders.

Global Developmental Disorders: Infantile Autism/Mental Retardation

As early as 1943, in his first article summarizing 11 observations of autistic children, Kanner stressed sleep anomalies in three children. Two of them could remain conscious for many hours without speaking; the third had an agitated and fragmented sleep.

Subsequently, Wing (1966), Hoshino et al . (1984), and Coleman and Gillberg (1985) summarized the sleep disorders of the autistic child as a temporal and behavioral irregularity in the consciousness–sleep alternation. A more important finding appears to be sleep difficulties as early as the first months of life, i.e., well before the diagnosis of autism. Such difficulties occurred in 3 of 11 children in the series of Kanner (1943). Similarly, Sauvage (1984) reported that 16 of 23 children had sleep problems in the first 6 months of life. This was interpreted as one of the major early "symptoms," along with unusual eating habits.

Few authors have recorded the sleep of autistic children. Ornitz et al. (1969) reported that the total sleep time and its organization are normal, and that only eye movements during REM sleep have a particular mode of appearance. Among current hypotheses of infantile autism, that of a dysfunction of the dopaminergic systems is often advanced (Bruneau et al. in this volume). Segawa (1985) concluded that dopaminergic and noradrenergic neurons function as circadian rhythm modulators, which is why sleep anomalies are observed very early in these infants. This hypothesis is tempting when we consider that circadian rhythm pacemakers are hypothalamic and that the earliest symptoms encountered in autistic children are problems with eating, for which the hypothalamus is one of the most important centers.

In terms of mental retardation, sleep disorders may exist but they are often related to the etiology of retardation. Disregarding these etiological aspects, sleep disorders are no more frequent in retarded children than in the control population (Okawa and Sasaki 1981). The organization of sleep is not very different from that observed in normal children of the same age. Shibasaki et al. (1987) found that the time of appearance of REM sleep was slightly later in retarded children. Petre-Quadens and Jouvet (1966) reported that total REM sleep duration was shortened in retarded children. This is surprising, since it is known that total REM sleep duration decreases during the first year.

Attention Deficit Disorder

Attention deficit disorder is a behavioral disorder combining lack of attention, impulsiveness, and hyperactivity. It is ten times more frequent in boys than in girls, occurring in 3% of all children in the United States (Golden, 1987). In more than 80% of the cases, agitated sleep is combined with frequent awakening. The combination of consciousness disorders, lack of attention, and hyperactivity is often found in psychopathology. Some authors have explained hyperactivity as a compensating behavior for a basic disorder represented by defective regulation of states of consciousness.

Disorders Influenced by Sleep

In this last section we will examine functional enuresis; this is classed by certain authors among the parasomnias, but this is not an unanimous decision. We will then briefly discuss the relationships between epilepsy and sleep.

Functional Enuresis

The DSM III (1980) recognizes the following as diagnostic criteria for functional enuresis:

—Repeated involuntary micturition, diurnal or nocturnal
—Occurrence at least twice monthly in children between 5 and 6 years of age, and at least once monthly in older children
—Absence of a causative physical disorder such as diabetes or epilepsy

In the context of this definition, the age criterion is unanimously agreed upon. The notion of frequency, on the other hand, is far from universally recognized. Thus, the DSM III gives twice monthly as the frequency criterion, while McLain (1979) proposed one to two times weekly, and Nino-Murcia and Keenan (1987) excluded any notion of frequency, basing their diagnosis instead on how the enuresis is supported by the child and the family.

Functional enuresis is more frequent in boys than in girls and almost 75% of the cases have a family history of bedwetting. Enuresis is primary when the child was never toilet trained, and secondary in the opposite case. But here again, some authors consider that enuresis is secondary when the child was trained for 3 consecutive months, while for others this period may be as long as 1 year. These differences in the definition explain the variations in frequency reported for primary functional enuresis: 7% according to the DSM III, 15% by Golden (1987), and 3%–30% by Keener and Anders (1985). The latter authors noted that frequency varies as a function of socioeconomic conditions, family milieu, race, family history, etc.

Organic causes are frequent for secondary enuresis but are rare for the primary disorder. Again, however, the reported percentages vary considerably from one author to another. Causes may be multiple, but they can be divided into three main classes: those related to a problem in the maturation of the urogenital system, those related to a systemic disorder, e.g., hyperthyroidism, diabetes, and urinary infection, and finally those related to a psychopathological problem. Ritvo et al. (1969) classified cases of enuresis into two groups: In cases without intercurrent psychopathology the child is not awakened by the need to urinate and the cause is probably related to a delayed maturation of the urogenital system. By contrast, in cases combined with a psychopathological problem the child may be awakened by the need to urinate but does not respond to this need.

The sleep organization of the bedwetter is normal and although some authors consider that the episode always occurs during SWS and in the first part of the night, the work of Mikkelsen et al. (1980) showed that enuresis may occur at any moment during the night.

It can thus be seen that there is no unanimity on either the diagnostic criteria or the pathophysiological processes of enuresis, and it is our opinion, in agreement with Nino-Murcia and Keenan (1987), that there is not one type of enuresis, but rather several, and that some are intimately related to states of consciousness while others appear to be independent. We will conclude this section by considering the influence of serotoninergic metabolism on functional enuresis.

Influences of the Serotoninergic Metabolism

Weizman and his colleagues (1985) carried out an interesting experiment to attempt to identify the possible organic causes of functional enuresis. By reference to the fact that animal studies demonstrated a change in ^{3}H-imipramine binding on human blood platelets which reflected a similar binding change on nerve cells, they examined 16 enuretic children and 22 healthy controls aged 9–17 years on the basis of this property linked to the serotoninergic system. In this regard, the report of Träskman-Bendz (1983) presented evidence of a possible relationship between functional enuresis and low serotoninergic metabolic rate. This led to the hypothesis that reduced serotoninergic activity could play a role in nocturnal enuresis.

Weizman's research group found a significant decrease in the number of ^{3}H-imipramine binding sites, whereas the dissociation constant on the blood platelets did not differ significantly between children with or without enuresis or between enuretic children with or without a genetic predisposition to enuresis. The results may indicate that a deficit in serotoninergic metabolism (at the peripheral or central level) is part of the pathophysiology of functional enuresis.

Given the finding that a vasopressin derivate (Minirin) is an efficient short-term treatment for bedwetting at night, and considering further that it at the same time stimulates the hippocampal serotonin secretion, which, in turn, relates to the antiamnesic effect of the medicament, the interesting hypothesis should be raised that an intact and functionally fully active serotoninergic system represents an important condition for memory processes. For this reason, one can speculate that this system hosts mechanisms which are highly important for perceiving, memorizing, and evaluating bladder signals that inform the central nervous system to exert voluntary control because of insufficient involuntary control. The methods of behavior modification, especially the conditioning techniques, most likely help the child to develop the attention necessary to successfully process and respond to the afferent signals spontaneously sent by his/her body (Rothenberger 1988).

Sleep and Epilepsy

Considering that sleep deprivation favors the onset of fits and considering the action of barbiturates on sleep and on epilepsy, it appears obvious that there is a close relationship between sleep and epilepsy. We will only summarize this aspect, since an exhaustive treatment would require discussion of the classificaation of the types of epilepsy and examination of the relationship with states of consciousness, and also determination of the impact of epilepsy on the sleep of the child.

In the young child, outbreaks of generalized slow waves at three cycles per second, characteristic of typical absence (petit mal) seizures, preferentially appear during phase 2 of sleep and are related to the appearance of K complexes. In addition, Janz (1962) reported that temporal lobe epilepsy occurs during sleep, and that grand mal epilepsy is more frequent in the adolescent. The latter appears preferentially during the REM sleep at the end of the night.

Hyperpyretic convulsions, flexion spasms of the West syndrome, and the fits of the Lennox syndrome or of progressive myoclonic familial epilepsy are not affected by sleep. There is only a decreased frequency of fits during SWS.

Finally, the influence of epilepsy on the sleep of the child is incontestable, but its magnitude varies depending on the causes of the epilepsy and on associated signs.

Conclusion

Table 4 shows the possible dominant pathophysiological mechanisms for each sleep disorder described in this part. It is interesting to note that it is not difficult to determine a dominant pathophysiological mechanism for each of the disorders discussed: disorders in circadian rhythms, anomalies in the organization of sleep, or retarded maturation. It thus appears important to continue research in these three particular areas of the pathophysiology of sleep.

Table 4. Possible dominant pathophysiological mechanisms for each sleep disorder

	Circadian rhythms	Sleep structure	Maturation
I. *Specific sleep disorders*			
Individual variations	+ + +	0	0
Nightmares	0	+	+ +
Pavor nocturnus	0	+ + +	+ +
Sleepwalking	0	+ + +	+ +
Narcolepsy	0	+ + +	+
Kleine-Levin syndrome	0	+	+ +
Sudden infant death syndrome	0	+	+ + +
II. *Sleep difficulties and psychopathology*			
Current problems	+ + +	+ +	0
Psychoaffective disorders	+ + +	+ +	
Global developmental disorder	0	+	+ + +
Attention deficit disorder	0	+ +	+
III. *Disorders influenced by sleep*			
Enuresis	0	+ +	+ +
Epilepsy	0	+ +	+

General Conclusion

We have seen that many sleep disorders in children are influenced or even dominated by exogenous factors, including upbringing, and the emotional/material environment. It is important to note the vulnerability of the sleep of children and the effects of the disorders on the future of the child. In many cases, however, these disorders—even when minimal—could be avoided or at least reduced in duration if simple professional advice were given, rather than pharmacological treatment. In this respect, information programs for families, doctors, attending physicians, pediatricians, child psychiatrists should be expanded in order to avoid erroneous patient management.

The pathological aspect of certain particularities of sleep such as individual variations in circadian rhythms or parasomnias should be minimized. In most cases, such particularities are normal or subnormal physiological events related to a well-defined period in the development of the individual. In this context, acquaintance with current data on changes in sleep and circadian rhythms is necessary, but it remains the case that knowledge of circadian rhythms and the changes in sleep during development are still fragmentary. For example, answers to the following questions are lacking: Why does the neonate live according to a 4-h ultradian rhythm? How does the ultradian rhythm become a circadian rhythm? What factors are responsible for the changes in the organization of sleep?

Finally, other disturbances in sleep may be considered early signs of a specific disorder of sleep, e.g., narcolepsy, or predictive of a syndrome of which the sleep difficulties are only an epiphenomenon. This is the case with general developmental problems in which sleep anomalies, albeit multiple, are most often combined with alimentary anomalies and should alert the physician.

More thorough knowledge of the maturational aspects of circadian rhythms and of the organization of sleep should lead to a substantial decrease in the interpretation of certain individual variations as pathological and to greater precision in the use of pharmacological treatment for sleep disorders.

Acknowledgment. This study was supported by grants from INSERM no. 859014, Fondation pour la Recherche Médicale, Fondation H. Langlois, Sécurité Sociale, and Conseil Régional de la Région Centre.

References

American Psychiatric Association (1980) DSM III, Diagnostic and Statistical Manual of Mental Disorders: Washington, DC

Anders TF (1978) State and rhythmic processes. J Am Acad Child Psychiatry 17:401–420

Anders TF, Keener M (1985) Developmental course of nighttime sleep-wake patterns in full-term and premature infants during the first year of life. Sleep 8:173–192

Anders TF, Emde R, Parmelee A (1971) A manual of standardized terminology, techniques and criteria for scoring of states of sleep and wakefulness in newborn infants. Los Angeles: UCLA Brain Information Service/BRI Publication Office

Aschoff J, Wever RA (1976) Human circadian rhythms: a multioscillatory system. Med Proc 35:2326–2332

Benoit O, Goldenberg-Leygonie F, Lacombe J Marc ME (1978) Sommeil de l'enfant présentant des manifestations épisodiques du sommeil: Comparaison avec l'enfant normal. Electroencephalogr Clin Neurophysiol 44:502–512

Benoit O, Lacombe J, Bouard G, Foret J and Marc ME (1981) Durée habituelle de sommeil, modification du rythme veille-sommeil et éveils nocturnes. Rev Electroencephalogr Neurophysiol Clin 11:89–95

Blurton-Jones N, Rossetti-Ferreita MC, Farquar Brown M, McDonald L (1978) The association between perinatal factors and later night waking. Dev Med Child Neurol 20:427–434

Browman CP (1980) Sleep following sustained exercise. Psychophysiology 17:577–580

Carey W (1974) Night waking and temperament in infancy. J Pediatr 84:756–758

Carskadon MA, Keenan S, Dement WC (1987) Nighttime sleep and daytime sleep tendency in preadolescents. In: Guilleminault C (ed) Sleep and its disorders in children. Raven Press, New York, pp 43–52

Coble P, Kupfer DJ, Reynolds CF and Houck P (1987) EEG sleep of healthy children 6 to 12 years of age. In: Guilleminault C (ed) Sleep and its disorders in children. Raven Press, New York, pp 29–41

Coleman M, Gillberg C (1985) The biology of the autistic syndromes. Praeger Publishers, New York

Coons S (1987) Development of sleep and wakefulness during the first 6 months of life. In: Guilleminault C (ed) Sleep and its disorders in children. Raven Press, New York, pp 17–28

Coons S, Guilleminault C (1982) The development of sleep wake patterns and non-rapid eye movement sleep stages during the first six months of life in normal infants. Pediatrics 69:793–798

Dekaban A (1970) Developmental anatomy and physiology of the central nervous system from birth to six years of age. In: Dekaban A (ed) Neurology of infancy 2nd edn, Williams and Wilkins, Baltimore, pp 1–48

Ferber R (1987-a) Circadian and schedule disorders. In: Guilleminault C (ed) Sleep and its disorders in children. Raven Press, New York, pp 165–175

Ferber R (1987-b) The sleepless child. In: Guilleminault C (ed) Sleep and its disorders in children. Raven Press, New York, 141–163

Ferber R, Boyle MP (1983) Persistence of a free running sleep wake rhythm in a one year old girl. Sleep Res 12:364

Golden GS (1987) Text book of pediatric neurology. Plenum, New York and London

Guilleminault C (1987) Disorders of arousal in children: somnambulism and night terrors. Guilleminault C (ed) Sleep and its disorders in children. Raven Press, New York, pp 243–252

Guilleminault C, Silvestri R (1982) Disorders of arousal and epilepsy during sleep. In: Sterman MB, Shouse M and Passouant P (eds) Sleep and epilepsy. Academic Press, New York, pp 513–531

Hallstrom R (1972) Night terrors in adults through three generations. Acta Psychiatr Scand 48:350–352

Hameury L (1986) Troubles du sommeil du nourrisson et du jeune enfant (0–3 ans). Rev Med de Tours 20, 8:525–528

Hobson, JA (1985) The neurobiology and pathophysiology of sleep and dreaming. Discussions in Neurosciences, vol II, No. 4. Fondation FESN, Geneva

Honda Y, Doi Y, Juyi T, Satake M (1985) Positive HLA-DR2 finding as a prerequisite for the development of narcolepsy. Folia Psychiatr Neurol Jpn 39:203–204

Horne JA, Ostberg O (1979) Individual differences in human circadian rhythms. Biol Psychol 5:179–190

Hoshino Y, Watanabe H, Yashima Y, Kaneko M, Kumashiro H (1984) An investigation on sleep disturbance of autistic children. Folia Psychiatr Neurol Jpn 38:45–51

Jacobson A, Kales JD, Kales A (1969) Clinical and electrophysiological correlates of sleep disorders in children. In: Kales A (ed) Physiology and pathologia symposium. Lippincott, Philadelphia, pp 109–118

Janz D (1962) The grand mal epilepsies and the sleeping-waking cycle. Epilepsia 3:69–109

Jouvet M (1969) Biogenic amines and the states of sleep. Science 163:32–41

Jouvet M (1984) Mécanismes des états de sommeil. In: Benoit O (ed) Physiologie du sommeil, Masson, Paris, pp 1–18

Kales J, Jacobson A, Kales A (1968) Sleep disorders in children. Progr Clin Pathol 8:63–74

Kales A, Soldatos C, Caldwell A (1980) Somnambulism. Arch Gen Psychiatry 37:1406–1410

Kanner L (1943) Autistic disturbances of affective contact. Nerv Childn Dis 2:217–250

Keener M, Anders T (1965) New frontiers of sleep disorders medicine in infants, children and adolescents. In: Michels R (ed) Psychiatry vol 2, Basic Books, New York, pp 1–12

Killduff TS, Bowersox S, Kaitin K, Baker T, Ciaranello RD, Dement WC (1986) Muscarinic cholinergic receptors and the canine model of narcolepsy. Sleep 9:102–106

Klackenberg G (1982) Sleep behavior studied longitudinally. Acta Paediatr Scand 71:501–506

Krieger DT (1980) Ventromedial hypothalamic lesions abolish food-shifted circadian adrenal and temperature rhythmicity. Endocrinology 106:649–654

Lacombe J (1980) Les rythmies du sommeil de l'enfant. Neuropsychiatrie de l'Enfance et de l'Adolescence 28:213–228

Martinius J (1986) Respiratorische Affektkrämpfe und Parasomnien: Symptom oder Diagnose? In: Gross-Selbeck, G (Hrsg.): Das anfallskranke Kind, Bd. 4. Edition M + p, Hamburg, pp 77–83

Mc Ginty DJ, Harper RM (1976) Dorsal raphe neurons: Depression of firing during sleep in cats. Brain Research 101:569–575

Mc Lain IG (1979) Childhood enuresis. Curr Probl Pediatr 9:4–36

Mefford IN, Baker TL, Boehme RE, Foutz AS, Ciaranello RD, Barchas JD, Dement WC (1983) Narcolepsy: Biogenic amine deficits in an animal model. Science 220:629–632

Mikkelsen E, Rapoport J, Nee L (1980) Chilhood enuresis: Sleep patterns and psychopathology. Arch Gen Psychiatry 37:1139–1144

Monod N, Pajot N (1965) Le sommeil du nouveau-né et du prématuré. Analyse des études polygraphiques chez le nouveau-né à terme. Biol Neonnate 8:281–307

Moore-Ede MC, Sulzman FM, Fuller CA (1982) The clocks that time us. Harvard University Press, Cambridge and London

Moore T, Ucko LE (1957) Night waking in early infancy. Arch Dis Child 32:333–342

Navelet Y, Anders TF, Guilleminault C (1976) Narcolepsy in children. In: Guilleminault C, Dement WC, Passouant P (eds) Narcolepsy. Spectrum, New York, pp 171–177

Nino-Murcia G, Keenan S (1987) Enuresis and sleep. In: Guilleminault C (ed) Sleep and its disorders in children. Raven Press, New York, pp 253–267

Okawa M, Sasaki H (1981) Disorders of sleep waking rhythm: A report of three cases. Shinkei Kenkyu No Shinpo 29:346–365

Ornitz EM, Ritvo ER, Brown MB (1969) The EEG and rapid eye movements during REM sleep in normal and autistic children. Electroencephalogr Clin Neurophysiol 26:167–175

Petitjean F, Buda C, Janin M, Sakai K, Jouvet M (1978) Patterns of sleep alteration following selective raphe nuclei lesions. Sleep Research 7:40

Petre-Quadens O, Jouvet M (1966) Paradoxical sleep and dreaming in the mentally retarded. J Neurol Sci 3:608–612

Ragins N, Schachter S (1971) A study of sleep behavior in two year old children. J Am Acad Child Psychiatry 10:464–480

Rechtschaffen A, Kales A (1968) A manual of standardized terminology, techniques and scoring system for sleep stages of human subjects. US Government Printing Office, Washington, DC

Ritvo ER, Ornitz EM, Gottlief F, Poussaint AF, Maron BJ, Ditman KS, Blinn KA (1969) Arousal and nonarousal enuretic events. Am J Psychiatry 126:77–84

Roffwarg HP, Muzio JN, Dement W (1973) Ontogenic development of the human sleep-dream cycle. In: Stone LJ, Smith HT, Murphy LB (eds) The competent infant. Tavistock, London, pp 273–282

Rothenberger A (1988) Enuresis nocturna. Was bedeutet "funktionell" beim nächtlichen Einnässen? Der Kinderarzt, 19:1351–1353

Rothenberger A (1989) Somnambulismus und Pavor nocturnus. Knotenpunkte in einem Verhaltensspektrum. Pädiat. Prax., 38:232–240

Sakai K (1981) Some anatomical and physiological properties of pontomesencephalic tegmental neurons with special reference to the PGO waves and postural atonia during paradoxical sleep in the cat. In: Brazier M (ed) The reticular formation revisited. Raven Press, New York

Salzarulo P, Fagioli I, Salomon F, Ricour C, Raimbault G, Ambrosi S, Cicchi O, Duhamel JF, Rigoard MT (1980) Sleep patterns in infants under continuous feeding from birth. Electroencephalogr Clin Neurophysiol 49:330–336

Sauvage D (1984) Autisme du nourrisson et du jeune enfant (0–3 ans). Masson, Paris

Schulz H, Salzarulo P, Fagioli I, Massetani R (1983) REM latency: Development in the first year of life. Electroencephalogr Clin Neurophysiol 56:316–322

Segawa M (1985) Circadian rhythm in early infantile autism. Shinkei Kenkyu No Shinpo 29:140–153

Shibasaki M, Kiyono S, Takeuchi T (1987) REM sleep latency during nocturnal sleep in mentally retarded children. Electroencephalogr Clin Neurophysiol 66:512–514

Träskman-Bendz L (1983) CSF 5-HIAA and family history of psychiatric disorder. Letter, Am J Psychiatry 140:1257

Wehr TA, Wirz-Justice A (1982) Circadian rhythm mechanisms in affective illness and in antidepressant drug action. Pharmacopsychiatr 15:31–39

Weitzman ED, Czeisler CA, Coleman RM, Spielman AJ, Zimmerman JC, Dement WC (1981). Delayed sleep phase syndrome: a chronobiologic disorder with sleep onset insomnia. Arch Gen Psychiatry 38:737–746

Wever RA (1972) The circadian system of man. Springer, Berlin, New York 276

Williams RL, Karacan I, Hursch CJ (1974) Electroencephalography of human sleep: Clinical applications, Wiley, New York

Wing L (1966) Early childhood autism: Clinical, educational and social aspects. Pergamon, New York

Yoss RE, Daly DD (1957) Criteria for the diagnosis of the narcoleptic syndrome. Proc Mayo Clini 32:320–328

Weizman A, Carel C, Tyano S, Rehavi M (1985) Decreased high affinity ^{3}H-imipramine binding in platelets of enuretic children and adolescents. Psychiatry Research 14:39–46

Zulley J, Wever RA (1982) Interaction between the sleep wake cycle and the rhythm of rectal temperature. In: Aschoff J, Daan S, Groos G (eds) Vertebrate circadian systems. Springer, Berlin Heidelberg, pp 253–261

Diagnostic and Therapeutic Aspects

Biofeedback: Evaluation and Therapy in Children with Attentional Dysfunctions

B. Rockstroh, T. Elbert, W. Lutzenberger, and N. Birbaumer

Introduction

Most of the problems in learning disabled (LD) and hyperkinetic (HK) children arise from their deficiency in attending selectively and/or in sustaining attention. This conclusion is based on experimental studies, psychological tests, neurological soft signs (John 1977), and electrophysiological investigations. Various electrophysiological indices point at impaired *attentional processes* in these children, for example smaller amplitudes of event-related potential (ERP) components that are related to attention such as N2 and P3 (Lovrich and Stamm 1983; Loiselle et al. 1980; Stamm et al. 1982) or P2-N2 amplitudes with longer latencies (Grünewald-Zuberbier and Grünewald 1982). Lubar et al. (1987) reported that the P3 component was found to have a somewhat higher amplitude for gifted children (IQ > 130) than for "normal controls" but was clearly reduced in amplitude for LD children, these differences being most pronounced for a semantic task. In a continuous performance test, Lutzenberger et al. (1986) found error rates and reduction of the P3 amplitude to be similar in children with attentional problems (identified by their school teachers) and children diagnosed by child psychiatrists as hyperactive (without overt neurological deficits). Furthermore, ERPs in anticipation of signaled or self-induced responses were found to be smaller in children with poor ability to concentrate as compared with well concentrating children. Examples are the CNV (contingent negative variation; Walter 1964), a slow cortical potential (SCP) that typically develops whenever a stimulus signals that a second, task-related event will be presented a few seconds later, or the Bereitschaftspotential (BP) that develops under voluntary response conditions. CNV and BP amplitude were reduced and the BP onset was delayed in poorly concentrating boys (Grünewald-Zuberbier and Grünewald 1982). In addition, enhanced slow activity in the EEG power spectra—in the theta range—has often been reported for LD or HK children as compared with controls (Hughes 1971; Stamm et al. 1982; Lubar and Lubar 1984; Lubar et al. 1987). Grünewald-Zuberbier and Grünewald (1982) observed higher baseline alpha amplitudes in HK children as compared with controls, as well as less alpha reduction in response to tasks. They interpreted this finding as "lessened ability to sustain 'attentional effort'" (p. 299).

Many results have suggested that attentional dysfunctions (as, for example, in HK children) be regarded as an inability to inhibit response tendencies. As hypothesized by Stamm and Kreder (1979), these attentional deficits could reflect an impaired functioning of the frontal lobes. On the basis of a physiological model of the interaction between excitatory cortical (frontal) and inhibitory thalamic and basal ganglia (caudate nucleus) functions, Stamm and Kreder assumed that "dysfunctions in prefrontal cortex would lead to inadequate excitation in caudate neurons and, consequently, to impairments in inhibitory control of physiological processes and motor responses" (p. 145). This view is in line with other models on the regulation of SCPs and attentive behavior (Skinner and Yingling 1977; Elbert and Rockstroh 1987). We have assumed that negative SCPs—which reflect on a neurophysiological level depolarization of apical dendrites within cortical neuronal networks—indicate the regulation of excitability thresholds within these cortical neuronal networks and that these excitability thresholds are regulated via feedback loops that comprise thalamo-cortical circuits, especially the mediothalamic-frontocortical system, the mesencephalic reticular formation, and the paleostriatum.

The functional significance of the basal ganglia for attentive, intentional, and goal-directed behavior has been demonstrated in animal experiments as well as in studies on Parkinson's disease. Since output from the motor cortex depends upon subcortical input from the basal ganglia (which project via the contralateral thalamus and supplementary motor area (SMA) to the motor cortex), a dopamine deficit would result in a failure to dampen adequately striatal excitation, which in turn would inhibit cortical activity. Animals become akinetic and their behavior becomes dependent upon external stimuli (stimulus bound) after blocking of dopaminergic activity (Schmidt 1983) or after lesions in the ventrolateral thalamus (Canavan 1986). For HK children Schmidt (1987 personal communication) speculates that the "paradoxical" effect of stimulant drugs on hyperactive behavior might be due to overactivity of excitatory neurotransmitters, leading to stereotypic behavior.

Whenever thresholds of excitation are set in advance, e.g., are reduced in anticipation of an expected stimulus input or performance, attentional processes can be tuned and efficient performance can be prepared for. If SCPs are associated with attentive behavior in such a way, inattentive behavior—poor concentration or a short attention span as observed in HK children—may be connected to a dysfunction in SCP *regulation*. On the one hand, this may show up in reduced CNV and BP amplitudes. The reduced SCP shifts indicate that cortical excitability is not tuned effectively in advance. The composition of the EEG power spectra, on the other hand, is also believed to indicate cortical arousal, lower excitation being connected with increases in sensorimotor rhythm (SMR; Sterman 1984) or increase in power in the theta range (Lubar 1984). Increased theta activity as found in LD and HK children might therefore be interpreted as sign of generally insufficient cortical excitation, whereas the deviant phasic SCPs point at insufficient *regulation* of excitation.

The abilities for self-regulation and the functional meaning of SCPs can be investigated within the biofeedback paradigm. It is well established by now (Elbert 1978; Elbert et al. 1980; Rockstroh et al. 1984a, b; Ueda et al. 1985;

Trimmel 1986) that healthy human subjects can learn to modify their SCPs "upon command", i.e., upon discriminative stimuli, if adequate feedback is given. Following an appropriate training time subjects are also able to maintain this control over their SCPs even if feedback is withdrawn. The self-controlled change in SCPs has been found to affect various behavioral responses. It was repeatedly demonstrated that self-induced increased negative SCPs facilitate efficient performance (for an overview of results see Rockstroh et al. 1982a, b, 1984a, b). Furthermore, we were able to demonstrate that patients with bilateral frontal lobe lesions of traumatic origin cannot maintain SCP modulation when continuous feedback was withdrawn (Lutzenberger et al. 1980). This finding confirms the hypothesis of a significant frontocortical contribution for SCP regulation (Skinner and Yingling 1977; Rockstroh et al. 1982a). It therefore seems possible that attentional problems can result from disturbances in the fronto-thalamic pathway. It is the very same circuit which has been associated with the distribution of preparatory negativity among cortical neuronal networks (Elbert and Rockstroh 1987). If hyperactivity and attentional dysfunctions are due to an impaired prefrontal–thalamic functional connection (Stamm and Kreder 1979), then an impaired ability to regulate their SCPs might be expected for HK children, comparable to that observed in patients with frontal lobe lesions.

Consequently, we have applied the *biofeedback paradigm to HK children*. Questions addressed by this study included: (a) Do these children exhibit problems in *learning* to control their SCPs? (b) Do they exhibit problems in *maintaining* the control without feedback and during attention-demanding "frontal" tasks? This may lead us to a further question of practical significance: If SCP regulation connected to the control of attentive behavior is impaired, then extended training by means of a biofeedback paradigm might improve SCP self-regulation, and this might affect attentional processes. Electroencephalographic biofeedback training has been used in LD children, e.g., by Lubar and Lubar (1984), Lubar et al. (1987), and Tansey (1984, 1985; Tansey and Bruner 1983) as an "internal cerebral exerciser," i.e., with the intention of activating "auto-stimulation of the brain's neural processes" which "results in increased functional efficacy on the part of the cerebro-cortico system undergoing such stimulation" (Tansey 1984 p. 163). The question of therapeutic effectiveness will be discussed after the presentation of the first feedback study.

SCP Biofeedback Methodology

Twenty-six 9-year-old boys from regular school classes with normal or higher intelligence were selected. First, teachers identified boys with attentional problems (n = 31) and boys with acceptable attention (n = 25). From these groups, 13 boys were assigned to the group with attentional problems (AP) and 13 boys were assigned to the control group with good attention according to their performance in a continuous performance test (for a description of this investigation, see Stamm et al. 1982). None of the subjects was under current medication.

The feedback procedure was adapted from one which had proven useful in earlier studies (Elbert et al. 1980). Continuous visual feedback of SCPs was provided by the outline of a rocket ship that moved from left to right across a TV screen, taking 6 s to do so. Bars marked two goals. Subjects were asked to direct the ship into one of the goals depending upon which of two discriminative acoustic stimuli was presented. Reaching the required goal was rewarded by a win point, whereas a loss point indicated inadequate performance. Additional verbal "social" reinforcement was provided in that the words "good" or "bad" appeared above win or loss points. The position of the rocket at any time point 't' during its 6-s flight was a linear function of the integrated EEG. Reaching one goal required a shift toward increased generation of cortical negativity; reaching the other goal required suppression of negativity or a positive shift referred to the mean of a 4-s pretrial baseline. The "negativity" goal was reached when the mean SCP shift from baseline exceeded $-12 \mu V$, hitting the "positivity" goal required no deviation from baseline or a positive shift. Artifact control procedures during the baseline interval and during the feedback interval were designed to prevent an influence of eye movements, electrode displacements, or muscular artifacts on the rocket's flight, i.e., on feedback and reinforcement (see Elbert et al. 1980, or Rockstroh et al. 1982a, for detailed description). Two blocks of 40 *feedback* trials each were followed by 20 *transfer* trials. In transfer trials only the discriminative stimulus but no feedback was presented. These transfer trials were scheduled to test for conditioning effects. In the second experimental session, 40 transfer trials followed 40 feedback trials. A choice reaction task including "matching-to-sample" was added to test for behavioral effects of the self-induced SCP shifts (see Fig. 1). This task was adapted from Stamm and colleagues (Stamm 1984; Born et al. 1982) as requiring frontocortical processing (see also Stamm 1969).

The EEG was recorded from frontal (Fz), central (Cz), and parietal (Pz) locations referenced to shunted earlobe electrodes, using a 30-s time constant. The vertical EOG was recorded from locations about 1 cm above and below the left eye. Data were sampled at a rate of 100 Hz and digitally filtered to 100-ms points. Trials were excluded from data analysis if the absolute mean of the VEOG exceeded $70 \mu V$ or those of any EEG channel $50 \mu V$. An average of 25% of trials were rejected because of artifacts. There was no significant group difference with respect to trial rejection. Voltage levels of SCPs and VEOG were determined for successive 0.2-s intervals during a 1-s pretrial baseline, 6-s feedback interval, and 5-s reinforcement or task interval. The SCP shift was scored early and late in the feedback interval, the "early CNV" being scored as maximum negativity during the first 3 s, the "late CNV" as the mean shift during the 6th second.

The EEG power spectra of the precentral and parietal recordings were analyzed by Fast Fourier transformation for the intervals 1.0–3.56 s and 3.54–6.0 s of the 6-s interval, using a Tuckey-Hanning window. Experimental effects were statistically evaluated by means of an analysis of variance with the between-

Display

Correct
lever movement

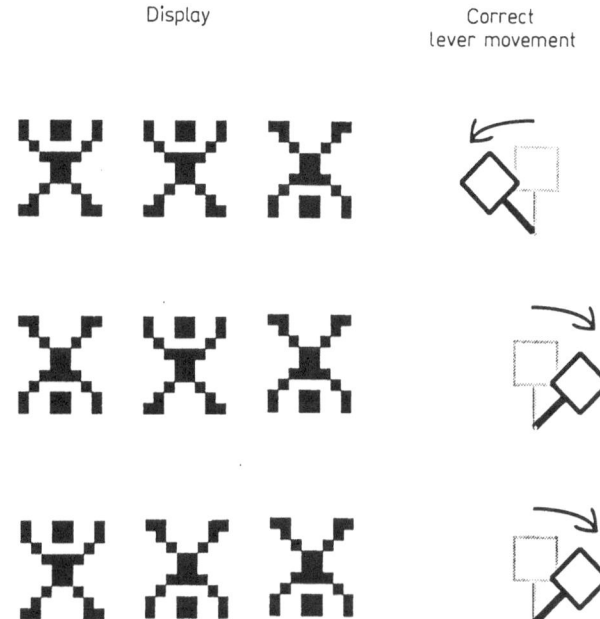

Fig. 1. Schematic drawing of the task that was presented at the end of each transfer trial: three little sketched men appeared on the TV screen. Either one or two of the three men were in an upright position, the other was/were upside down. The direction into which a lever had to be moved with the thumb of the dominant hand indicated whether two men were standing upright or upside down. This choice reaction task involving matching-to-sample is assumed to involve frontocortical functioning (Born et al. 1982; Stamm 1984)

subject factor Groups (children with attentional problems, AP, vs controls) and the within-subject factors Required Shift (negativity increase vs negativity suppression), Type of Trial (feedback us transfer/task), and Electrode (Fz, Cz, Pz).

SCP Biofeedback Results

The AP children achieved SCP differences in the required directions under *feedback* conditions in both sessions. Under *transfer* conditions without feedback, however, AP children proved unable to modify systematically their SCPs. In contrast, control subjects slowly learned the task of SCP self-regulation across the sessions: they developed differentiation in SCPs only during the second session, this differentiation being most pronounced during transfer trials. Fig. 2 illustrates these different developments for arbitrarily selected subjects from both groups. In the second session, AP children produced a mean differentiation of $10 \mu V$ during feedback but $0 \mu V$ during transfer/task trials, while control subjects increased SCP differentiation from $2 \mu V$ during feedback to $5 \mu V$ during

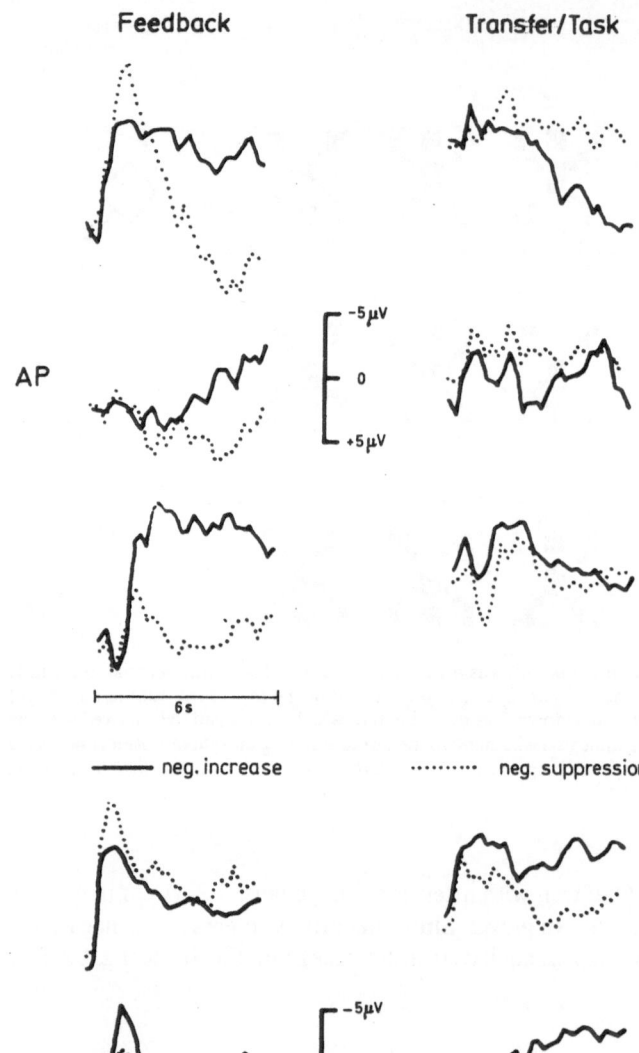

Feedback   **Transfer/Task**

AP

−5μV
0
+5μV

6s

——— neg. increase   ·········· neg. suppression

C

−5μV
0
+5μV

6s

transfer/task trials. [This group-specific development is confirmed by the ANOVA interaction of Groups × Required Shift × Type of Trial in the second session, $F(1, 24) = 7.6$, $P < 0.05$, for the SCP shift during the last second of the 6-s interval.] In other words, AP children were not able to maintain the SCP control under transfer conditions, although they did learn the task when feedback was provided. SCP regulation in these children proved to depend on *immediate, contingent* feedback.

Within the 6-s interval two SCP components were scored as "early"[1] and "late CNV." Averaged across all recordings and types of trial, the late CNV turned out to be around zero in AP children as compared with controls [$-5 \mu V$; Groups $F(1, 24) = 5.0$, $P < 0.05$]. This group difference was especially pronounced during transfer trials, in which the late CNV increased in amplitude to $-8.5 \mu V$ in controls while no late negative shift occurred in AP children [Groups × Type of Trial, $F(1, 24) = 4.8$, $P < 0.05$]. This result indicates an *impaired ability to generate negative SCPs contingent on task anticipation* in AP children. Closer inspection of the scalp distribution of the late CNV revealed that control children exhibited a frontoparietal negativity on transfer/task trials (with parietal negativity even increasing from feedback to transfer trials by $8 \mu V$). AP children, on the other hand, showed smaller late CNV amplitudes at all locations [with parietal amplitude even decreasing by $4 \mu V$ from feedback to transfer/task trials (see Fig. 3[2]); these group differences give rise to an interaction Groups × Electrode × Type of Trial, $F(2, 48) = 4.0$, $P < 0.05$].

Finally, the P3 to the acoustic S^D as well as to the second stimulus (S^D offset and task in transfer trials) was significantly reduced in AP children as compared to controls [$F(1, 24) = 8.0$, $P < 0.01$]. Furthermore, a positive deflection with maximum during the first 500 ms following S^D presentation was smaller in AP children than in controls, especially during task trials [Groups × Type of Trial, $F(1, 24) = 6.2$, $P < 0.05$].

Furthermore, AP children exhibited enhanced activity in the theta range (4–8 Hz) as compared with controls (see Fig. 4; Groups in the second session, $F(1, 24) = 7.8$, $P < 0.05$), this being especially pronounced at Pz and under conditions of required negativity suppression [Groups × Electrode × Required

---

[1] The early CNV exhibited central predominance in the first and frontal predominance in the second session. It turned out to be somewhat larger on transfer/task trials with required negativity increase and smaller on trials with required negativity suppression (Required Shift × Type of Trial, $F(1, 24) = 4.2$, $P < 0.05$). No group differences were found for the early CNV.

[2] Transfer/task conditions—as compared with feedback trials—produced the larger frontocentral late CNV amplitudes in all children.

◀───

Fig. 2. Slow potential shifts in μV during the second session averaged for three arbitrarily selected children with attentional problems (*AP*) and three control children with adequate attention (*C*). SCP *differentiation* is illustrated by averages across trials with required negativity increase (*solid lines*) and trials with required negativity suppression (*dotted lines*) separately for feedback (*left*) and transfer (*right*) conditions

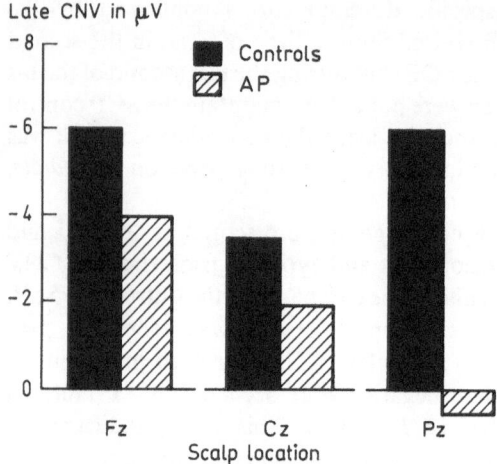

Fig. 3. Scalp distribution of the negative SCP during the last second of the 6-s feedback interval (late CNV in μV, *ordinate*) at frontal (Fz), central (Cz), and parietal (Pz) sites for controls and children with attentional problems (*AP*) *Columns* represent group means under transfer/task conditions of the second session

Shift, $F(2, 48) = 4.6$, $P < 0.05$]. Controls exhibited the expected alpha peak in their EEG power spectra (8–12 Hz), which is known to be more pronounced over posterior locations (Pz) than frontocentrally. Fast (10–12 Hz) synchronized EEG activity was nearly absent in AP children. This result is in support of *enhanced slower frequency activity* as found, for example, for LD children (Stamm and Kreder 1979; Hughes 1971; Lubar et al. 1987; Klorman et al. 1979; Grünewald-Zuberbier and Grünewald 1982).

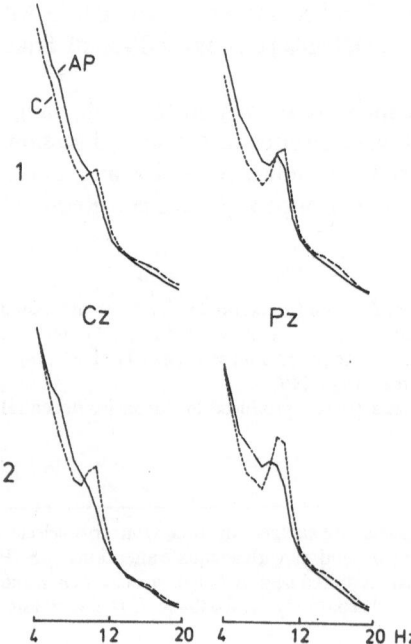

Fig. 4. EEG power spectra between 4 and 20 Hz for the first (*1*) and second (*2*) experimental sessions, averaged separately for controls (*C, dashed*) and children with attentional problems (*AP, solid*)

In the second session, the effects of self-induced SCP shifts were evaluated on *performance* in a choice reaction task (see Fig. 1). AP children responded incorrectly in 4.9 out of 20 trials (24%), while the error rate for controls was 1.7 (8%) [Groups, $F(1, 24) = 6.3$, $P < 0.05$]. Thus, *AP children exhibit poorer performance in a task that is supposed to require frontocortical processing.* Control children tended to respond faster (mean response latency, RL, 1.52 s computed for trials with correct responses) than AP children [mean RL 1.73 s; the group difference approaches significance with $F(1, 24) = 3.0$, $P < 0.1$]. When both RL and errors were added to obtain an overall measure of performance, the group difference became pronounced: $F(1, 24) = 10.7$, $P < 0.01$.

Discussion

Children with attentional problems exhibited a specific impairment in SCP regulation, namely an inability to modulate their SCPs depending upon the discriminative stimuli, as soon as immediate and continuous feedback was withdrawn. This result is reminiscent of observations in patients with frontal lobe lesions (Lutzenberger et al. 1980). These patients, too, achieved SCP differentiation under feedback conditions but were unable to maintain this SCP regulation when feedback was withdrawn (see Fig. 5). This similarity of results between the two groups—AP children and frontal lobe lesioned patients— strengthens the assumption of a mild prefrontal lobe dysfunction underlying attentional and learning problems (Stamm and Kreder 1979).

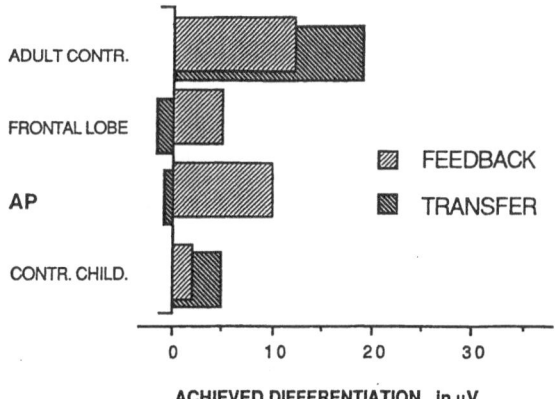

Fig. 5. Average SCP differentiation (in μV) achieved during the second training session in healthy adults (from Rockstroh et al. 1982b), patients with frontal lobe lesions (from Lutzenberger et al. 1980), and children with attentional problems (*AP*) and controls from the present study. Note the comparability between AP children and patients with frontal lobe lesions in their inability to maintain SCP control without feedback (*transfer*). Children demonstrate less differentiation than adults, but both exhibit the more pronounced differentiation during transfer trials without feedback

The possibility of a frontocortical dysfunction in these children is further supported by the finding that the late CNV amplitude in anticipation of the task was smaller in AP than in control children also over the frontal cortex (see Fig. 3). Relating the position of figures and the direction of lever movement is known to involve frontocortical functioning (Stamm 1969, 1984). On the other hand, the most pronounced group difference in late CNV amplitude was found for the parietal recording. The task is visually presented and, therefore, probably involves parietal activation. The motoric lever switching response, finally, should activate the areas below Cz. Consequently, control children show a widespread negativity in anticipation of the task. Negativity is even more pronounced frontally and parietally than at the Cz electrode (see Fig. 3). AP children exhibit a reduced negativity at all locations, which is virtually absent over Pz. Given the assumption that the frontal cortex is mainly involved in the regulation of SCPs and attention (Skinner and Yingling 1977; Elbert and Rockstroh 1987), a failure to *regulate and distribute* SCPs adequately can be attributed to inadequate frontocortical-mediothalamic functioning. The inadequate parietal tuning could, however, also result from a parietal deficit or inadequacies in other parts of the SCP regulation loop (e.g., basal ganglia). The inadequate SCP distribution confirms previous results (e.g., by Grünewald-Zuberbier and Grünewald 1982). Furthermore, ERP components that show parietal predominance and can be associated with the processing of stimuli— such as P3 or LPC—turned out to be smaller in AP children than in controls, which also supports previous findings (Grünewald-Zuberbier and Grünewald 1982; Lutzenberger et al. 1986). An inadequate processing of the signal stimuli, together with impaired preactivation of cortical networks, may have contributed to the poorer performance of AP children in the choice reaction task.

SMR Biofeedback

If attentional dysfunctions and performance deficits may be at least partially attributed to inadequate SCP *regulation*, as indicated by impaired SCP *self*-regulation, the next question would be whether or not the SCP regulation can be improved by means of training of SCP self-regulation. In fact, EEG feedback techniques have been applied in the treatment of HK and LD children. This approach so far has been restricted to the feedback of activity in EEG frequency bands, namely the SMR. Specifically, it has been argued that biofeedback and subsequent conditioning of increased amplitudes of a 14-Hz neural discharge pattern over the central Rolandic area (SMR) will result in "increased bilateral sensorimotor transactions yielding substantive remediation of the learning disabilities" (Tansey 1985 p. 86), due to internal exercising and by way of recruitment of neuronal activity within sensorimotor networks. The stimulation of the sensorimotor cortex by means of SMR feedback and conditioning should increase cognitive functions in a similar way to enrichment of environmental stimuli. Indeed, Tansey (1984, 1985; Tansey and Bruner 1983) reported a significant increase in SMR due to feedback training, which was accompanied by significant increases in verbal and performance IQ indices. Similarly, Lubar and Lubar (1984) found increased SMR activity in LD children with hyperkinetic symptoms after training periods of 10–27 months. Parallel improvement in school performance validated the training. Since SMR is associated with motoric quiescence (Sterman 1984), SMR feedback may furthermore improve performance by means of reducing hypermotoric behavior. Lubar et al. (1987) trained children to first increase SMR (12–15 Hz) and then to increase beta activity (16–20 Hz), while simultaneously suppressing slow activity (4–8 Hz). Sixty-nine LD and 34 control children were trained twice weekly for 4–8 months. Lubar et al. (1987) report "clear increases in school performance" as measured by various achievement tests.

Conclusion

The practical significance of EEG biofeedback training for LD children has been demonstrated, and it seems likely to be of similar significance for HK children. The question of which *specific* EEG biofeedback procedure might be most adequate for a child with a given diagnosis is an urgent challenge for future clinical research. If it is the more "tonic" aspect of cortical arousal which constitutes the malfunctioning, then a modification of EEG frequency bands might be helpful in order to "normalize" EEG power spectra. If the "phasic" regulation of excitability underlying attentive behavior is disturbed, SCP biofeedback training might be chosen first: Whenever an event is anticipated or a performance is prepared for, it is advantageous for the organism that thresholds for excitation are reduced in advance, so that task- or performance-related stimuli are presented to an easily excitable neuronal tissue. Efficient performance would be the consequence. Such a process may be assumed to underlie attentive behavior. Attentional dysfunctions, as stated above, might result from inadequate regulation of cortical excitability. Strengthening the threshold regulation by means of SCP biofeedback might improve attentive behavior.

Acknowledgments. The research reported here was supported by the Deutsche Forschungsgemeinschaft and the Max Kade Foundation (grant to the second author). The reported experiment was supported by Drs. John S. Stamm (SUNY), Jan Born, and Peter Schlottke. We would like to acknowledge especially the stimulating and supportive assistance of John Stamm during his stay at our Department.

References

Born J, Whipple S, Stamm J (1982) Spontaneous cortical slow-potential shifts and choice reaction time performance. Electroencephalogr Clin Neurophysiol 54:668–676

Canavan AGM (1986) Functions of the basal ganglia in man and monkey. Dissertation, University of Oxford

Elbert T (1978) Biofeedback langsamer kortikaler Potentiale. Minerva, Munich

Elbert T, Rockstroh B, Lutzenberger W, Birbaumer N (1980) Biofeedback of slow cortical potentials. Electroencephalogr Clin Neurophysiol 48:293–301

Elbert T, Rockstroh B (1987) Threshold regulation—a key to the understanding of the combined dynamics of EEG and event-related potentials. J Psychophysiol 4:317–333

Grünewald-Zuberbier E, Grünewald G (1982) Event-related EEG changes in children with different abilities to concentrate. In: Rothenberger A (ed) Event-related potentials in children. Elsevier, Amsterdam, pp 295–316 (Developments in neurology, vol 6)

Hughes JR (1971) Electroencephalography and learning disabilities. In: Myklebust HR (ed) Progress in learning disabilities. Grune and Stratton, New York

John ER (1977) Functional neuroscience, vol 2, Wiley, New York

Klorman M, Salzman L, Pass H, Borgstedt A, Dainer K (1979) Effects of methylphenidate on hyperactive children's evoked responses during passive and active attention. Psychophysiology 16:23–29

Loiselle DL, Stamm JS, Maitinsky S, Whipple SC (1980) Evoked potential and behavioral signs of attentive dysfunctions in hyperactive boys. Psychophysiology 17:193–201

Lovrich D, Stamm SJ (1983) Event-related potentials and behavior correlates of attention in reading retardation. J Clin Neuropsychol 5:13–37

Lubar J (1984) Applications of operant conditioning of the EEG for the management of epileptic seizures. In: Elbert T, Rockstroh B, Lutzenberger W, Birbaumer N (eds) Self-regulation of the brain and behavior. Springer, Berlin Heidelberg New York, pp 107–125

Lubar L, Lubar J (1984) Electroencephalographic biofeedback of SMR and beta for the treatment of attention deficit disorders in a clinical setting. Biofeedback Self Regul 9:1–23

Lubar L, Lubar J, Gross D (1987) Neurodiagnosis and EEG biofeedback treatment for attention deficit disorder in children. 1st international conference on biobehavioral self-regulation and health, Honolulu

Lutzenberger W, Birbaumer N, Elbert T, Rockstroh B, Bippus W, Breidt R (1980) Self-regulation of slow cortical potentials in normal subjects and patients with frontal lobe lesions. In: Kornhuber H, Deecke L (eds) Motivation, motor and sensory processes of the brain. Electrical potentials, behavior and clinical use. Progr Brain Res 54: pp 427–430

Lutzenberger W, Rockstroh B, Schlottke P, Birbaumer N, Elbert T, Stamm J (1986) Event-related potentials during a continuous performance test in children with attentive problems. In: McCallum WC, Zappoli R, Denoth F (eds) Cerebral psychophysiology: studies in event-related potentials (EEG Suppl 38). Elsevier, Amsterdam, pp 126–128

Rockstroh B, Elbert T, Birbaumer N, Lutzenberger W (1982a) Slow cortical potentials and behavior. Urban and Schwarzenberg, Munich

Rockstroh B, Elbert T, Lutzenberger W, Birbaumer N (1982b) The effects of slow cortical potentials on response speed. Psychophysiology 19:211–217

Rockstroh B, Birbaumer N, Elbert T, Lutzenberger W (1984a) Operant control of EEG and event-related and slow brain potentials. Biofeedback Self Regul 9:139–160

Rockstroh B, Elbert T, Lutzenberger W, Birbaumer N (1984b) Operant control of slow brain potentials: a tool in the investigation of the potential's meaning and its relation to attentional dysfunction. In: Elbert T, Rockstroh B, Lutzenberger W, Birbaumer N (eds) Self-regulation of the brain and behavior. Springer, Berlin Heidelberg New York, pp 227–239

Schmidt W (1983) Involvement of dopaminergic neurotransmission in the control of goal-directed movements. Psychopharmacology 80:360–364

Skinner J, Yingling C (1977) Central gating mechanisms that regulate event-related potentials and behavior. In: Desmedt J (ed) Attention, voluntary contraction and event-related cerebral potentials. Karger, Basel, pp 30–69

Sterman MB (1984) The role of sensorimotor rhythmic EEG activity in the etiology and treatment of generalized motor seizures. In: Elbert T, Rockstroh B, Lutzenberger W, Birbaumer N (eds) Self-regulation of the brain and behavior. Springer, Berlin Heidelberg New York, pp 95–106

Stamm J (1969) Dorsolateral frontal ablation and response processes in monkey. J Comp Physiol Psychol 67:535–546

Stamm J (1984) Performance enhancements with cortical negative slow potential shifts in monkey and man. In: Elbert T, Rockstroh B, Lutzenberger W, Birbaumer N (eds) Self-regulation of the brain and behavior. Springer, Berlin Heidelberg New York, pp 199–215

Stamm J, Kreder S (1979) Minimal brain dysfunction and neurophysiological disorders in hyperkinetic children. In: Gazzaniga (ed) Handbook of behavioral neurobiology, vol 2. Plenum, New York, pp 119–150

Stamm J, Birbaumer N, Lutzenberger W, Elbert T, Rockstroh B, Schlottke P (1982) Event-related potentials during a continuous performance test vary with attentive capacities. In: Rothenberger A (ed) Event-related potentials in children. Elsevier, Amsterdam, pp 273–294 (Developments in Neurology, vol 6)

Tansey M (1984) EEG sensorimotor rhythm biofeedback training: some effects on the neurologic precursors of learning disabilities. Int J Psychophysiol 1:163–178

Tansey M (1985) Brainwave signatures—an index reflective of the brain's functional neuroanatomy: further findings on the effects of EEG sensorimotor rhythm biofeedback training on the neurologic precursors of learning disabilities. Int J Psychophysiol 3:85–100

Tansey M, Bruner RL (1983) EMG and EEG biofeedback training in the treatment of a 10-year-old hyperactive boy with a developmental reading disorder. Biofeedback Self Regul 8:25–37

Trimmel M (1986) DC potentials of the brain. In: Martin I, Papakostopoulos D (eds) Clinical and experimental neuropsychology. Croom Helm, London, pp 312–338

Ueda M, Furumitsu I, Kakigi S (1985) Self-regulation of contingent negative variation (CNV) using immediate feedback. Jpn J Physiol Psychol Psychophysiol 3:1–9

Walter WG (1964) The contingent negative variation: an electrical sign of significance of association in the human brain. Science 146:434

Is the Use of Psychotropic Drugs Helpful Within Child Psychiatry?

J. Martinius

Introduction

Specialized child psychopharmacology dates back to 1937, when Bradley introduced amphetamines into the treatment of disturbed children. Since then there have been attempts to deepen our knowledge about psychostimulant drug action and to widen the spectrum of psychoactive substances for use in children. As regards the psychostimulants, remarkable progress has been made. For other drugs, mostly those in wide use in adult psychiatric pharmacotherapy, research has remained sporadic, leaving knowledge rather restricted. A period of relative enthusiasm in the late 1960s for employing antidepressant medication in children suffering from a large variety of complaints believed to represent depressive "equivalents" was quickly terminated by a change in general attitudes toward the use of psychoactive drugs, particularly in children. Public opinion in general is against using drugs to treat behavior problems, and even among child psychiatrists opinion is divided: Those oriented toward biological understanding who are dealing with children and adolescents presenting with severe and acute or chronic disorders, namely the psychoses, have to rely on pharmacotherapy and would wish to see knowledge improved and systematized. Others, primarily committed to a psychological approach and therefore, perhaps unknowingly dealing with different patients, see little need for the use of drugs and repudiate them.

Besides this controversy ethical issues have been raised which have to be taken seriously. Diverse opinions, be they well founded or not, have to be given due consideration, and the matter remains highly sensitive. In this context it is very suggestive to ask whether psychoactive drugs are a help and if so, whom do they help, the children, the parents, the doctor, the school, or all of them?

General Principles of Psychotropic Drug Action in Children

Basically, the effects exerted by psychotropic drugs in the child are those known from the adult, notwithstanding the fact that the physical properties of the child's organism differ from those of the adult, in some respects greatly. Present knowledge of the absorption, distribution, and chemical action of drugs in the mature organism therefore cannot be directly applied to the developing one. With lower gastric acidity, absorption may be slower; hepatic metabolism, on the other hand, is usually faster and the blood-brain barrier is more permeable. Some factors reinforce drug action and others inhibit it comparatively more than in the adult, their gross effect making drug action seemingly similar. In detail, however, it may be different. The pharmacology and pharmacokinetics of psychoactive drugs in children are not well understood and virtually nothing is known about sensitivity, binding, and alteration of central nervous receptors in children. From animal studies it is known that receptors vary in their developmental rate (Boreus 1982), meaning that this is most probably the case in human development also, and that at a given time each child has an individual receptor status responsible for an individual sensitivity. Children readily react with dystonic effects to comparatively low doses of a neuroleptic, perhaps due to a heightened receptor sensitivity or to a more easily perturbed receptor balance. On the other hand, tardive dyskinesia does not appear to be as much of a problem in children as in adults. As long as knowledge remains scanty one should, for practical purposes, titrate dosages individually and carefully monitor behavioral effects.

With the exception perhaps, of psychostimulants, basic mechanisms of psychotropic drug action in children are understood too little to lend convincing support to the idea of treating children with such substances. Child psychiatry largely has to rely on knowledge derived from adult psychiatry, used analogously. In view of this, the acuteness and severity of a problem often have to justify the decision for drug treatment. There remains a need for research.

Psychological Effects of Drug Treatment

When medications are given to children little attention is generally paid to the child's feelings, expectations, and reactions "to taking pills." Yet children do react psychologically. A strong placebo effect has been demonstrated (Barkley 1977), meaning that positive expectations frequently accompany the initiation of treatment. Usually children are not in favor of having to take drugs and neither are adolescents. On the other hand, pills are believed to offer a concrete form of help. Children who see their problem and suffer from their experience with themselves and their environment can be convinced to try drugs and may even express strong interest in taking them. It is of utmost importance to work with a child, combatting resistances or heightening interest prior to initiating therapy. Parents view psychotropic drug treatment of their child with increasing reluctance, thereby consciously or unconsciously determining the child's rejecting attitude and the failure of treatment. Securing solid backing from parents, the teacher, and others is part of any therapeutic regime, in itself being part of the help to be given.

The need to take medication signals to parents and to the child that there is a physical illness, namely of the brain. This attitude can help to reduce guilt feelings.

The initiation of pharmacotherapy and observed success come as a relief which in turn may improve family interactions. If this happens, drug treatment of the child will have helped both the child and his environment. Unfortunately, the identification of drug treatment with brain damage can lead to overt or unconscious rejection of the child and to projecting all problems onto the child, avoiding necessary modifications of parental behavior and of traumatic family interactions. Psychological advantages and disadvantages have to be weighed carefully and individualized. A maximum of help can be obtained if psychopharmacotherapy, family counseling, and psychotherapy are not viewed as mutually exclusive means of intervention but as complementary to each other. Pharmacotherapy can prepare the ground for introspection, reflection, and insight, for realization of feelings, and for facilitating expression in gesture, play, and language, thereby opening access to psychotherapy. Psychotherapy in turn can help pharmacotherapy, particularly where the development of the self-image has to be detached from taking drugs. Again, present knowledge is based on more or less anecdotal reports, not on systematic work, the importance of which is obvious.

Formally Established Indications for Psychotropic Drug Treatment in Children

Psychosis

Psychotic symptoms occur in children in the form of agitation, bizarre behavior, and disordered perception and thinking without clear evidence of delusions and/or hallucinations. The resulting syndromes often are suggestive of schizophrenia although the picture typically known in adults is rarely seen before the age of 10. In the acute stage pharmacological treatment is mandatory, using antipsychotic substances, notably the butyrophenones, thioxanthenes, or phenothiazines. These so-called neuroleptic drugs have been shown to have blocking actions, among others, upon the dopamine receptor. They are apt to control agitations, to reorganize thought processes, and to suppress hallucinations and delusions, should they be present. The younger the child, the more psychosis tends to be atypical and to have an "organic" basis. The benefit from antipsychotic drug treatment is obvious in acute psychotic episodes. Withholding it would contradict established knowledge and the obligation to help. Whether the long-term prognosis is improved in children is open to question. Furthermore, whether neuroleptic treatment helps in psychoses which take a primarily chronic course, presenting with negative symptoms such as apathy, remains to be shown. First reports about the treatment of adolescents with clozapine and our personal experiences are promising (Siefen and Remschmidt 1986; Martinius 1988).

Depression

Depression is certainly frequent in children and even more so in adolescents, but whether endogenous depression occurs before the age of 10 is still disputed. In the late 1960s the diagnosis of "masked depression" was popular in children, and antidepressant drug treatment was advocated (Frommer 1968). Since then attention has shifted towards more accurate diagnosis on a descriptive level and towards the development of depression scales analogous to those used in adults (Kovacs 1986). Thereby it has been shown that depression in childhood has much in common with depression in adulthood but that there are distinct behavioral differences. Brain mechanisms in childhood depression are far from being explored. There is one study on the dexamethasone suppression test (Poznanski et al. 1982) which is believed to have "some diagnostic validity." An important biological finding is the fact that more than 50% of children with major depressive disorder have a parent who has had major depression or an affective illness (Popper and Famularo 1983). Few controlled studies on the action of anti-depressant drugs in children have been performed so far, but among them is the study by Petti and Conners (1983) which demonstrated differential responses to imipramine. Individually there was a significant reduction of anger/hostility, incongruous behavior, and self-depreciation. The drug treatment enabled children to verbalize previously suppressed feelings and to relate feelings to others. If these findings can be confirmed they pose a strong argument for antidepressant drug treatment in children and for combining it with psychotherapy.

Tricyclic antidepressants are known to have other indications in child psychiatry, for instance in compulsive-obsessive disorders, particularly if there is a family history of major depression. Hyperkinetic children do sometimes respond favorably to tricyclic antidepressants, an effect which is not well understood. It is suggested that a subgroup of hyperkinetic children suffer from depression or from a basic disturbance which is related to it. Panic disorders are another indication for tricyclic antidepressants in childhood.

Affective psychoses in adolescents appear to improve under treatment with carbamazepine. At least the combination of carbamazepine with lithium appears to act in doses lower than when either is given alone (Poustka and Lehmkuhl 1983).

Mania

Childhood mania is an ill-defined clinical entity whereas in adolescents mania takes the picture known from adults. The cyclic occurrence of impulsiveness, aggressivity, and temper tantrums is sometimes classified under "mania," particularly if alternating with episodes of depressive mood. Among the children so classified is a lithium-responsive subgroup. These children have shown a biological marker in the form of altered neurophysiological responses. McKnew et al. (1981) found an augmentation of evoked potential measurements characteristic of adults with bipolar illness.

The area of depression and mania in children is undoubtedly one of great attraction to biological research and pharmacotherapy, and it is difficult to understand why relatively little research is being undertaken. Clearly the help children suffering from severe depression experience from antidepressant medication outweighs the risks of such treatment.

Hyperkinetic Syndromes

No other problem in child psychiatry has been investigated more intensely than the hyperkinetic/attention deficit syndrome. This holds true for clinical investigations of psychopathology and etiology and for psychological, neurophysiological, and pharmacological research. Over more than 50 years a large body of literature has accumulated, giving answers and leaving questions (Taylor 1986). The clinical description is standardized, yet there are still divided opinions as to classification. There is more than one effective drug treatment and besides this other modes of effective treatment have been established, including family counseling, behavior modification, individual therapies, and dietary regimes. Still the problem is not thoroughly understood. Theories exist about underlying brain mechanisms involving transmitter imbalance or deficits of transmitter function, but no one theory has been proven except for the general realization that harmful exogenous influences (toxic, physical), genetic factors, and pathogenic social interactions cause and maintain the problem in an individually variable composition. There is no single biological marker, and if there is anything new in this field it is the recognition of the fact that there never will be a one-dimensional explanation.

Over long periods the search for one cause, however, has guided research and hampered success. Overactivity is related to the level of arousal and/or the malfunctioning of inhibitory mechanisms acting in ways for which there should be a neurochemical and/or neurophysiological explanation. The search for explanations has been misled by the assumption that the hyperkinetic syndrome is one clinical entity (and likewise one biological entity). Findings have been contradictory or nonspecific.

Abnormal variations in arousal are part of the syndrome but not its origin. Central monoaminergic mechanisms are involved and are influenced by drug treatment (Shaywitz et al. 1984), but their role in causation remains unclear. Drug response itself has been viewed as a pharmacological "probe" for the investigation and understanding of underlying brain pathology. The study of peripheral noradrenaline (Focken et al. 1984) has produced nonspecific results. This fits in well with the observation of Rapoport et al. (1980) that the response to dextroamphetamine is markedly similar in hyperactive and normal boys. If research on the causes of hyperactivity/attention deficit wants to progress, it must abandon the idea that the "hyperkinetic syndrome" is a specific condition (Zametkin and Rapoport 1987).

The clinical diagnosis meanwhile remains descriptive, relying on behavior rating scales (Conners 1973) and on medical and psychological workup analyzing individually the possible causative factors and verifying accompanying problems. Barkley (1982) suggested that the following criteria be used for identification of hyperactive children:

1. Parental/teacher complaints of inattention, impulsiveness, restlessness, and poor compliance and self-control
2. A score on a standardized rating scale of hyperactive behavior of at least two standard deviations above the mean for normal children
3. Reported age of onset of symptoms less than 5 years
4. Duration of symptoms at least 12 months
5. Pervasiveness of symptoms across at least 50% of the situations on the home situation questionnaire
6. IQ estimate of at least 70
7. Exclusion of autism, psychosis, and gross neurological disease

Beyond any doubt stimulants, given in daily doses of 0.3--1 mg/kg body weight to hyperactive children, do improve ability to focus and sustain attention, decrease impulsiveness, and support social behavior in most cases. Experience has taught, however, that only one-third of the children show an effect promptly and consistently. An equal number show a less pronounced and inconsistent response and the rest get worse.

Short-term stimulant treatment is a help for many overactive children, and no less so for their parents and teachers. Stimulants do not promote or induce maturation. Follow-up studies have shown that psychopathological features persist and that in some children who have received stimulant treatment over the course of years, cessation of treatment provoked the immediate recurrence of all the original symptomatology. Besides the administration of stimulants other forms of treatment have to be offered: specific helps for learning, family counseling, and individual social education and therapy, to name only the more important ones.

There are other psychotropic drugs which are effective in calming hyperactive children. As a primary alternative or in those who are not responding to stimulants, tricyclic antidepressants can be used. And in some children presenting with hyperactive-impulsive and antisocial behavior low to medium dose neuroleptic treatment is indicated. Our own experience with the butyrophenone pipamperone has been convincing.

Learning Disabilities

There is no drug which can increase intelligence. Learning behavior, however, is a complex process involving nonspecific prerequisites such as attention and specific cognitive components such as discrimination and memory. Attention can be improved by stimulant medication. On the other hand, pharmacotherapy has little to offer to enhance specific cognitive processes. Many attempts in this direction have failed, not infrequently due to flaws in method. According to recent reports the situation appears to have changed somewhat. Carefully controlled treatment of dyslexic children with piracetam resulted in a marked improvement in reading comprehension skills (Helfgott et al. 1987). This finding is interesting from a clinical and theoretical point of view. In animal studies it could be demonstrated that piracetam enhances the transcallosal response (Giurgea and Moyersoons 1979), indicating that piracetam has a functional selectivity which no other psychoactive drug displays.

Multiple Tic Disease

A small but significant number of children seen in child psychiatry present with multiple involuntary movements and vocalizations (multiple tics, Tourette's syndrome) which can be helped in most cases by drugs blocking dopamine receptors. This effect so far has not been fully explained. Neurophysiological and neurochemical studies point towards a functional dopamine hyperactivity, particularly in striate structures (Cohen et al. 1978; Rothenberger 1984). Neuroleptic drugs such as haloperidol and pimozide have been found to be highly effective in most cases. More recently tiapride, another dopamine receptor blocking agent, has been introduced. Tiapride has the advantage of causing only minor side-effects and is therefore the drug of first choice.

Sleep Arousal Disorders

A number of behaviorally well-defined disorders of sleep in children have been found to have a neurophysiological basis. Among these, night terrors (pavor nocturnus) and sleep walking are the most frequent. They represent dissociations of sleep organization, leading to "incomplete" awakening. Treatment with imipramine has proven to prevent the occurrence of abnormally deep sleep. Whether this is the mechanism to prevent the occurrence of the symptom is still a matter of debate. Psychological stressors, often found in association with sleep arousal disorders, have to be dealt with in conjunction with pharmacotherapy.

Predicting a Response to Pharmacotherapy

Much would be gained by initiating psychopharmacotherapy and maintaining it if responses and side-effects were predictable on a quantitative scale. So far this is wishful thinking because biological markers allowing precise diagnostic classifications have not been found in children. Clinical descriptions are, on the other hand, still too vague (as in depressive syndromes) or too global, incorporating pathogenetically different subgroups under one diagnostic heading (as in hyperkinetic syndrome). Therefore research presently follows both lines: discovery of biological traits indicating vulnerability and improving the appropriateness and reliability of clinical descriptions, defining groups and subgroups. Very little biological research useful for predicting drug response has been performed within child psychiatry. Nevertheless, Halliday et al. (1984) reported variability of the visual evoked response to be the critical measure of stimulant medication effects in children. The same appears to apply to reaction time measurements: hyperkinetic children whose performance on a continuous performance test was most variable showed the best response to stimulant treatment (Martinius 1982). In addition, a combination of clinical features has been evolved for the purpose of predicting response to stimulant medication (Popper and Famularo 1983), although we in fact do not know what the reliable predictors of stimulant response are, except, possibly, severity. These features are:

1. Organic brain damage
2. A family psychiatric history of overactivity
3. A history of intrauterine overactivity
4. Neurological soft signs or minor physical anomalies
5. A history of agitation in response to sedatives

It remains to be stated that research into pharmacological response prediction is much needed in child psychiatry, especially for depressive syndromes, schizophrenia and other psychoses.

Risks

The help that can potentially or actually be obtained from treatment with psychoactive drugs needs to be measured against the potential risks or actual side-effects and untoward reactions. A hyperactive child who under stimulant treatment becomes very quiet and depressive suffers more from the drug than is justified by the gains in behavior. Side-effects and toxicity of psychoactive substances in children are well known and basically parallel those known from adults. Close surveillance of treatment will usually be sufficient to establish a situation in which an improvement outweighs the side-effects. In some respects children are more sensitive than adults, e.g., to neuroleptic medication, while in others they appear to be less reactive, e.g., fewer cardiovascular disturbances occur under treatment with antidepressant drugs.

Drug dependence does not appear to be a problem in children. Side-effects may be subtle yet significant, such as impairment of learning under treatment with chlorpromazine or related substances. In this context it is to be noted that too few studies on psychoactive drugs are done in children. The rules are clearly set, meaning that there is no justification for continuing to treat children with drugs which have not been properly studied in them.

Conclusion

The existing indications for treatment with psychoactive drugs in children are few but important. Schizophrenia and atypical psychoses, severe forms of depression, pervasive hyperactivity, agitation and aggression, and multiple tic disease demand pharmacological intervention for immediate help. In most cases improvement can be achieved without taking risks that are too high. Withholding such treatment would violate our professional mission and act against humanity. Naturally, if the child is helped the family and others in its environment may profit. However, it has to be remembered that psychoactive drugs are used to treat not causes but symptoms. Other forms of treatment have to complement the use of such drugs.

Basic and clinical research in child psychopharmacology is not reflecting the importance of this area in child psychiatry. If knowledge about the brain processes underlying disease is to be furthered and if treatment is to be improved, research needs to be intensified. This requires a change in public opinion as well as conscious recognition of the biological foundations of our own professional work by all child psychiatrists.

References

Barkley RA (1977) A review of stimulant drug research with hyperactive children. J Child Psychol Psychiatry 18:137–165

Barkley RA (1982) Specific guidelines for defining hyperactivity in children (attention deficit disorder). In: Lahey B, Kazdin A (eds) Advances in clinical child psychology. Plenum Press, New York, pp 137–175

Boreus LO (1982) Principles of pediatric pharmacology. Churchill Livingstone, New York

Cohen DJ, Shaywitz BA, Caparulo D, Young JG, Bowers MB (1978) Chronic, multiple tics of Gilles de la Tourette's disease CSF acid monoamine metabolites after probenecid administration. Arch Gen Psychiatry 35:245–253

Conners CK (1973) Rating scales for the use in drug studies with children. Psychopharmacol. Bull 9:24–84

Focken A, Rossel E, Wellistein H, Appel E, Costa D, Palm D (1984) Wirkungen von Methylphenidat bei hyperkinetischen Kindern mit minimaler zerebraler Dysfunktion-Beeinflussung pyschologischer, physiologischer und biochemischer Parameter. Z. Kinder-Jugendpsychiatr 12:235–249

Frommer B (1968) Depressive illness in childhood. Br J Psychiatry 2:117–136

Giurgea C, Moyersoons F (1979) The pharmacology of callosal transmission. A general survey. In: Russel Steele H, van Hof D, Berlucci M (eds) Structure and function of cerebral commissures. Mac Millan, London, pp 282–298

Halliday R, Callaway E, Rosenthal JHI (1984) The visual ERP predicts clinical response to methylphenidate in hyperactive children. Psychophysiology 21:114–121

Helfgott I, Rudel G, Korlewicz H, Krieger J (1987) Effect of piracetam on reading test performance of dyslexic children. In: Bakker D, Wilsher C, Debruyne H, Bertin N (eds) Develpmental dyslexia and learning disorders. Karger, Basel, pp 110–122

Kovacs M (1986) A developmental perspective on methods and measures in the assessment of depressive disorders: the clinical interview. In: Rutter M, Izard CE, Read PB (eds) Depression in young people. Guilford, New York, pp 435–465

Martinius J (1982) Psychopharmakologisch—experimentelle Studien. In: Steinhausen HC (ed) Das konzentrationsgestörte hyperaktive Kind. Kohlhammer, Stuttgart, pp 43–51

Martinius J (1988) Psychopharmakotherapie der Psychosen im Kindes- und Jugendalter. Acta Paedopsychiatr 51:188–195

McKnew DH, Cytrin L, Buchsbaum MS, Hamovit J, Lamour M, Rapoport JL, Gershon ES (1981) Lithium in children of lithium responsive parents. Psychiatry Res 4:171–180

Petti TA, Conners CK (1983) Changes in behavioral ratings of depressed children treated with imipramine. J Am Acad Child Psychiatr 22:355–360

Popper CW, Famularo R (1983) Child and adolescent psychopharmacology. In: Levine MD, Carey WB, Crocker AC, Cross RT (eds), Developmental behavioral pediatrics. Saunders, Philadelphia, pp 1138–1159

Poustka F, Lehmkuhl G (1983) Kombinationsbehandlung mit Lithium und Carbamazepin bei affektiven Psychosen im Jugendalter. Z Kinder-Jugendpsychiatr 11:388–398

Poznanski EO, Carroll BJ, Banegas MC (1982) The dexamethasone suppression test in prepubertal depressed children. Am J Psychiatry 139:321–324

Rapoport J, Buchsbaum M, Weingartner H, Zahn T, Ludlow C, Bartko J, Mikkelsen EJ (1980) Dextroamphetamine: cognitive and behavioral effects in normal and hyperactive boys and normal adult males. Arch Gen Psychiatry 37:903–913

Rothenberger (1984) Bewegungbezogene Veränderungen der elektrischen Hirnaktivität bei Kindern mit multiplen Tics und Gilles de la Tourette Syndrom. Postdoctoral thesis, University of Heidelberg

Shaywitz SE, Shaywitz BA, Cohen DJ, Young JG (1984) Monoaminergic mechanisms in hyperactivity. In: Rutter M (ed) Developmental neuropsychiatry. Churchill Livingstone, Edinburgh, pp 330–347

Siefen G, Remschmidt H (1986) Behandlungsergebnisse mit Clozapin bei schizophrenen Jugendlichen. Z Kinder-Jugendpsychiatr 14:245–257

Taylor EA (ed) (1986) The overactive child. Clinics in developmental medicine no 97. Blackwell, Oxford

Zametkin AJ, Rapoport JL (1987) Neurobiology of attention deficit disorder with hyperactivity: Where have we come in 50 years? J Am Acad Child Adol Psychiatry 26:676–686

A Call for Reliable and Valid Psychophysiological Indices in Daily Practice

M. Timsit-Berthier

Introduction: The Difficulties of Psychiatric Practice

In daily practice, diagnosis and treatment in psychiatry are usually difficult, ambiguous, doubtful, and more related to "art" than to "applied science." Four reasons for these difficulties can be stated:

1. The theoretical knowledge which is at the basis of psychiatric practice is constantly being extended but it is also in a state of fragmentation, with many conflicting viewpoints. A list of the different trends in contemporary psychiatry is sufficient to prove the truth of this assertion: psychoanalysis, learning theory, behavioral schools, cognitive, gestalt, and other versions of individual psychology, and the many schools of social psychiatry.
2. The forms of therapy used in psychiatry display the same diversity: psychotherapy, pharmacotherapy, biofeedback, family therapy, ergotherapy, etc. In most cases it is difficult to assess the efficacy of these methods insofar as, much more than in any other field of medicine, one cannot fail to observe spontaneous recovery irrespective of medical intervention and solely due to changes in life conditions.
3. Accurate assessment of psychiatric symptoms by means of structured interviews and newer diagnostic systems has not improved the situation as much as had been hoped, because of the absence of correlations between symptoms and pathogenesis. Accordingly, it is well known that the same symptom may be related to opposite psychopathological conditions; it can be a direct expression of a primary deficit but it can also be the result of a secondary reaction to overactivity of the same function. A good example of this is provided by motor hyperactivity, which may be related either to excessive dopaminergic activity, as has been described in some psychoses, or to a decrease in the arousal level, as has been described in hyperkinetic children. Thus in biological psychiatry, as well as in psychoanalytic theory, it is often difficult to know whether the symptoms with which we are confronted express the "direct effect of morbid processes" or a "compensation."

   Moreover, the symptomatic form taken by dysfunctional states is closely related to socioeconomic and cultural factors, as well as to family history, and it is easy to understand that the "same" disease may be differently expressed by an immigrant worker who does not easily speak the language of his psychiatrist and by a well-known politician caring about his own image.

4. Last but not least is the general problem of distinguishing with clear-cut criteria between "normal" and "pathological" behavior. In fact, in contrast to the rules which describe general physiological functioning devoted to the central task of regulation of body processes and maintenance of life (like respiration, digestion, and blood circulation), it is impossible to define only one model of psychological functioning. On the contrary, the mental processes which allow the human being to interact with a complex environment may be organized in several different ways, and each of them might be appreciated in the light of short-term results but also long-term consequences. Thus, even if they induce "abnormal" behavior which may, at times, embarrass family members and other people, intensive stress reactions in response to traumatic events may finally lead to sustained coping and a feeling of health and well-being. Such a statement may be illustrated by psychophysiological research on parachute training: this showed that the stronger the initial physiological reaction and subjectively reported fear, the shorter the training and the better the coping with sky jumps (Ursin 1980). Therefore, the value of the behavioral adaptation is difficult to appreciate with objectivity and it is most often judged according to philosophical or ethical criteria. These difficulties may explain extreme abuse of psychiatric interpretations by politicians in totalitarian countries.

Finally, whether the progress realized in the field of nosology by the development of structured interviews (such as PSE and SADS) and of diagnostic criteria (RDC, DSM III), it seems clear that psychiatric theory is incomplete, psychiatric practice is imperfect, and both are sometimes perverted by philosophical and social interests. Thus, there is a large body of reasons to call for valid and reliable indices in psychiatry.

Event-Related Potentials in Psychiatry: Hope and Disappointment

In such a perspective, event-related potentials (ERPs) seem to be excellent candidates for providing the clinician with valid and reliable information: First of all, from a *strictly practical point of view*, ERPs recording is a noninvasive method, easy to do, and may be repeated without the slightest problem for the patient. These explorations have been made possible by the technical developments in electronics and the progress in data processing. Most of the problems raised by physiological and/or technical artifacts have been solved, and ERPs data are now accurately measured with the help of powerful statistical methods which are able to take into account the EEG signal variability. Moreover, substantial methodological progress has been made, first with standardization of the experimental paradigms allowing the recording of CNV, P300, mismatch negativity, and processing negativity, then with the collection of normative data among healthy, normally functioning children and adults. A good example of such standardization is the International Pilot Study of CNV in Mental Illness carried out in Belgium, England, and The Netherlands (Abraham et al. 1980; Timsit-Berthier et al. 1984; Verhey et al. 1986).

From a *theoretical point of view*, the recording of ERPs offers the possibility of examining several aspects of brain activity directly in the intact subject, who is able to report his experience: so this technique may be considered as a window onto the working brain, and as an intermediate level of analysis between biochemistry and behavioral phenomena. It is therefore not surprising that W. Grey Walter, understanding very early the potential value of ERPs for the psychiatric clinician, after observing concordance between electrophysiological and psychiatric assessments in only a few neurotic patients claimed "I suggest that the CNV is an objective indicator of some aspect of mentality that reflects psychopathology in a very direct and trustworthy way.... the operational value of such observations is already considerable in our clinic. With a small computer and suitable programs, all the essential data can be collected, computed, tabulated, plotted and labeled in less than one hour, providing the clinician with a set of objective observations, in less time than it takes him to conduct a neuro-psychiatric examination" (Walter 1975).

But, only 3 years after this statement and the first publications of lack of statistical correlations between ERP measures and nosological categories in cross-sectional studies, the section related to ERPs and psychopathology of the IVth Congress of Event Related Potentials (EPIC IV) was introduced by the provocative question "Why is slow potential research in psychiatry unproductive?" and the contributors pointed out that there had been eight papers directly addressing psychopathology at EPIC II and EPIC III, but only six papers at EPIC IV (Knott and Tecce 1978).

Six years later, Roth et al. (1984) further pointed out that ERPs "have failed to play a significant role in psychiatric clinics..." and they illustrated this statement by adding that "even at psychiatric research meetings, ERPs only grace a few posters in the wings, while center stage is occupied by biochemical studies...." How can we understand such a disappointement? In our opinion, it arises from the secret desire that continues to guide much research in psychiatric electro-physiology: finding distinctive brain wave patterns for the major functional psychiatric disorders, some "signature" waveform of mental disease, similar to the "spikes and waves" pattern in epilepsy. It is the same desire which guided the first research of Hans Berger, a psychiatrist in the clinic in Jena (Germany), who wanted to discover an objective marker of mental diseases. Such a wish is in itself clearly justified since, in the absence of laboratory tests, it is difficult to assess the validity of diagnostic categories. But we have to confess that in electrophysiology as well as in the other fields of biological psychiatry, up to now no specific neurobiological pattern has ever been described for a specific diagnosis, and as put by Shagass: behind "twisted thoughts" there do not lie "twisted brain waves" (1977). The absence of diagnostic "signature" waveforms has led investigators to search for quantitative rather than qualitative deviations from normal in the brain potentials of psychiatric patients. Thus, cross-sectional studies have been done extensively in order to search for statistical differences in ERPs measures between patients presenting specific symptoms or syndromes and controls paired by age and sex. Significant deviant patterns of latency and amplitude ERPs have been discovered in psychiatric populations, but it must be confessed that these changes are not universal within a diagnostic group nor specific enough to be used as a biological marker. For example, CNV and P300 amplitude decreases have been described in schizophrenia as well as in depression and dementia (review in Roth et al. 1986). Within child psychiatry, similar deviant patterns were found in autistic children, in hyperkinetic children, and in children with attention disorders (reviewed by Rothenberger 1982). In summary, we have to realize that, as far as psychiatric nosology is concerned, ERPs do not constitute valid and reliable indices.

Event-Related Potentials in Psychiatry: Their Actual Value

There is, however, an alternative approach which consists in giving up the strict medical point of view which tries to use ERPs as a simple diagnostic tool and in adopting a psychopathological perspective which aims at a better understanding of the psychophysiological processes which may be altered in an individual subject suffering from psychiatric problems. In this perspective, the question consists in wondering about the nature of information that ERPs are able to offer psychiatrists. In our opinion, ERPs are of great interest for the clinician, because they are able to provide information on three main topics:

1. First, ERPs provide information about *the most elementary part of mental functioning*, which comprises the detection of a stimulus, its categorization, and then the selection of a simple motor program. The timing of an elementary mental process may be provided by the latency of N200 and P300. Such an electrophysiological study of mental chronometry may have clear clinical applications since it allows the microanalysis of the different stages of information processing and the detection of cognitive deficits. A good example of such an application might be in patients who pose diagnostic problems, for example dementia vs depression: simultaneous measurements of P300 latency and reaction time would help to discriminate those patients suffering from organic brain damage, who display an increase in both these measures, from those suffering from a functional disease and displaying only an increased reaction time. These methods may be also useful for the objective evaluation of posttraumatic neuropsychiatric sequelae. Moreover, ERPs would be able to provide information about the amount of compensatory brain processes in the early stage of dementia or in slightly brain-damaged patients. In such subjects the study of the relationship between performances, attentional cost, and ability to sustain attention may be of great interest before undertaking a

rehabilitation program. For example, performance output may be maintained for a simple task, but at the price of a high energy cost, and measurement of ERP amplitudes such as CNV and P300 during cognitive performance allows the assessment of this subtle attentional loss (Parasuraman and Nestor 1986).

2. But the study of ERPs permits much more than simple testing of information processing; it also allows *the assessment of controlled processing and behavioral regulation*, which are complex operations, incorporating ideas of flexibility and dynamism of the brain. The simplest example of such an interpretation is provided by measurement of the CNV amplitude during easy tasks. The progressive reduction in CNV amplitude when prolonging the task and repeating the measurements seems to express the transition from a controlled and effortful to an automatic mode of processing. Experiments showed that some subjects were more involved than necessary in easy tasks, dealing with them as if they were highly demanding or stressful, but displayed a response deficit when facing high demands. This functional particularity, which has been observed in migraineurs (Timsit et al. 1986) and in anhedonic subjects (Pierson et al. 1987), may be the expression of psychological "vulnerability." Moreover, ERPs are able to provide information about the processes which regulate the transition between sensorimotor and cognitive modes of processing, as defined by Paillard (1986). This transition between the objective and subjective worlds is a source of adjusting difficulties, and it is often the field containing most of the problems raised by the psychopathological condition. A good index of the quality of the transition is given by the study of CNV resolution and by the postimperative negative variation (PINV). The PINV seems to be an index of processing mechanisms devoted to coping with external stimuli, and it appears specifically whenever the subject experiences lack of control, i.e., when he cannot find a time-locked relation between a given response and the outcome of this response, or, equivalently, between an outcome and his expectation about this outcome (Bolz and Giedke 1981). PINVs have been specifically described in severe psychopathological conditions (such as schizophrenia and manic-depressive illness) (Roth et al. 1986) and seem to express a "dissonant" state of the brain, i.e., a lack of concordance between the sensorimotor and the cognitive system (Timsit-Berthier et al. 1987a; Rockstroh et al. 1979).

3. Last but not least, ERPs (in particular the CNV) may give information about the functional reactivity of the cortical catecholaminergic and cholinergic systems. Indeed, some significant correlations have been evidenced between CNV amplitude and the results of neuroendocrine challenge tests exploring the aminergic systems. These "dynamic" neuroendocrine methods provide an indirect index of central neurotransmission. They consist in measuring the hormonal response (growth hormone, GH) to a specific eliciting factor [apomorphine, a dopamine (DA) agonist, or clonidine, a norepinephrine (NE) agonist], thus giving a neurochemical assessment of catecholaminergic receptor reactivity. We observed in control as well as in depressive patients that the higher the GH secretion in response to apomorphine, the higher the

CNV amplitude (Timsit-Berthier et al. 1986, 1987b). Such results are consistent with the neurochemical models of CNV proposed by Marczynski (1978) and Libet (1984), who hypothesized that negative shifts result from the mediation of the cholinergic system and would be mediated by DA while positive shifts would be controlled by NA neurons. Their hypotheses were also supported by pharmacological studies in animals and humans (review in Rebert et al. 1986). Such results have found a practical application in the study of headache sufferers: First, it has become possible to differentiate migraineurs from subjects with tension headache (Maertens et al. 1986). Second, the measure of CNV amplitude allows the detection, among migraineurs, of "responders" to a given therapy: the higher the amplitude of the CNV, the more adequate the response to beta-blocker treatment (Schoenen et al. 1986). Similarly, our daily practice revealed that among patients showing an acute anxiety state, those with very high CNV amplitudes were most likely to respond well to neuroleptics.

Conclusions: Realizations and Prospects

As far as one can overcome the deception caused by the absence of clear-cut correlations between nosological categories in psychiatry and electrophysiological abnormalities, one may see that the methods of functional exploration of the CNS constitute an interesting source of information about cerebral functioning in its elementary mental operations as well as in its dynamic capacities of adaptation. In addition such methods yield items of knowledge about the functional reactivity of the cortical catecholaminergic systems, which are not directly accessible in human clinical research. However, to become of practical interest to the clinical psychiatrist, during the coming years the study of ERPs should fulfil three fundamental objectives:

First, the method of functional exploration should comprise *different experimental paradigms* allowing the study of various ERPs (processing negativity, N2, P3, CNV, and PINV) in order to study the electrophysiological correlates of the different processes of information processing, decision making, regulation, and control. In this way the problems of the relationships between all the different ERPs will no longer be put in terms of rivalry, but in terms of complementarity. It will no longer be necessary to decide which is the best index for detecting schizophrenia or depression; rather it will be a matter of selecting the more suitable battery of ERPs for adequate evaluation of the particular psychopathological problem under study.

Another, quite fundamental necessity is that the *results be obtained quickly*, as for any other examination in medical practice, and put into relationship with each individual case in order to cast more light on clinical findings and aid decision making. Normative data should be established in every laboratory and for each particular ERP as a function of the chosen experimental paradigm.

Finally, it will be necessary to set up a *presentation of the results* which will allow the opening of a dialogue between the neurophysiologist and the psychiatrist on the one hand, and between the psychiatrist and the patient on the other. Such an integration of biological data into the patient–psychiatrist dialogue could bring an intermediate aspect into their relationship, resulting in a diminution of the psychiatrist's ascendancy over the patient. The basic problem which is raised here is to know how to integrate in our daily practice the data yielded by functional explorations, and how to enter biological aspects into the network of words which is the cornerstone of the psychiatrist's action. For this, there is no option other than innovation and invention, with all the risks they entail.

Acknowledgements. We wish to thank H. Mantanus for helpful discussion and comments, A. Gerono for preparing the manuscript, and M. Salmon for correcting the English version.

References

Abraham P, Docherty TB, Spencer SC, Verhey FH, Lamers TB, Emonds PM, Timsit-Berthier M, Gerono A, Rousseau JC (1980) An international pilot study of CNV in mental illness. In: Kornhuber HH, Deecke L (eds) Motivation, motor and sensory processes of the brain. Prog Brain Res 54:535–542

Bolz J, Giedke H (1981) Controllability of an aversive stimulus in depressed patients and healthy controls: a study using slow brain potentials. Biol Psychiatry 15:441–452

Knott JR, Tecce JJ (1978) Event-related potentials and psychopathology: a summary of issues and discussion. In: Otto DA (ed) Multidisciplinary perspectives in event-related brain potential research. US Environmental Protection Agency. Washington DC, pp 347–354

Libet B (1984) Heterosynaptic interaction at a sympathetic neuron as a model for induction and storage of a postsynaptic memory trace. In: Lynch G, Mc Gaugh JL, Weinberger N (eds) Neurobiology of learning and memory. Guilford, New York, pp 405–430

Maertens de Noordhout A, Timsit-Berthier M, Timsit M, Schoenen J (1986) Contingent negative variation in headache. Ann Neurol 19:78–80

Marczynski TJ (1978) A parsimonious model of mammalian brain and event related slow potentials. In: Otto DA (ed) Multidisciplinary perspectives in event-related brain potential research. US Environmental Protection Agency, Washington DC, pp 626–634

Paillard J (1986) Cognitive versus sensorimotor encoding of spatial information. In: Ellen P, Thinus-Blanc C (eds) Cognitive processes and spatial orientation in animal and man. Nijhoff, The Hague, pp 2–35

Parasuraman R, Nestor O (1986) Energetics of attention and Alzheimer's disease. In: Hockey GR, Gaillard AW, Cols MG (eds) Energetics and human information processing. NATO ASI Sens D, No 31. Nijhoff, pp 395–407

Pierson A, Ragot R, Ripoche A, Lesevre N (1987) Electrophysiological changes elicited by auditory stimuli given a positive or negative value: a study comparing anhedonic with hedonic subjects. Int J Psychophysiol 5:107–123

Rebert CS, Tecce JJ, Marczynski TJ, Pirch JH, Thompson JW (1986) Neural anatomy, chemistry and event-related brain potentials: an approach to understanding the substrates of mind. In: McCallum WC, Zappoli R, Denoth F (eds) Cerebral psychophysiology: studies in event-related potentials (EEG Suppl 38). Elsevier, Amsterdam, pp 343–373

Rockstroh B, Elbert T, Lutzenberger W, Birbaumer N (1979) Slow cortical potentials under conditions of uncontrollability. Psychophysiology 16:374–380

Roth W, Tecce J, Pfefferbaum A, Rosenbloom M, Callaway E (1984) ERPs and psychopathology. I. Behavioral process issues. In: Karrer R, Cohen J, Tueting P (eds) Brain and information: event-related potentials. Ann NY Acad Sci 425:496–522

Roth W, Duncan C, Pfefferbaum A, Timsit-Berthier M (1986) Applications of cognitive ERPs in psychiatric patients. In: McCallum WC, Zappoli R, Denoth F (eds) Cerebral psychophysiology: studies in event-related potentials (EEG Suppl 38). Elsevier, Amsterdam, pp 419–438

Rothenberger A (ed, 1982) Event-related potential in children. Elsevier, Amsterdam. (Developments in Neurology, vol 6)

Shagass C (1977) Twisted thoughts, twisted brain waves? In: Shagass C, Gershan S, Friedhoff A (eds) Psychopathology and brain dysfunction. Raven, New York, pp 353–378

Schoenen J, Maertens de Noordhout A, Timsit-Berthier M, Timsit M (1986) Contingent negative variation and efficacy of beta-blocking agents in migraine. Cephalalgia 4:229–233

Timsit M, Schoenen J, Timsit-Berthier M (1986) A psychophysiological approach to psychosomatic headache. In: McCallum WC, Zappoli R, Denoth F (eds) Cerebral psychophysiology: studies in event-related potentials (EEG Suppl 38). Elsevier, Amsterdam, pp 455–459

Timsit-Berthier M, Gerono A, Rousseau JC, Mantanus H, Abraham P, Verhey FHM, Lamers T, Emonds P (1984) An international pilot study of CNV in mental illness, 2nd report. In: Karrer R, Cohen J, Tneting P (eds) Brain and information: event-related potentials. Ann NY Acad Sci 425:629–637

Timsit-Berthier M, Mantanus H, Poncelet M, Marissiaux P, Legros JJ (1986) Contingent negative variation as a new method to assess the catecholaminergic systems. In: Gallai V (ed) Maturation of the CNS and evoked potentials. Elsevier, Amsterdam, pp 260–268

Timsit-Berthier M, De Thier D, Timsit M (1987a) Electrophysiological and psychological aspects of the derealization state induced by nitrous oxide in nine control subjects. Adv Biol Psychiatry 16:90–101

Timsit-Berthier M, Mantanus H, Ansseau M, Devoitille JM, Dal Mas A, Legros JJ (1987b) Contingent negative variation in major depressive patients. In: Johnson R, Rohrbaugh JW, Parasuraman R (eds) Current trends in event-related potential research (EEG Suppl 40). Elsevier, Amsterdam, pp 762–771

Ursin H (1980) Personality activation and somatic health. In: Levine S, Ursin H (eds) Coping and health. Plenum, New York, pp 259–280 (NATO conference series)

Verhey F, Lamers T, Timsit-Berthier M, Mantanus H, Rousseau JC, Gerono A, Abraham P, Mumford J, Spencer S, White G (1986) An international pilot study of CNV in mental illness. III. In: McCallum WC, Zappoli R, Denoth F (eds) Cerebral psychophysiology: studies in event-related potentials (EEG Suppl 38). Elsevier, Amsterdam, pp 458–460

Walter WG (1975) The contingent negative variation as an aid to psychiatric diagnosis. In: Kietzman M, Sutton S, Zubin J (eds) Personality and psychopathology. Experimental approaches to psychopathology. Academic, New York, pp 197–205

Authors' Index

Subject Index